Biodiversity, Environment & Sustainablility

Biodiversity, Environment & Sustainablility

Editor
Dr. Jagbir Singh

MD Publications Pvt Ltd
New Delhi
www.mdppl.com

Proceedings of the International Conference on Biodiversity, Environment and Sustainability Challenges for Future held on 4 to 6 September, 2008 at School of Environment Studies, University of Delhi (North Campus), Organised by Swami Shraddhanand College, University of Delhi, Alipur, Delhi -110036 and supported by the following agencies:

- UGC, New Delhi
- All India Foundation for Peace and Disaster Management, New Delhi.
- DST
- Delhi Pollution Control Committee
- Ministry of Environment, Delhi
- CSIR, New Delhi
- Maharastra Bank.
- ICCR
- MD Publications Pvt Ltd, New Delhi.

Published by :

M D Publications Pvt Ltd
"M D House", 11, Darya Ganj,
New Delhi-110 002
Phone : +91-11-41563325 (Customer Service)
+91-11-41562846 (Editorial)
E-mail (orders) : **contact@mdppl.com**
Website : **www.mdppl.com**

ISBN : 978-81-7533-148-8

Price : Rs. 550.00

Published and Printed by Mrs. Rajni Gupta on behalf of
M D Publications Pvt Ltd at Printext

Preface

Life on Planet Earth has reached a crisis point. The very word "bio" means life and almost synonymous with it is energy and having polluted the world so badly in the last 200 years, we now need to look at ways of using clean renewable forms of energy. Water is the great sustainer of life in all its forms and we need to clean up the rivers of the world.

The International Conference on Biodiversity, Environment and Sustainability being held at Delhi University, India from September 4-6, 2008 will be presenting the newest and best ideas on an holistic approach in preserving the treasured information of the ancient use of Herbs grown in the Himalayas, used in agriculture and animal husbandry, together with the problem of water, soil, energy, tourism and terrorism, urbanization and air pollution.

This Conference promises to be the new nucleus of ideas for the future sustainability of our planet and we have to make governments realize their responsibility, and to remind them of the saying

"Not to know is bad - not to wish to know is even worse."

The world belongs to every living creature in it.

September, 2008

Dr. Jagbir Singh
Convener-ICBESCF-2008
&
Head, Department of Geography
Swami Shraddhanand College,
University of Delhi

Acknowledgements

The impetus of many encouraging minds from all over the world, as well as here in India, have merged to bring the most modern ideas together with almost a Tsunami wave of Knowledge to offer solutions to the world's most pressing problems involving Biodiversity, the Environment and its Sustainability. These are our challenges for the future and for the success of this Conference to be held at the school of Environmental Studies, University of Delhi Sept. 4-6-2008. I would like to acknowledge with gratitude the following people. Firstly I would like to express my heartfelt thanks to Dr. J.L. Bhat, Principal, Swami Shraddhanand College, who ones again has given me a great inspiration and encouragement in making this Conference possible.

My thanks go to Mr. J.K. Dadoo, Secretary to the Minister for the Environment who has taken a very keen interest in this conference and for providing funds for the conference, I am very grateful to him. I am also indebted to Prof. M.S.S. Rawat, Prof. Harjeet Singh, Ms. Joan S. Gilmour and also my students Ms. Babita, Rekha, Nidhi, Manju, Meenakshi, Mr. Sunil and Amandeep. Without the great enthusiastic team effort of all concerned, this conference would not have been possible.

DR. JAGBIR SINGH
Convener-ICBESCF-2008
&
Head, Department of Geography
Swami Shraddhanand College,
University of Delhi

Acknowledgements

Contents

1

The Future Possibility of Geothermal Energy Resources in Australia and how Australia can help India

PROF. JOAN SCHREIJAEG-GILMOUR

Introduction

Biodiversity is the thread which holds the ecology of the environment in its place. Planet earth is about 4.55 billion years old and life has survived on earth for nearly 4 billion years. Geologists and biologists from all over the world have carried out scientific research on rocks studying the decay of radioactive forms of elements of rocks.

The Greek world "isos" meaning "equal" and "topos" meaning "place" has formed the word "isotope" used in chemistry to explain each of two or more forms of the same element that contain equal numbers of protons but different numbers of neutrous in their nuclei. The decay rates of isotopes found in rocks are precise, meaning that nowdays we can state precisely how long rocks have existed. It has been proven that moon and meteorite rocks also prove the 4.55 billion years age of our solar system corresponding to the age of Planet Earth.

In Greenland, the oldest rocks are 3.8 billion years old, South Africa 3.6 BILLION and Australia 3.36 billion years old. Rocks which were 3.4 billion years old contained microfossils of single celled life and younger rocks contained more diversity. All this suggests that Planet Earth is very conductive to life. The sun is an external energy source near enough to the Earth to prevent freezing but not close enough to evaporate and destroy everything. These conditions allow a carbon based chemistry to flourish. Carbon is abundant and reactive without being dangerous and the basis for the most complex and varied chemistry. For 3.1 billion years, 80% of the history of life, biodiversity was dominated by bacteria and other life forms now dismissed as microbes and the biodiversity of the Earth is the story of the bacteria. Although human beings have destroyed and polluted much of the planet we have not threatened the overall survival of life itself.

Oxygen is the waste product from photosynthesis splitting water to obtain hydrogen ions.

such as calcium and iron in the sea. As this iron and other sinks for the oxygen were consumed, atmospheric oxygen increased. The first Ice Age may have been created by the oxygen replacing the carbon dioxide methane rich greenhouse atmosphere. For many prokaryotes, which are members of a large group of organisms (including bacteria and blue green algae) that lack true nuclei in their cells, oxygen was a deadly poison. Global ecosystems were destroyed and remnant communities were banished to habitats beyond oxygen's reach such as waterlogged mud and the deep sea. For others this was an opportunity, as aerobic creatures in an environment with a metabolic process requiring oxygen to release energy, took over. The origin of atmospheric oxygen was the result of prokaryote photosynthesis as even the smallest; most ancient life form was capable of modifying the planetary environment. A global biodiversity catastrophe due to oxygen toxicity and climate change occurred and even though life was not wiped out, oxygen tolerant prokaryotes took over and new cellular organisms developed from an intricate association of prokaryotes. Since the advent of oxygen, a further global catastrophe took place called the Vendian Epoch. This was round 670 million years ago and in its aftermath, large animals appeared with startling speed. At 570 million years, the earth underwent an almost global glaciation.

Geological strata contain huge glacial deposits from what were sea levels, equatorial regions. Iron rich deposits suggest a marked decline in atmospheric oxygen and carbon. This massive glaciation was overlain with carbonate rocks that required carbon dioxide and rich warm seas for their deposition.

About 770 million years ago, most of the land masses were small fragmented and equatorial. There was high rainfall and geochemical processes stripping carbon dioxide from the atmosphere. Variations in planetary tilt and an ever increasing reflection of solar energy from the sun resulted in a glaciation one kilometre thick covering most of the Earth's core and even volcanic peaks did not freeze over. With biological systems closed down or banished to the ocean, carbon dioxide from volcanic sources began to increase. By 575 million years ago, deep sea animals recolonized the empty seas. The ability of biodiversity to regulate the global environment was ever whelmed by the unusual combination of circumstances.

At the end of the Paleozoic Period 251 million years ago a global catastrophe threatening the very existence of life on Earth, coincided with a period of ferocious lava eruptions called flood-volcanism. The outpouring of the lava lasted for over 1 million years, dumping 2 million cubic kilometres across the landscape. From this global impact arose huge additions of atmospheric greenhouse gases causing a 6°C warming. Changes to levels of oxygen and carbon isotopes in the rocks suggest a catastrophic breakdown of biochemical systems as biodiversity's ability to drive global geochemical systems was overwhelmed. Recovery from the Permian extinctions was the beginning of a contest between different groups of vertebrates to dominate terrestrial habitats. About 179 million years later at the end of the Mesozoic Period, this contest was resolved with the extinction of the Dinosaurs.

We benefit from biodiversity in many ways.

In 2001 the United Nations launched the millennium Ecosystem Assessment to explore how changes to ecosystem services may affect our well being. It began with core ecosystem services such as nutrients, soils and primary production. These produced 3 classes of direct service.

1. Provisioning services – food and water.
2. Regulating services – climate and disease.
3. Cultural services – spiritual and educational ones.

These were followed by 4 constituents of well – being.

1. Security – clean and safe shelter.
2. Basic material for a good life (e.g. resources for livelihood).
3. Health (e.g. nourishment).
4. Good social relations (e.g. cultural expression)

Biodiversity is fundamental to freedom and choice and all of these 4 constituents need to be fulfilled to create freedom and choice. We are reliant on biodiversity to keep the planet healthy and safe and living systems are drivers of environmental services vital to human health and security.

In 1972, the first United Nations Conference on the Human Environment met in Stockholm, Sweden. It was a dialogue of the deaf. Developed countries asked poorer countries to clean up environmentally destructive development and developing countries wished for economical growth even if pollution and degradation were the price.

In 1987 the Brundtland Report promoted the use of sustainable development

In 1992 at the Rio Earth Summit the developed world focussed on climate change, destruction of tropical forests and species loss.

By September 2002, the Rio Earth Summit was calling out People Planet Prosperity, Greenpeace compared what was required with what was achieved.

Biodiversity is the richness and variety of life on earth and the priorities for the world's poor are Water Agriculture Energy and Health and all form the foundation for Sustainable Development.

The flowers and insects and bacteria and forests and coral reefs are biodiversity.

We have had life in some form or another on this planet for nearly 4 billion years.

Within the last five years there has been an enormous awakening in the Consciousness of Man that the speed at which the environment and ecology is rapidly deteriorating everywhere is signaling to the scientists that the time has come when we have to rethink our attitudes to our immediate environment- The Clean House Effect, the suburb in which we live, the town or city in which we live and the country in which we live. It is now each country's responsibility to exchange ideas and technical knowledge for the betterment of the planet as whole. Each of us is challenged daily by how much we consume, how much we recycle and how much we waste.

At the International Conference held at Delhi University in March 2005, I uttered the words that "the farmers are the backbone of any country." It is very encouraging to read about all of the scientific research and experiments which are being carried out by the brilliant scientists here in India on water quality, soil quality, genetic strains of seeds, pollution problems, environmental problems, microbiology, biotechnology, mosquito control and water treatment, urbanization and earthquake impacts.

This all connects to biodiversity and health and for modern man, these two factors of biodiversity and health have graduated to an economical shock which is causing our minds to explode.

We realize now in 2008 that our present levels of coal and oil consumption cannot continue. We are being forced to find alternative forms of energy and find them quickly because we have seen what damage has been done up until now from the extreme air pollution problems caused by industry, the water pollution problems in the rivers, lakes and oceans of the world caused by negligent chemical factories and the degradation of forests on land in all parts of the world for urbanization and wrong farming methods leading to soil deprivation and erosion.

Much interest worldwide is for the commercial development of alternative sources of energy namely Wind, Solar, Hydroelectricity and Nuclear Power.

It is a well known fact that Geothermal Power using water heated by volcanoes has been used for heating and electricity generation, in both Iceland and New Zealand for the best part of sixty to eighty years. India needs new forms of energy. Despite recent strong economic growth, about half of the Indian population – more than 500 million people – does not yet have access to electricity.

India's Prime Minister, Dr. Manmohan Singh has said that India's economical development must be energy efficient based on non- fossil fuels and on renewable sources of energy. Geothermal Power is carbon free energy available round the clock and in a world which aspires to cut greenhouse gases, it may be the answer to India's power problems. Solar power cannot provide it and neither can wind.

Global shifts in climatic zones have produce natural shifts in agricultural productivity the spread of tropical diseases and with other abnormal extremes of weather and undersea activity, we have witnessed first hand the Sumatran Tsunami of 2004 as well as Hurricane Katrina in New Orleans and the current enormous flooding in Ohio, America of the Mississippi River. On Sydney Television recently in June 2008 an OHION sugar farmer was shown to be standing by his farm watching the raging Mississippi flow by. Visible were only the tips of leaves from his acres of a huge sugar plantation which he had purposely planted to produce Ethanol for Car fuel to save the use of petrol from polluting the atmosphere. Now, his farm was all part of the Mississippi.

Nature will win over man every time. Our awareness of current threats to biodiversity and prospects for our future survival arises in past because we know of animals and plants that have become extinct in the past. The protection of the Himalayan eco-system must be protected at all costs.

The future possibility and application of the Geothermal Energy Reserves in Australia is exactly how our knowledge and research to date can help India.

In many parts of the world including Europe, India, China, the United States and Australia there is rock strata which is found at least 5 kilometres below the earth's surface and in Australia many companies are seeking to capture Australian's underground heat to prove that deep-earth geothermal power is commercially viable.

In Australia at a town called INNAMINCKA, about 1100 kilometres north of Adelaide, work is going on to tap the resources of deep earth geothermal power which will give carbon free energy to a country which is aspiring to cut greenhouse gas emissions. The power supplied would be far greater than what solar power or wind can provide at the moment and while Nuclear Power is on the table, many are seeking to capture Australian's underground heat to prove that deep-earth geothermal power is commercially viable.

In Australia at a town called INNAMINCKA, about 1100 kilometres north of Adelaide, work is going on to tap the resources of deep earth geothermal power which will give carbon free energy to a country which is aspiring to cut greenhouse gas emissions. The power supplied would be far greater than what solar power or wind can provide at the moment and while Nuclear Power is on the table, many Australians are opposed to it. The problem, here, of course, is that there is no possibility of disposing of the rods.

According to the chief executive of the firm Geodynamics, there is enough heat in the rocks of the Cooper Basin, an area on which the town of Innamincka sits, to replace all the coal-fired power stations in Australia for more than 250 years.

The Innamincka granites are buried under 3 kilometres of sedimentary rock. About 5 kilometres below the earth's surface the rocks's temperature is 250 Degree. Heat is generated by the radioactive decay of elements in the granite, but it cannot escape. The original plan was to pump water deep underground under high pressure in order to crack the granite and create a path for the water to flow but there was no need to add water as it was already there, trapped underground for the past 3million years. Already one well just 4 kilometres down is sufficiently strong and hot to run a 1 Megawatt power station by the end of 2008. This will be enough to power the drilling camp and the town of Innnamincka. More wells are planned to produce 500 Megawatts by 2015, enough to replace all the coal fired power stations in Australia for more than 250 years.

Geothermal power will become economical once coal and gas plants have to pay for their carbon emissions, a trading scheme which begins in 2010. Once the water is pumped out hot enough and fast enough to drive the turbines on the surface, it can keep doing it for ages.

Australia has some of the hottest rocks in the world and it has been estimated that 1 cubic kilometre of hot granite has about the same stored energy as 40 million barrels of oil. With several thousand cubic kilometres of these granites, Australia has enough heat to last millennia.

Carbon Credits do not stop Droughts, Earthquakes, Typhoons, Floods, Bushfires, Tornadoes, Hurricanes or Tsunamis and every country must realize that to find new, alternative sources of power is to conserve the biodiversity of the future. ' When Mahatma Gandhi said that "The earth provides enough for everyone's need but not everyone's greed" little did he know that his words are day would be applied to the politics of greenhouse gas emissions.

References

1. Jain, A.K. (1990), *The Making Of Metropolis: Planning And Growth Of Delhi*, National Book Organization, Delhi.
2. Jairaj, B. (1995), *Regulating Ground Water Of Delhi: Scope For Legal Intervention* Spatio-Econimic Development Record, Vol.2, No.6, pp, 42-46 New Delhi.

5. Kundu, A. (1993), *In the Name of Urban Poor: Access to Basic Amenities*, Sage Publication, New Delhi.

6. Kumar, Sheel, (2003), *Drinking Water Problems in Delhi Metropolitan Region*, an Unpublished M. Phil. Thesis, submitted to Department of Geography, University of Delhi, Delhi.

7. Nagpaul, H. (1996), *Modernization And Urbanization In India: Problems And Issues*, Rawat Publication New Delhi.

7. Postel, S, (1985), *Conserving Water: The Untapped Alternative,* World Water Paper-67, September.

8. Reddy, R. (1998), *Urban Water Crisis: A Case Study of Jaipur,* Rawat Publication, Jaipur.

9. Sinha, A.K. (1991), *Water Management: Crucial role of Minor Irrigation,* Survey of Indian Agriculture, The Hindu, Chennai.

10.. Shiklomanov, I.A. (1999): *World's Fresh Water Resources*. In Gleick. P.H. (ed): Water in Crisis. Oxford University Press, New York.

11. Sundersen, P. (1995), Water *Demand and Supply,* Water Resource Research Vol. 26, No.6, June 1995.

12. Sikka, V.M. (1995), *Ground Water Resource Of Delhi State; An Overview,* in Spatio-Economic Development Record, Vol II No. 6, New Delhi, pp 31-38.

13. Trivedi, P.R.(2000), *Water Supply Management*, Indian Institute of Ecology and Environment, New Delhi.

14. Zerah, M.H. (2000), *Water Unreliable Supply in Delhi,* Manohar Publisher, New Delhi.

2

Studying the Effect of Various Densities on the trend of growth, Yield and component Yield of Three Varieties of Soybean in Climatic Conditions of Kermanshah

KEYVAN SHAMSI AND SOHEIL KOBRAEI

Introduction

Producing sufficient foodstuff is considered as one of the most momentous human issues in today's globe. In many developing countries produced food does not suffice consumption; so it will find larger dimensions again in the future, as projected.

Studies show that 90% of calories and 80% of proteins are needed for humans and are directly supplied by plant sources.

In this sense, oily grains are important as one of the main crops with their various products to supply a part of human community needs. Among oily grains, soybean plays an important role to provide calorie and protein needed for humans. Soybean importance in the agricultural industry relies on much oil (20%) and plenty of protein (40%) of the grains. (1)

Given to studies previously performed, planting density seems to have meaningful effects on, as one of important agronomical factors, the growth process, yield components, and ultimately on the yield of different soybean cultivars. Among these studies we can point to experiments by Majidi, (5) Egli, (8) Taqizade. (2) These researchers believe that although yield components, per plant decrease with the increase of planting density, the reduction of yield components can be compensated for by the rising number of plants per area unit. Therefore, the yield increases.

Materials and Methods

This experiment was carried out on the Mahidasht research field, Kermanshah, in May 2000. To perform this research, a factorial test was employed in the form of a complete random block layout with 4 repeats in which a cultivar factor at3 levels including Williams (V1), Zan (V2),

Clark (V3) and a density factor at 3 levels including 3 spacing of plants on rows (D1) 3 cm, (D2) 5 cm, (D3) 7 cm were examined. Thus, the experiment contained 9 treatments placed on 36 test plots. Storage operations were timely and routinely done such as the campaign against pests and diseases, fertilization, weeding and campaign against weeds, and irrigation.

To investigate the soybean cultivars growth process at various densities, sampling was conducted one time on every 16th day that is, 20 days after planting, to physiological maturity (R8) Sampling area on each plot was 0.3 m2. Plants were harvested by shears from the soil surface. Having measured the leaf area, we attempted to isolate various parts of many plants and noted dry weights of any one separately.

An area equal to 3.6m2 was removed from the midst of all plots, with eliminating two marginal rows and 1.5 m from ends of rows in order to calculate the final yield, the biological yield as well as the harvest index.

Also, in order to access the yield components at the final harvesting time, 5 plants were randomly taken out from the harvesting area of any plot and the specifications, as well as the morphological parameters, were measured and recorded for each plant.

Analyzing data variances were performed based on factorial tests in the shape of a complete random block layout.

The Duncan method was employed to compare averages, and the Harvard GRAPHIC, STATG, MSTATC programmes were used to analyze statistical data.

Results and discussion

To study changes of dry weight (1) and leaf area index,(2) mathematical equations with different degrees were used, but the best equation obtained was quadratic, as follow : $y = \mathrm{Exp}(a + bx + cx2)$. Results showed that the dry weight of plant shoots decreased with the increase of the density so that the minimal accumulation of total dry matter of a plant was observed with the density D1.

On the other hand, with increasing the density, the leaf area index rose and the greatest leaf area index pertained to density D1 among different densities.

Among various cultivars, Clark enjoyed the largest rates of the dry weight of shoots and leaf area index. This cultivar possessed the maximum duration of the leaf area.

Shibles et al. (11), Paruez et al. (10), and Ganjali (4) have reported similar results during their experiments.

For cultivars V1 and V3, the crop growth speed was nearly the same and equal, but V3 reached the maximal quantity of CGR with more time lag from planting.

For all cultivars, the crop growth speed became negative at the end of the growth season due to leaves shedding and the consequent reduction of the accumulation of dry matter.

Among different densities, D2 had the most and D3 had the least CGR. In this experiment, the enhancement of the density from D3 to D2 caused the CGR to increase but with the further increase in density, the quantity of CGR decreased slightly.

The results of comparing the changes trend of RGR showed that the Williams cultivar among others and density D3 possessed the maximum quantity of RGR.

RGR quantity was lower and its reduction trend was faster for D1 compared with 2 other densities.

Enyi (7) and Shojaii Noferest (3) also reported that with the increase in density, the amount of RGR is reduced so that they measured the highest and lowest amounts of RGR in the minimum density and the maximum density, respectively.

The experimental results also showed that with the increase in density the following items were enhanced: plant height, spacing of the formation of the first subbranch from soil surface, the length of mid nodules, the number of gnarls on the major stem, the number of grains per pod per sub-branches per plants, grain yield per area unit and physiological yield; and the following items are reduced : the number of sub- branches, the number of gnarls per sub-branches per plant, the number of pods per sub-branches per plant, the number of grains per sub-branches per plant, grain dry weight per sub-branches per plant.

In this experiment, protein percentage; oil percentage harvest index ; 100- grain weight per major branches, per sub-branches, and per plant ; the number of grains per pods on major- sub- branches were not affected by planting density.

These results show some features are further affected by genetic factors rather than environmental factors. For low densities, the number of sub-branches and hence their shares per yield are raised due to less competition.

In any case and according to obtained equations, the most significant components of the yield which entered the model before others were the number of plants per area unit and the number of nodes per plants.

Leumman and Lambert (9) have tested the effect of density on the yield and its components at row spacing of 50, 100 cm and planting spacing on rows of 7.5, 3.75, 1.87, 1.25 cm and concluded that with an increase in density higher yields are obtained while the number of sub-branches and the number of pods per plant are reduced.

Also, Ablett et al. (6) suggested that soybean yield components are reduced with an increase in density.

During his experiment, Majidi (5) stated that although the reduction of planting density could improve grain yield per plant, it failed to compensate the yield reduction caused by a deficiency of plants per area unit.

Conclusions

The results of this research ultimately showed that among different treatments, V1D1 treatment (Williams cultivar and 3 cm plant spacing on row) produced the highest yield per area unit.

Williams cultivar had the largest spacing of formation of the first sub-branch from the soil surface, on the other hand, which can facilitate its mechanized harvesting.

It also had the higher harvest index in comparison with other cultivars. The reduction of yield components is compensated for by increasing the plants per area unit; therefore, the yield increases, although with high densities the enhancement of the density leads to the reduction of the yield components per plant style.

References

1. Alyari. H, Shekari, F. 1998. "Oily grains" (phyisoagro). Amidi press.
2. Ta Qizade. M 1992. "Evaluating the effects of different seed to plant density ratios in mixed farming on the yield, its components, and qualitative specifications of soybean cultivars". Thesis of agronomy MS. Ferdawsi university of Mashhad. Iran.
3. Shojaii No ferest. K, 1993. "Studying the effects of plant density on physiological properties, of water use efficiency, grain yield and its components of two limited unlimited growth soybean cultivars." Thesis of agronomy MS. Agriculture college of Esfahan industrial university. Iran.
4. Ganjali, A. 1992. "Examining the effects of different plant density patterns on soybean yield and photosynthetic potential in Karaj region". Thesis of agronomy MS. Tehran teacher training university. Iran.
5. Majidi, Ganjali. A. 1997. "The effects of planting and plant density patterns on the yield, its components, and superficial features of soybean Williams cultivar in Karaj". Magazine of sapling and seed. No2, 15th pp 142-155 year. Iran.
6. ablett, G. R. G. C. Schleihauf, and A. D. Maclaren. 1984. Effect of row width and population on soybean yield in south western Ontario. Can. J. Ptant Sci. 64 : 9-15.
7. Enyi, B. A. C. 1973. Effect to plant population on growth and yield of soybean (Glycin Max). J. Agric. Sci. Carb. 81 : 130-138.
8. Egli, D. B. 1988. Plant density and soybean yield. Crop Sci. 28 : 977- 981.
9. Lehman, W.F., and J. W. Lambert. 1960. Effect of spacing of soybean plant between and within rows on yield and its components. Agron., 52 : 84- 86.
10. Paruez, A. Q. F. P. Gardner. And K. J. Boote. 1989. Determinate and indeterminate type soybean cultivar responses to pattern, density and planting date. Crop Sci. 22: 150-157.
11. Shibles, R., I. C. Anderson., and A. H. Gibson. 1975. Soybean p. 151-190. int. evans (ed). Crop physiology. Cambridge university press, Cambridge.

3

Traditional uses and Distribution of plants in Cold Desert of Trans- Himalaya

K.N. Singh, Pankaj Jamwal and Brij Lal

Introduction

The dependence of mankind on the plant resources is well known since the evolution of the human race on this planet earth. From ancient times till today people healed themselves with the traditional herbal medicines. Over sixty per cent of the world population, eighty per cent in the developing countries, depends directly on plants for their medical purposes (Shrestha & Dhillon, 2003) and of these eighty five per cent use plants or their extracts as the active substance (Sheldon et al, 1998). Traditional medicine is still recognized as the primary health care system in many rural communities because of its effectiveness, but lack of modern medical facilities in remote areas and of course, cultural preferences, with traditional herbal medicines. (Taylor et al, 1995; Tabuti et al, 2003). In an Indian context, the use of plants as medicine is mentioned in the ancient Indian literature, "*Rig Veda*" and "*Charak Samhita*", and describes the Himalayas as the best habitat for medicinal plants (Samant et al, 1998). Presently, about 2500 wild plant species are reported in the use for medicinal purposes in the Indian sub continent, of which about 300 taxa are used in 8000 pharmaceuticals in India (Ahmad, 1993) on the basis of ethno-information, based on tribal knowledge through centuries. Most of the herbs are in use by the local inhabitants through generations who still continue the traditional practice of herbal cure (Jain, 1991).

The state of Himachal Pradesh with varying land forms from sub-tropical to alpine cold deserts, shares 23.52% of 17000 flowering plant species in India. With the wide range of elevation, ecological landforms and endemic plant species, the state occupies an important area of prime biodiversity concern in the Himalayas. Himachal Pradesh is situated in the lap of the western Himalayas and is considered a veritable emporium of medicinal plants, which occupy an important place in the "Vedic" treatises. It is quite evident through literature that "*Ayurveda*", the ancient science of medicine has its origin in Himachal Pradesh (Chauhan, 2003). A good account of work on the ethno botany particularly on the medicinal uses of plants can be extracted from the publications of Chauhan (1988); Chauhan and Chauhan (1988); Sharma et al. (2003); Sood et al. (2001). Though efforts are in progress, still no scientific and systematic inventory of available resources of medicinal plants with regard to their specific distribution, uses and recepies/ mode of preparation, quality etc. is maintained (Chauhan, 2003).

Of 3500 known species, more than 1000 species have been documented as medicinal and aromatic from the state. Around 500 medicinal and 150 aromatic plants are used in India. The alpine cold desert, usually above 3800 m remain under snow cover for 4-8 months and is remarkable for the variety of beautiful coloured flowers forming a rich storehouse of the medicinal plants (Chauhan 1999; Chauhan, 2003).

In this paper, the authors have especially emphasized on comprehensive first hand information of the medicinal uses and distribution of flowering plants of the high altitude cold desert of the Spiti valley, which are in use traditionally for the last many generations. But unfortunately, this traditional wisdom is vanishing rapidly due to the modern facilities. It is therefore of great importance to document the traditional knowledge systematically before it is lost forever. We are sure that this primary information of an ethno ecological survey will serve the national database on flowering plants in relation to their medico-ethno botany and ethno ecological characteristics.

2. Study area

Spiti valley, an area of 12,210 sq. km. lies in the northeastern corner of Himachal Pradesh between 31°42'33"-32°58'N latitude and 77°21'-78°35'E longitude. It represents the characteristics of a typical cold desert in the district of Lahaul-Spiti in Himachal Pradesh. It encompasses rich trans Himalayan flora and fauna, which is bounded by Tibet on the northeast, Kinnaur in the southeast, Kullu in the west and Ladakh in the north with an average elevation of 4270 m (Fig.1). The climate is dry and cold where the temperature goes down to as low as -45°C and receives heavy snowfalls during winter and has an annual rainfall of 17mm (Kapadia, 1996). The vegetation of the area is broadly categorized as a dry alpine steppe (Champion and Seth, 1968), which comprises various types of flowering herbaceous flora, of which many plant species are used as medicines. A total of 23 rare and endangered medicinal plants have been reported from the Spiti valley (Kala, 2000), which are in use medically to cure various diseases and are therefore considered under high human pressure.

"Mane" a natural glacial lake at an elevation of 4575 mts and the spiti river, which originates from the heights of the "*Kunjum Pass*" are the main sources of water in the valley. Lush green natural patches of *Hippophae* species and *Myricaria squamosa* are present on both sides of the Spiti river with a few small patches of the *Dactylorhiza hatagirea* in the marshy areas. Spiti is also famous for its fossil rich rocks. (District Census Book, Lahul & Spiti, 1991). A Buddhist community, the largest tribe in the valley, mainly dominates the region. The tribal people believe on the " *Tibetan System of Medicine*" and are mostly dependent upon traditional healers, locally called "*Amchis*" to cure various diseases as they have done for generations. A large number of medicinal plants, their local nomenclature, distribution and associated indigenous knowledge of the study area, need proper documentation because of the rapid depletion of the resources with the passage of time. For instance, few studies have been carried out in relation to the status and conservation of medicinal plants of this area. Perusal of available literature indicates that no comprehensive medico-ethno botanical work has been published so far on flowering plants and in particular, in the Spiti valley. Therefore, in the present paper an attempt has been made to document certain flowering plants, which are used by the tribals for their medicinal purposes, distribution, mode of preparation and administration and parts used.

3. Methodology

Ethno-ecological surveys were conducted during 2003 in the entire Spiti valley covering more than forty hamlets in twenty-two different localities with the elevation ranging between 3300 m to 4800 m above msl. Ethno-medical information on medicinal flowering plants in a variety of natural landscapes of the valley was collected by way of developing close contacts, through interviews and discussions with elderly people and local herbal practitioners (*Amchis*) following Martin (1995). Field visits were made with the "*amchis*" for identification and verification of the plant species in use. Well-structured questionnaires were used to document the medicinal and ecological information of the plants. The data is arranged in alphabetical order of the family name. The scientific name of each plant along with its herbarium number, local name, life form, part used, flower colour, locality, distribution, altitude, use and route of administration is given in tabular form (Table 1). New claims of unknown uses for the plants are indicated by an asterisk (*) mark. The range of distribution and geographical location of the plant resources were recorded by the Global Positioning System (GPS). Plant specimens of the herbs in use were subsequently identified and the nomenclature given by Aswal and Mehrotra (1994) was followed, at a Botanical survey of India (Northern Circle), Dehra Dun. The specimen collected were properly pressed, processed, mounted and after labelling, they were finally deposited in BSI (Dehra Dun) in an herbarium of the Institute of Himalayan Bioresource Technology (CSIR), Palampur, Himachal Pradesh with a unique voucher number.

4. Results and discussion

The ethno ecological survey identified 50 flowering plant species, belonging to 25 families used for medical purposes by the natives of the Spiti region. These are sparsely distributed in the

Table 1: Medicinal flowering plants in cold desert of Spiti valley

Scientific name	Local name	[a]Life form	[b]Part used	Flower colour	Locality, Distribution, Altitude (m)	Uses and route of administration
Apiaceae*Bupleurum falcatum* Linn (4275)	Thonpu	H	L,Fl	Yellow	Telimg 31°59′27″N-78°04′18″E 3704	Fresh parts are collected and dried under shade to prepare powder that is used to cure stomach pain
Apiaceae*Chaerophyllum villosum* Wall.ex DC. (4278)	Shigu zera	H	S,L	White	Sagnum 32°01′25″N-78°02′56″E 3851	Seeds and leaves are consumed as vegetable to cure cold and cough*. Also cures stomach pain due to cold
Apiaceae*Ferula jaeschkeana* Vatke (4300)	Thunak	H	S,R	Light yellow	Bhar 32°03′19″N-78°05′08″E 3908	Powder prepared from the dried parts is consumed daily to cure chest pain*
Apiaceae*Pleurospermum stylosum* Benth. ex Cl. (4310)	Yumo-dijen	H	Wh	Yellow	Sagnum Tud 32°01′25″N-78°02′56″E 3851	Tablets prepared from the powder extracted from dry parts help are consumed to ease the child birth and to reduce delivery pain*
Asteraceae*Scorzonera divaricata* Turcz. (4203)	Thunpu	H	Sh,L	Pale	Langza 32°16′27″N-78°04′26″E 4252	Decoction prepared at low temperature is used to cure jaundice and dysentry*
Asteraceae*Saussurea bracteata* Decne. (4309)	Pang-Poy	H	L,Fl	Pink	Sagnum Tud 32°01′25″N-78°02′55″E 3857	Juice of aromatic fresh parts as well as dry powder is used an ingredient in the medicine to cure food poisoning and it also acts as cooling agent*

AsteraceaeLactuca *rapunculoides* (DC.) Clarke (4315)	Gonpu	H	L	Purple	Sagnum 32°01′21″N-78°02′52″E 3900	Juice extracted from young leaves is taken with water to cure severe stomach pain and scrams*
Asteraceae*Artemisia scoparia* Waldst. and Kit (4334)	Phur-Mang	H	L,S	Pale	Lari 32°05′04″N-78°24′39″ E3345	Dried leaves and seeds are crushed slightly and boiled in water. Prepared decoction is used to cure ear and tooth ache and is applied with mustard oil*
Asteraceae*Erigeron borealis* (Vierch.) Simmons (4342)	Mathok-Lugmik	H	L	Light pink	Kaza-Tud 32°15′05″N-78°03′16″E 3703	Dry powder is used as an ingredient in the medicine to cure food poisoning*
Asteraceae*Taraxacum officinale* Wigg. (1513)	Sarchen-Metok	H	Wh	Yellow	Humna 32°15′50″N-78°01′28″E 3700	Powder prepared from the above ground parts is orally consumed to cure joint pains* and kidney problems
Asteraceae*Artemisia maritima* Linn. (1560)	Shoma	H	R	Yellow	Kaza 32°15′49″N-78°01′29″E 3650	Juice extracted from the roots is applied to cure burned skin*
Asteraceae*Erigeron multiradiatus* Benth. (4247)	Luk-Mik	H	L	Purple	Shagtal-gete 32°19′11″N-78°00′23″E 4197	Juice extracted from fresh leaves is taken with water to get relief from burning sensation in the stomach
Asteraceae*Tanacetum falconeri* Hook.f. (4284)	Khanpa	H	Wh	Light yellow	Mud 31°57′33″N-78°01′44″E 3808	Whole plant is dried and powdered to use in the medicines to cure joint pains and for blood purification*

Asteraceae*Lactuca macrorhiza* Hook.f. (4292)	Gonpu	H	Wh	Blue	Mud 31°57′37″N-78°01′44″E 3832	Powder prepared from the dry parts is used as ingredient in the medicine to cure stomach pain*
Asteraceae*Crepis flexuosa* (DC.) Benth (1534)	Homa Sili	H	Wh		Rangrik-Humna 32°15′32″N-78°01′28″E 3697	Juice extracted from the fresh parts of the plant cures jaundice*
Asteraceae*Cousinia thomsoni* C.B. Clarke (4208)	Tulse	H	R	Purple	Kibber 32°19′09″N-78°00′25″E 4300	Root powder about 2-2.5 g is taken regularly to cure swelling and joint pains
Boraginaceae*Arnebia euchroma* (Royle) Jonston (4205)	Khamed	H	R	Purple	Kibber 32°19′27″N-78°00′18″E 4102	Sap cures cough* and also acts as blood purifier and natural dye
Boraginaceae*Lin delofia anchusoides* (Lind.) Lehm. (4256)	Showarag	H	L	Dark Purple	Chang-Lung 32°15′53″N-78°03′48″E 4022	Fresh leaves are burned and then applied on cuts and wounds for immediate/ early healing
Brassicaceae*Aphragmus oxycarpus* (Hk.f. &Th.) Jafri (1520)	Chhuruk	H	Wh	Pink	Humna 32°15′49″N-78°01′29″E 3650	Powder extracted from the dry parts is used as ingredient to cure the urinary problems and to regulate the urine flow*
Brassicaceae*Lepidium latifolium* Linn. (4226)	Chulti	H	L,Fl	White	Rangrik 32°15′32″N-78°01′28″E 3697	Powder is consumed twice a day with water to cure joint pains

Capparidaceae *Capparis spinosa* Linn. (4330)	Martokpa	H	Sh	White	Tabo 32°05′41″N- 78°22′26″E 3366	Powder extracted from the dried green shoots is taken orally to cure urinary problems and liver infection.
Chenopodiaceae *Chenopodium botrys* Linn. (4301)	Sha	H	Sh, L	Red	Sagnum 32°01′25″N- 78°02′56″E 3851	Vegetable prepared from tender shoots and leaves cure severe headache
Crassulaceae*Sedum tibeticum* Hk.f. andTh. (4262)	Sholo	H	Wh	Pink	Khongme-Tud 32°13′57″N- 78°07′45″E 4800	Juice is used as an ingredient in the medicine to cure trouble (pain) in heart and lungs*
Elaeagnaceae*Hippophae rhamnoides* Linn. (4268)	Cherma	S	Br	Orange	Rangrik 32°15′10″N- 78°02′23″E 3620	Berries are used to prepare herbal tea to cure tuberculosis. It also cures cancer, jaundice and helps in blood purification*
Elaeagnaceae*Hippophae tibetana* Schlecht (4348)	Chhr-Tuan	S	Br	Orange	Takcha 32°26′53″N- 77°41′32″E 4123	Herbal tea is prepared from the berries which cures cough, congestion, jaundice and purifies the blood*
Ephederaceae*Ephedra gerardiana* Wall.ex Stapf (4232)	Tse	H	Sh	Red	Kee 32°15′20″N- 78°02′43″E 3658	Juice is used as an ingredient in the medicine to cure joint pains
Ephedraceae*Ephedra regeliana* Florin (4333)	Thak-Tse	H	Sh	Red	Lari 32°05′04″N- 78°24′39″E 3345	Juice extracted from fresh leaves and berries cures joint pains and also considered as general tonic for sick and weak people*

Fabaceae*Astragalus thomsonianus* Benth. (4210)	Satkar	H	Fl,L	Light pinkish	Langza 32°16′27″N-78°04′26″E 4252	Powder prepared is used to cure gastric trouble due to cold. Also reduces joint pains*
Fabaceae*Thermopsis inflata* Camb. (4213)	Tapa	H	S	Light Yellow	Hikkim 32°14′24″N-78°06′24″E 4516	Powder prepared from the dry ripened seeds cures joint pains and taken as general tonic by the weak people after long sickness*
Fabaceae*Oxytropis lapponica* (Wahl) Gay (1840)	Chhushin Darm	H	Wh	Bright Purple	Humna 32°15′49″N-78°01′29″E 3650	Used as ingredient in the medicine to cure joint pains*
Fumariaceae*Corydalis moorcroftiana* Wall. (4263)	Tongre-sewa	H	Fl,L	Yellow	Khongme-Tud 32°13′57″N-78°07′45″E 4800	Powder extracted from the dried parts is taken orally to cure bone marrow diseases*
Gentianaceae*Gentiana moorcroftiana* Wall. (4218)	Santik	H	Fl,L	Blue	Rangrik Humna 32°14′58″N-78°02′14″E 3661	Sap extracted is used to cure jaundice and for blood purification*
Gentianaceae*Gentiana tubiflora* (D.Don) Griseb (1514)	Chatik	H	Wh	Purple	Humna 32°15′50″N-78°01′28″E 3700	Juice extracted from the fresh parts is taken orally in the morning hours to cure jaundice*
Gentianaceae*Gentianopsis detonsa* (Rttb.) Ma. (1844)	Chatik	H	L,Fl	Blue	Humna 32°15′50″N-78°01′28″E 3740	Juice extracted from fresh leaves and flowers is taken orally to cure jaundice which also acts as blood purifier*

Geraniaceae *Geranium pratense* Linn. (4285)	Podh-Lo	H	Wh	Dark Purple	Mud 31°57′33″N-78°01′44″E 3808	About 2 g of powder extracted from the plant after drying under shade is taken orally with water to cure cough, jaundice and stomach disorders
Iridaceae*Iris ensata* Thunb. (4336)	Thema	H	S	Purple	Lari 32°04′37″N-78°25′33″E 3306	About 10 gms of seed powder is consumed orally to kill the worms in the stomach*
Lamiaceae *Dracocephalum heterophyllum* Benth. (4201)	Shim-thingli	H	L, Fl	White	Lanza 32°16′27″N-78°04′26″E 4252	Sap extracted from fresh leaves and flowers is used as an ingredient to cure eye diseases
Lamiaceae*Hyssopus officinalis* Linn. (4271)	Tiyanku	H	Fl,L	Dark Purple	Telimg 31°59′27″N-78°04′18″E 3704	Powder is used to cure cough, cold and an ingredient in the medicine to purify the blood
Liliaceae*Allium carolinianum* DC. (4207)	Lahud	H	Fl,L,Rh	Light pinkish	Lanza 32°15′60″N-78°03′56″E 4114	Paste prepared from the fresh parts is taken orally with water to cure stomach pain. Vegetable is also prepared to cure stomach disorders*
Orchidaceae*Dactylorhiza hatagirea* (D.Don.) Soo. (4217)	Wangpo-lakpa	H	Rh	Red	Rangrik Humna 32°14′58″N-78°02′14″E 3661	Powder of the tubers dried under shade is boiled in milk and taken with empty stomach to increase spermatogenesis* and considered as general tonic
Polygonaceae*Rheum emodi* Wall. ex Meissn (4313)	Tukshu	H	L,R	Pink	Sagnum-Tud 32°01′26″N-78°02′58″E 3832	Young leaves are crushed to extract juice and is taken with water to cure stomach pain due to internal injury

Polygonaceae*Rumex orientalis* Bernh. ex Schult (4293)	Shomang	H	L	Light red	Mud 31°57′35″N- 78°01′40″E 3836	Juice extracted from the fresh leaves is applied to cure burned skin due to hot water*
Polygonaceae *Polygonum tortosum* Boiss. (4215)	Nayalo	H	Fl,L	Pinkish White	Hikkim 32°01′25″N- 78°02′56″E 3851	Used as ingredient in the medicines to cure stomach and intestinal troubles. Also acts as cooling agent and taken orally to cure jaundice*
Ranunculaceae *Aconitum rotundifolium* Kar. and Kir (4264)	Bonkar	H	Wh	Purple	Khongme-Tud 32°13′57″N- 78°07′45″E 4800	Juice extracted from fresh parts is used as medicine to cure jaundice and also helps in blood purification*
Rosaceae*Rosa webbiana* Wall.ex Royle (4233)	T-siya, Seva	S	Fl, fr	Pink	Kee 32°15′20″N- 78°02′43″E 3658	Juice is orally consumed regularly to cure jaundice. Also considered as rejuvenating tonic
Rubiaceae*Rubia tibetica* Hook.f. (1893)	Thak-Thokpa	H	Br,S	Yellow	Kee 32°15′20″N- 78°02′43″E 4000	Ripened fruits along with seeds are used as ingredient in the medicines to purify blood*
Saxifragaceae *Bergenia stracheyi* (Hk.f.& Th.) Engl. (2728)	Gatikpa	H	R	Pink	Sagnum-Tud 32°01′25″N- 78°02′55″E 3857	Paste is applied in mouth to cure ulcer/ blister
Scrophulariaceae *Pedicularis punctata* Decne (4245)	Lungru-serpo	H	Fl,L	Purple	Gete 32°18′25″N- 78°01′38″E 4415	Powder prepared from the dry parts is used as an ingredient to prepare medicine to control blood pressure*
Solanaceae*Hyoscyamus niger* Linn. (4344)	Thuglang	H	S	Yellow	Kee-Humna 32°15′13″N- 78°02′48″E 3625	Seed smoke is used to cure toothache which probably acts as wormicide
Tamaricaceae*Myricaria squamosa* Desv. (4305)	Humbu	S	Sh,L	Light Pink	Bhar32°03′19″N- 78°05′08″E3908	Juice extracted from fresh young shoots and leaves is used to prepare medicine to cure joint pains*

[a]Part used: Wh-Whole plant, L- Leaves, Fl- Flowers, R- Roots, S- Seeds, Rh-Rhizome, Fr- Fruits, Sh- Shoots, Br-Berries
[b]Life form: H- Herb, S- Shrub
New medical remedies reported by *

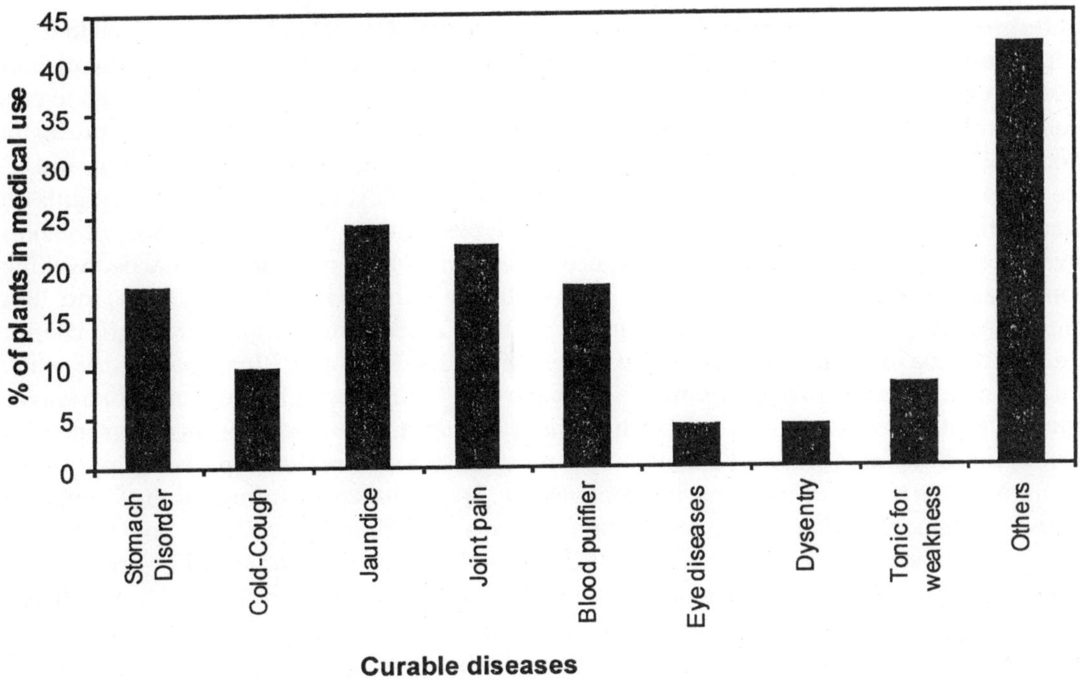

Fig. 1. Percentage of plants used in various medical remedies

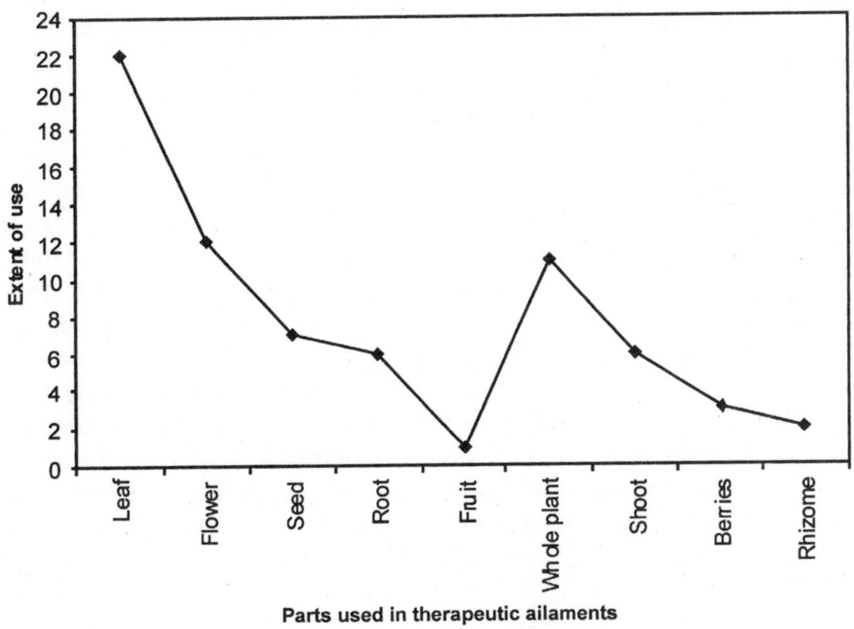

Fig. 2 Plant parts used in medical remedies.

altitudinal range of 3300 m to 4800 m above mean sea level. The data obtained is given in table-1. It is observed that the maximum number of species, which contributes to the ethno medicines, belongs to family asteraceae. Of 50 documented medicinal plants, about 18% are used to cure stomach disorders, 10% cure colds and coughs, 24% are used to cure jaundice, 18% to purify the blood, 22% to cure joint pains, 4% are applied to cure eye diseases and dysentery each, and 8% are taken orally as a general tonic to check weakness (Fig. 2). About 42% of the total plants are used by the "*amchis*" or local healers to cure various types of other diseases viz., kidney problems, fever, headache, food poisoning, burns, cancer, chest pain, blood pressure, toothache, swelling, bone marrow disorders, tuberculosis, child delivery pain, urinary, heart, liver and lung complaints, cuts and wounds. Most of the species have more than a single therapeutic use and are found to be of an herbaceous nature except *Myricaria, Rosa,* and *Hippophae* species which are perennial woody shrubs. Among various parts of the medicinal plants used like leaves, flowers, roots, seeds, rhizome, fruits, shoots and berries; leaves and flowers are used in the maximum of the cases and are considered the most efficient and effective parts, mainly due to the high concentration of bioactive ingredients (Fig. 3). The recorded information of medicinal remedies was compared with the earlier published work of ethnobotany in the Spiti valley and it was found that the present study throws light on 35 new claims (indicated by an asterisk * mark) in Indian ethnomedicine to treat different diseases. The information collected from different sources regarding the traditional medicinal use of plants, was re-examined by consulting the important works pertaining to Indian medicinal plants and ethnobotany such as the Wealth of India (Anonymous, 1948-1992), Indian Medicinal Plants (Kirtikar and Basu, 1975), Dictionary of Indian folk medicine and ethnobotany (Jain,1991), Flora of Lahaul-Spiti (a Cold Desert in North West Himalaya) (Aswal and Mehrotra, 1994), Medicinal and aromatic plants of Himachal Pradesh (Chauhan, 1999) and Ethnobotany of cold desert tribes of Lahoul-Spiti (N.W.Himalaya) (Sood et al., 2001).

5. Conclusion

The present study indicates that the study area harbours a remarkable diversity of flowering plants with medicinal properties. It is recommended that photochemical and pharmacological studies need to be carried out to confirm the validity of properties attributed to these species. It was observed that it was mostly the elderly people and the healers (*amchis*) who possess substantial knowledge of the plants and their uses in comparison with the younger generation because of gradual socio-cultural changes. Therefore awareness is very important among the young masses regarding the importance of the herbal wealth of the region. Rapid documentation of the plant resources and their uses is needed immediately because of the fast erosion of the vital ethnic knowledge regarding the traditional uses of the local flora, before its is lost forever. Similar kind of work should be carried out throughout the Himalayan region that definitely will contribute to maintain the systematic database on medico-ethnobotany and related ecological features of the diverse and unique bioresource of the nation.

References

1. Adhikari, R and K Adhikari (eds.) 2003. Farmers' Rights to Livelihood in the Hindu-Kush Himalayas. SAWTEE, Kathmandu, Nepal.
2. Anon. 2003. Framework for an Action Plan on 'Wetland Conservation and Wise Use in the Himalayan high Mountains Ramsar and ICIMOD, Kathmandu, Nepal Available at: http://www.ramsar.org/key_himalayan_plan.htm
3. Bhatnagar, Y. V, V. B. Mathur, T. McCarthy. 2002. A regional perspective for snow leopard conservation in the Indian Trans Himalaya. Envis Bulletin on Wildlife and Protected Areas 2002(1):57-76.
4. Chundawat, R. S and G S Rawat. 1994. Indian Cold Deserts: A Status Report on Biodiversity. Wildlife Institute of India, Dehradun.
5. Gupta, A. 2000. Governing Biosafety in India: The Relevance of the Cartagena Protocol. Belfer Center for Science and International Affairs (BCSIA) Discussion Paper 2000-24, Environment and Natural Resources Program, Kennedy School of Government, Harvard University.
6. Kala, C. P. 2001. A Study of Traditional Health Care & Medicinal Plants. In Anon. (2001)
7. Conserving Bio-Diversity in the Trans-Himalaya: New Initiatives of Field Conservation in Ladakh.
8. First annual Technical Report (1999-2000), Wildlife Institute of India, International Snow Leopard Trust and US Fish and Wildlife Service.
9. Maheshwari, G. 2000. A survey of bio-indicator community of Chironomidae (Diptera) of North-West Himalayan lakes. National consultation on conservation of high altitude wetlands in Ladakh. WWF-India at Leh, July 2000.
10. Mani, M. S. 1962. Introduction to High Altitude Entomology. Methuen and Co., Ltd., London.
11. MoEF. 2006. National Environmental Policy. Ministry of Environment and Forests. Government of India. Approved by the Union Cabinet on 18 th May 2006.
12. Rodgers, W. A and H. S. Panwar. 1988. Panning a Protected Area Network in India. Wildlife Institute of India, Dehradun. Government of India. 341 pages.
13. Wikramanayake, E et al. 2006. The WWF Snow Leopard Action Strategy for the Himalayan Region. WWF, Bhutan.

4

Public - Private Partnership in Forest Seed Systems: Key to rapid improvement of biodiversity, environment and sustainability

Dr. Preeti Rajpal Singh and Dr. Antonia Agmata Paliwal

Introduction

Today, more than ever before there is a need of addressing the issue of forestry germplasm systems, the pillars of biodiversity, environment and sustainability. The unavailability of forest germplasms is a major bottleneck in any assisted plant colonization (APC), assisted greening of the environment (AGE) and enhanced plant productivity (EPP) efforts to combat rapid loss of genetic base, degradation of environment and unsustainable practices.

Forestry Germplasm Systems (FGS) are more complex than Agricultural and Horticultural Germplasm Systems. The majority of forestry species are undomesticated, hence, their FGS are less understood, underdeveloped and neglected. The people dealing on FGS in trade are unorganized, secretive and unidentifiable. The main reasons behind these are: 1) FGS vary according to species and there are too many to deal with in the forest ecosystem, 2) FGS are located in remote areas, 3) FGS are inaccessible due to administrative restrictions (seed sources are publicly owned), and 4) uncertainty of monetary returns in dealing with FGS. Consequently, we are confronted with the perennial problems of dubious forestry germplasm quality, unreliable germplasm supply of desired species, erratic pricing and smuggling/illegal trading in the developing countries. These are major bottlenecks in our efforts to accelerate biodiversity conservation, rehabilitation of degraded environments and sustainability.

Nevertheless, we need forest germplasms urgently irrespective of the problems associated with FGS. By forging partnership between private and public sectors on Forest Germplasm Systems (FRS) these challenges can be addressed. As a result, the availability and accessibility of the right species, quality and quantity of forest germplasm are assured wherever and whenever they are needed. The cooperation of these two sectors, leads to a win-win situation. One fundamental factor to ensure the success of partnership between private and public sectors in

FGS however, is the development and adoption of Forest Reproductive Material Certification standards in a country. In the Certification standard, categories of reproductive materials are identified and the private sector can choose which category/categories with which they would like to be involved. This paper will outline the successful Certification Standard in developed countries and the proposed one in India.

Characteristics of Forest Germplasm Systems (FGS)

The majority of the forestry species have a long life cycle. Most of them are undomesticated. Their reproductive biology and propagation technologies are less understood. Research and development on FGS requires the prioritization of the species' use due to the vastness of species present in the forest ecosystem and the limited time and other resources we have. R & D on FGS can be tailored for biodiversity conservation, environment rehabilitation or sustainability. Presently, the private sector involvement is leading the R & D of FGS with a few species such as *Poplars, Eucalyptus and the few others* to increase productivity and quality in India.

The ownership and control of FGS are generally absent in the private sector because the parent material (seed sources) are owned by the government which prohibit their collection and use, unlike the agriculture or horticulture GS where farmers own the parent material and make decisions as to what to do with the germplasm. There are three major methods employed in FG procurement: 1) institutional (Forestry Staff), 2) contractual (Forestry Departments award contract to a group) and 3) individual or group with or without a permit.

The majority of the collected forest germplasm is used by the Forestry Department themselves. In India, annually 80% of the collected seeds in Method 3 were used by the Forestry Department. The remaining 20% were sold to NGOs and private farmers (Paliwal, 2007) irrespective of quality.

Forestry seed (forest germplasm) technology is quite advanced in India, Malaysia and Thailand. However, no developing countries in Asia have as yet adopted a Forest Reproductive Material Certification standard. Many of these countries however, have drafted the Certification Framework with the assistance of DANIDA in the late 1970s. Through the FORTIP (Forest Tree Improvement Project) in 1995 funded by the UNDP and FAO, eleven countries in South and Southeast Asia were assisted to improve their forest germplasm by establishing clonal/seed/seedling orchards, seed production areas, seed stands. Many of these established seed sources are still in existance and managed today (Pers. Comm.). Unfortunately, Certification of Forest Reproductive Materials has not been integrated in the FORTIP project. As a result, the 11 countries were not able to benefit fully from the reproductive materials derived from the seed sources they have improved. Had Certifcation been adopted by these countries, their governments would have been able to create private partnership in various categories of forest reproductive materials.

What is the Certification of the Forest Reproductive Material Scheme and why is it important in building public-private partnership in FGS ?

The object of the certification of tree seed and plants, is to maintain and make available to the public sources of seeds, plants and other propagating materials of superior provenances and cultivars so grown and distributed as to ensure the genetic identity and high quality of the seed and plants (Mathews *et al.*, 1964). Superior provenances are identified by means of provenance tests. The duration of these tests depends on the rate of growth and the length of the economical rotation for each species but, in general, it must be accepted that provenance tests are essentially long-term experiments. The need to provide safe prescriptions for the movement of seed and plants so that expensive losses in wood production can be avoided has led foresters in countries with large variations of site and climate to devise regions of provenance or zoning schemes defined in terms of latitude or distance and elevation. The rapid progress made with selection and breeding in the genus *Populus* has been largely due to the easy propagation of many species and hybrids by means of stem cuttings. The efforts of forest geneticists and tree breeders will be largely nullified unless certification procedures are available to maintain the genetic purity and quality of the seed and plants.

In the developing countries, the forest seed trade which can be promoted by the private sector is almost non-existent. This is primarily due to the lack of access to the seed sources and the practice of distributing seedlings free of cost by the government. Forest seed trade however, can be a driver to accelerate our work in biodiversity, environment and sustainability, if given recognition. There are the same advantages in a greater development of the private forest tree seed trade as there are in private trade in agricultural seed or in any other business enterprise. There would seem little doubt that the healthy and successful competition between private and state-enterprise organizations, in countries where these exist, in forest tree seed, must depend on the adoption and careful adherence to a universally acceptable seed certification system.

An Outline of National Forest Certification Scheme

The elements of a comprehensive national certification scheme for forest seed and plants are as follows:

- Inspection of the seed source by a qualified professional forester before pollination of the seed crop. At this time (which is from 6 to 21 months before seed collection) the quality of the seed trees, incidence of inferior trees and the effectiveness of the isolation can be checked. Many forest trees do not produce seed every year and, after the first inspection, subsequent re-inspections are made only in years when the seed is to be collected.
- Assessment, if possible, of the cone or fruit crop by a qualified professional forester at a stated time (about 90 days) before collection begins.
- Collection of the cones or fruits by a registered seed collector; extraction, cleaning and packaging of the seed at a registered seed extraction plant; and storage of the seed at a registered seed store. The records of collection, processing and storage are made available for inspection and the labels must conform to the minimum requirements.
- Testing, under the rules of the International Seed Testing Association, of an adequate sample of the seed at an official seed testing station.
- Sowing of the seed in a registered nursery where labels and records satisfy minimum requirements. Inspection of the seedlings and transplants should be made by a qualified professional forester before they are lifted and dispatched.

Categories of Reproductive Material

Indo-Danish Project recognized the following four broad categories of forest reproductive materials:

Source-Identified Forest Reproductive Material

The two requirements of this category are (i) the seed zone where the reproductive material is collected and the origin of the basic material (which may be indigenous or non-indigenous) shall be defined and registered as per guidelines notified by the Designated Authority (ii) the seed shall be collected processed and stored and plants shall be raised under the control of a Designated Authority.

Selected Forest Reproductive Material

The three requirements of this category are (i) the provenance or seed zone where the reproductive material is collected and the origin of the basic material (which may be indigenous or non-indigenous) shall be defined as per guidelines notified by the Designated Authority (ii) the reproductive material shall be derived from basic material which conforms to the requirements as may be notified by the Designated Authority; (iii) the seed shall be collected, processed and stored and plants shall be raised as per guide lines notified by Designated Authority.

Forest Reproductive Material form Untested Seed Orchards

The three requirements of this category are (i) the seed zone from which the components of the seed orchard came and the origin of these components (they may be indigenous or non-indigenous) or breeding records shall be defined and registered as per guidelines notified by the Designated Authority. (ii) The reproductive material shall be derived from the basic material conforming to the requirements as may be notified by the Designated Authority; (iii) the seed shall be collected, processed and stored and plants shall be raised as per guidelines notified by the Designated Authority.

Tested Reproductive Material

The four requirements of this category are (i) the place where the reproductive material is collected and the origin of the basic material (which may be indigenous or non-indigenous) shall be defined and registered by the Designated Authority in the manner to be notified by it (ii) the genetic superiority of the basic material should have been proven or shall be proved by field tests. (iii) The results of the tests shall be recorded by the Designated Authority; (iv) the seed shall be collected, processed and stored and plants shall be raised as per guide lines notified by the Designated Authority.

List of Approved Basic Materials

The Designated Authority and the Regional FRM Certification Committees shall issue guidelines to enable each State Government to compile and furnish the following information to Designated Authority in respect of FRM sources:-

i. List of seeds available in different seed zones (source identified material).
ii. List of seed production areas
iii. List of plus trees.

iv. List of seed orchards.

v. Test results of progeny and provenance trials.

Minimum Requirements for Certification of Seed and Plants

The minimum requirements for certification are as follows (*Mathews et al.*, 1964):-

Previous history of the seed source

A seed stand or seed production area may be formed in a natural stand (established by natural regeneration) or plantation (established by planting or sowing) of indigenous, introduced or unknown origin.

Isolation from foreign pollen

The requirements for isolation of seed production areas and seed orchards from large stands of inferior trees of the same or closely related species and for the width of isolation strips vary somewhat, with species and country, but a generalized plan for seed production areas and seed orchards is: 1,000 meters from a large stand of inferior trees and 100 to 150 meters from the nearest inferior tree - thus fixing the width of the isolation strip.

Field inspection of seed trees, seed crop and plants

Inspection of seed sources, seed production areas and seed orchards is most often made by the state forest service or by qualified professional foresters nominated by the certifying agency. The standards for seed production areas are based on the number of vigorous, well-formed and healthy seed trees per hectare and the removal of inferior trees. Inferior trees are marked by the inspector and if possible felled while he is in the area, but the rate at which inferior trees are removed is often determined by local silvicultural requirements and re-inspection may be necessary when the fellings are completed. During seed collection and processing, the cones/ fruit and seed must be handled separately by seed lot.

Standards for germination and purity

The methods of determining purity, germination and the moisture content and the standards of health supervision with regard to seed-borne pests and diseases are generally based on those set by the Forest Tree and Shrub Seed Committee of the International Seed Testing Association (ISTA) Standards of germination and purity for purposes of sowing the seed and getting a good yield of plants are available for a very wide range of species.

Labels and certificates

In 1950, FAO and IUFRO (FAO, 1952) jointly prepared an international certificate of origin and quality for tree seed and plants and this carries a large amount of information. This certificate has great value in exchanges of small quantities of seed for provenance research, but something simpler is needed for bulk trading in tree seed and plants. Such a certificate would bear the following items (Mathews *et al.*, 1964):-

Certifying agency
Consignee (name and address)
Sender (name and address)
Species (Latin name)

Subspecies variety or cultivar name
Category (source-identified, selected, certified)
Provenance (reference number, place or region, elevation)
Gross weight of packages
Contents (seed or plants, quantity)
Disinfections treatment (date, place, method).

Reference numbers

Comprehensive yet simple methods of identifying the origin of seeds and plants by means of reference numbers or letters are of great assistance in administering a comprehensive certification scheme. FAO has adopted the universal decimal classification system by which each country is identified by a permanent number consisting of three digits The OECD has a similar system.

How biodiversity, environment and sustainability be promoted through public-private partnership and Certification Scheme

The demand for FG is unpredictable especially for biodiversity conservation and the greening of a degraded environment. A good estimate however, is available in India on some economically known species as shown on the tabulated figures below (Table 1). It can be safely estimated that the country has an annual seed requirement of about 10,000 tonnes.

Table 1
Annual Seed Requirement of the Country for Planting Programmes

Species	Common Name	Quantity (Tonnes)
Abies and Picea spp.		8.0
Anacardium occidentale	Cashewnut	6.4
Acacia spp.		62.5
Bombax ceiba	Semal	1.2
Bamboos	Bans	68.7
Cedrus deodara	Deodar	7.5
Casuarinas equisetifolia		12.8
Dalbergia latifolia	Rose wood	3.6
D. sissoo	Shisham	11.8
Dipterocarpus Spp.	Gurjan	336.0
Eucalyptus Spp.		13.8
Gmelina arborea	Gamhar	29.7
Pinus Spp.	Pine	32.0
Quercus Spp.	Oak	29.0
Shorea robusta	Sal	846.0
Santalum album	Sandalwood	10.8
Tectona grandis	Teak	1760.0
Grasses		27.0
Other Species		6470.0
TOTAL		**9736.8**

Source: Khullar et al. (1992)

Through a Certification Scheme, the input of private sectors becomes valuable in the timely procurement of the above FG without sacrificing the quality of the material and destruction of the seed sources. Natural seed sources are not only scattered but also shrinking rapidly. Therefore, there is an urgent need to develop identified seed sources such as plus trees, seed stands, seed production areas and seed orchards and to establish central seed collection and distribution in each State in partnership with the private sector.

The Government of India during the seventies had sanctioned a project known as "Indo-Danish Project on Seed Procurement and Tree Improvement". After detailed deliberations and research, the project issued an agreed document for the implementation of forest seed certification by the states namely (i) Certification of forest productive material in India (Revised Scheme, 1979). Subsequently one more document i.e. Seed Zonation followed in India (Revised Scheme, 1979 was also issued). Though these two official documents were finalized in consultation with all the state forest departments for implementation, yet unfortunately, in the absence of any central interventions or a forest seed law, most of the forestry departments did not implement the same and were left far behind in the very important field of tree improvement and certification of the FRM. Current forestry trends in India also give high importance to non-wood forest products and medicinal / aromatic plants. The approach for producing reproductive material for such categories will also be more or less the same as for timber species with certain modifications.

Producers and users of FRM cannot be certifiers of the same if the FRM is meant for sale and trade. Further, as explained earlier, issues related to the production and certification of forest seed/ FRM are very different from seed for agricultural crops. Hence, we recommend an enactment of a separate legislation by the Parliament of India, and for the setting up of a suitable institutional framework for the certification of FRM. Another requirement is to frame appropriate rules for implementing the provisions of the proposed Law including the setting up of an autonomous Designated Authority for the purpose.

The Designated Authority shall have the overall responsibility of implementing the proposed FRM in accordance with the provisions of the said Bill and rules to be made thereunder. The Authority may constitute the desired number of Committees as it may deem appropriate to exercise such powers and perform such duties as may be delegated to them by the Designated Authority

Along with the Certification Scheme, Forestry Experts in India are recommending a creation of a National Board of Forest Genetic Resources at ICFRE, Dehradun with suitable mandate on the lines of NBPGR for various issues related to Forest Genetic Resources including ex situ conservation of endangered and rare species of forest flora, authentic seed samples, registration of clones and supply of basic clonal planting stock to prospective producers of certified clonal planting stock etc. This would allow private sectors to come forward and become registered producers and traders of FG.

Implications of the Seed Bill of 2004 in India in Public-Private Partnership

The Forestry Department supplies nursery-raised plants to the public on a large scale. For this the reproductive material will have to be certified according to the new law. The basic principle of seed certification or reproductive material is that it should be certified by an independent and autonomous certification agency that is different from the producer or seller of the seed. It will thus be appropriate to establish a certification agency that can deal with the forest seeds.

Adequate attention must also be given to poor farmers (kisan nurseries), whose requirements will usually be distinct from those of larger scale plantation programmes. In this context, the small bag approach developed under a DANIDA project in Nepal deserves a serious consideration. Small farmers cannot afford to travel long distances and need only small amounts of seed. Therefore, the genetically improved and high quality seeds need to be brought to the farmers. High quality seed produced by national/state tree seed improvement programmes and packed in small bags can be distributed using existing distribution networks for farm and horticultural commodities (Moestrup, Schmidt and Nathan, No date).

Enforcement of the bill and the Private Sector

According to the new seed bill no seed can be sold without certification. Some forestry reproductive material is produced / multiplied by the private parties. Among the species are nursery plants of poplar and eucalyptus clones and nursery plants of Casuarina, Acacia, Melia etc.. Private dealers in forestry seeds also sell forestry seeds in large quantities but the source of such seeds is normally not known and mostly these are procured from forests illegally and are of dubious quality. The new seed act will put an end to this illegal and unprofessional procurement, as the dealers will have to specify the source of procurement before certification. Seed sold by the private parties will also be a subject matter of inspection by seed inspectors appointed under the proposed Act. As discussed above, State Governments may designate suitable officers as Seed Inspectors. Similarly reproductive material of a medicinal and aromatic species of forest origin will also come under its preview. This will pave the way to availability and accessibility of FG to workers in biodiversity and environmental improvements.

A large number of forest seed species are grown in the private areas specially the seeds and the reproductive material of medicinal and aromatic plants. Minimum standards have not yet prescribed for these seeds. National institutes like the ICFRE, IBP GR, ICAR etc. may be assigned this task.

Opportunities for Private Sector

Once the Certification of Forest Reproductive Materials Scheme is adopted in a country or region, the public and private sectors can forge partnership in any FGS they are interested in such as:

FGS for Biodiversity

Private sectors composed of tribal groups would like to deal on the Source-Identified Forest Reproductive Material as they are closest to the area of collection and they have traditional knowledge of FGS. The materials in this category requires proper identification of source only. Hence, tribal communities can be engaged as private partners in source-identification, production and trading and the management of an area. The first step however, is that the private sector should recognize the rules set by the Certification Agency and register with the Agency to become a legitimate partner.

FGS for Environmental Use

The private sector can partner with respective government authorities in the production, trading and seed source management in two categories Source-Identified and Selected Forest Reproductive Material.

FGS for Sustainability

This is where the public-private partnership has been highly successful as an improvement of FGS is highly profitable. The private sector takes the initiative of the R & D categories of Untested and Tested Seed Orchards and multiplies them using advanced technology in tissue culture and Molecular Markings.

In India, an important role of the private sector in the Forest Germplasm Systems (FGS) is evident in the germplasm improvement of some species of ***Poplars, Eucalyptus, Acacias, Casuarinas and Leuceana.*** The FGS improvement initiative was started mainly by large wood-based corporations such as WIMCO, ITC, BILT and some Paper Mills to serve their own needs of quality planting materials. The high productivity of the improved germplasm of both poplar and Eucalyptus (20-50 m^3/ha/year compared with 4-10 m^3/ha/year) and better wood quality (Anonymous, ICFRE, 2007). These corporations are tied up with individual farmers using their improved germplasm. This paved the way to extensive commercial agroforestation and farm forestry plantations among small land holders in India. Soon the quality of germplasm of these species became uncertain as the germplasm production has been taken up by individuals outside these corporations. In the absence of germplasm the standard and a third party, designated to check the authenticity of these materials, has found that individual farmers have been cheated, which they realize only after several years (at least 3-5 years after planting).

In conclusion, the partnership of the private and public sector after the adoption of Certification standards of Forest Reproductive materials will decrease the genetic loss in its natural habitat as more quality germplasm is available for use in the biodiversity conservation and seed sources which will be documented and monitored as per the scheme. Similarly, the greening of the environment with an appropriate species becomes faster and easier if planting materials of good quality are readily available from a registered trader. More input in R & D of FGS is expected with the adoption of the Certification Standard as the private sector can freely trade/export its material across the globe, hence, promoting the sustainability of resources.

Note: Most of the materials on Certification of Forest Reproductive Material is from the Anonymous Paper Document, ICFRE, 2007 for India (as example) and OECD Document 2007 (as model).

References

1. Anonymous (ICFRE, 2007). Preliminary Report of Policy Sesearch Study on "Need for InstitutionalMechanism for Certification of Planting Stock Material of Forest Seed and Vegetative Origin"
2. Anonymous (1979) Indo-Danish Certification of Forest Reproductive Material in India.
3. Moestrup, S., L. Schmidt and I. Nathan (No date). *Guidelines for Distribution of Tree Seeds in Small Bags Small Quantities and High Quality*. Forest Landscape. Faculty of Life Sciences. University of Copenhagen.
4. Seeds Bill 2004. www.seednet.gov.in
5. FAO (No date). *Statistics on the production and trade of forest seeds and other forest plant material.*
6. Khullar P., R.C. Thapliyal, B.S. Behniwal, R.K Vakshasya and Ashok Kumar (1992). *Forest Seed*. ICFRE, Dehra Dun. 409 pp.
7. OECD (2007). *OECD scheme for the certification of forest reproductive material moving in International trade*. Organisation for Economic Corporation and Development. Trade and Agricultural Directorate. Paris, 20 June, 2007. www.oecd.org/agr/forest
8. Paliwal, A.A. (2007). Survey on Forestry Seed Trade in India. (Unpublished)

Annexure 1

The OECD Certification Scheme (As Model)

The OECD Scheme for the Control of Forest Reproductive Material Moving in International Trade, first adopted in 1974, is a certification tool to facilitate international trade in forest seed and plants. The Scheme reflects the wish for governments to have these materials correctly identified with a view to minimizing uncertainty in achieving successful afforestation.

The object of the OECD Scheme- Certification of Forest Reproductive Material Moving in International Trade, 2007 hereinafter called "the OECD Forest Seed and Plant Scheme" or "the Scheme", is to encourage the production and use of seeds, parts of plants and plants that have been collected, transported, processed, raised and distributed in a manner that ensures their trueness to name. Material covered by the Scheme is intended for use in a variety of forestry purposes (OECD, 2007). The scheme is open to all members of OECD as to well as any member of the United Nations, its Specialized Agencies or the World Trade Organization desiring to participate therein. Currently, there are 22 participating states. The list includes countries of Europe, U.S.A., Canada, Madagascar and Rwanda.

Afforestation may relate to wood production (which remains one of the major uses of forests), soil protection and fighting erosion, producing trees for ornamental or recreational purposes (Christmas trees); landscaping, etc. More than 250 species of trees are currently eligible for certification under the OECD Scheme which covers an area of more than 13 million hectares. Forest stands, tree seed orchards and tree clones are approved by the participating countries as the basic material for harvesting the reproductive material, i.e., seeds or grafts. Different OECD labels are used according to the category of material. The labelled material is then recognised internationally as quality guaranteed and of certified origin or source.

OECD Categories of Reproductive Material

Under the Scheme, reproductive material falls into categories. These categories are applied to the reproductive material, which is derived from approved basic material like the EU scheme basic. Basic material must have the approval of the Designated Authority.

There are two categories recognised in the Scheme under which reproductive material can be certified, namely: Source-identified this is the minimum standard permitted in which the location and altitude of the place(s) from which reproductive material is collected must be recorded; little or no phenotypic selection has taken place. Selected the basic material must be phenotypically selected at the population level.

There are two types of basic material recognised in the Scheme from which reproductive material can be collected, namely: Seed source and Stand.

Other categories, to be named "Qualified" and "Tested", that can involve other types of basic material (seed orchard, parents of family (ies), clone, clonal mixture) are under consideration for future extension of the Scheme. Many countries, especially from Africa and South America have expressed their interest in the OECD Scheme.

Approval of Basic Material

The unit of approval is the basic material. Each unit comprises a single entry in the National Register. Prior to approval, the basic material (except for basic material intended for the production of reproductive material to be certified as source-identified) will be inspected by the Designated Authority. When approved by the Designated authority, it will be maintained under its supervision until the approval is withdrawn. The approval of basic material shall be withdrawn if the minimum requirements are no longer fulfilled. RE-inspections must be made at intervals decided by the Designated Authority.

Registration of Approved Basic material

National Register

The Designated authority must establish and maintain a National Register in which each unit of approved basic material is recorded. The National register will contain full details of each unit including management, site and administrative details. A map or plan must be made available on request.

Inspection

The Designated Authority must control all categories of reproductive material, at least by random checking during collection, processing, storage, raising, labeling and sealing.

Method of Operation of the Scheme

Designated Authority

(a) The government of each country participating in the OECD forest seed and plant scheme will designate the authority to implement the scheme in that country.

(b) The name and address of the National Designated Authority and any changes in its designation will be circulated by the Secretariat of the OECD to all countries participating in the scheme.

Review and Co-ordination

(a) The operation and progress of the scheme will be reviewed as necessary at meeting of representatives of the National Designated Authorities. These meetings will report on the working of the scheme and make such proposals as are deemed necessary to the committee for agriculture of the OECD.

(b) The necessary co-ordination of the operation of the scheme at the International level will be ensured by the Secretariat of the OECD.

Annexure 2

Forest Seed Certification in India (An Example)

The Indo-Danish Project on Seed Procurement and Tree Improvement, Hyderabad, formulated a scheme entitled "Certification of Forest Reproductive Material in India". The scheme was based on OECD Standards, Scheme and Guides. It came out with a set of rules, which were to be strictly complied with, in the collection, transportation, processing, storage, sampling, labeling and sealing of seed for distribution.

The broad programme of seed certification, as formulated by the Project is discussed below.

Categories of Reproductive Material (Minimum Standards)

The scheme recognized the following four categories of stands for seed collection, which have been listed in accordance with increasing genetic control of seed quality. The term "Reproductive Material", in the text, refers to cones, fruits and seeds intended for the production of plants, parts of plants, stem, leaf and root cutting, scions and layers intended for the production of plants; plants raised by means of seeds or parts of plants; it also includes natural regeneration. Basic material has been defined as the stands and seed orchards for reproductive material produced by sexual means and clones for reproductive material produced by vegetative means (Khullar et al, 1992).

Source-identified Reproductive Material.

This category is defined and registered as reproductive material of indigenous, and non-indigenous, origin collected from (i) an area falling within the bounds of a Seed-Zone as demarcated on the seed map and approved by the Designated Authority and (ii) stand, within the control of State Forest Departments.

Selected Reproductive Material

Reproductive material approved for collection or rising as selected reproductive material is derived from selected stands and cultivars, which satisfy the minimum requirements for basic material.

Reproductive Material from Untested Seed Orchards

i) Reproductive material approved for collection or establishing as Untested Seed Orchard Reproductive Material, is derived from untested seed orchards, which satisfy the minimum requirements for basic material as specified under the head Minimum Requirements;

ii) Reproductive material derived from single species seed orchards, for conservation of gene resources or breeding by selection, for the criteria set out under the head 'Minimum requirement', is included in this category. Material from such orchards, that are from a single region or provenance, can be included in the category selected at the discretion of the Designated Authority;

iii) Reproductive material derived from seed orchards, established to produce species, hybrids or provenance hybrids, can only be included in this category when early tests show that the objectives of the orchards have been attained.

Tested Reproductive Material

i) Reproductive Material approved for collection or raising as Tested Reproductive Material, shall originate from seed orchards stands or cultivars whose genetic superiority to appropriate standards, in one or two characters important to forestry, has been proved by comparative tests conducted in specified environments. Environments for such comparative tests are set out under the head 'Minimum requirement';

ii) Tested reproductive material can also be obtained from trees grown from a representative sample of the original seed lot, which was used to establish the trials, provided that the Designated Authority is satisfied that this material is similar in genotype to the original basic material.

iii) Superiority can be certified only in terms of the environment in which the test has been carried out. Moreover, superiority revealed by a test can be certified as characteristic only upto, and including, the most advanced age at which it has been observed in the test.

iv) Before approval, a check must be made to ensure that the basic material is not significantly different from the source from which the reproductive material included in the test was obtained, adequate measures have been taken to reduce pollination from outside sources, and its boundaries are clearly marked in the field. Seed orchards must fulfill the essential requirements.

Approved Basic Material

In each state in India, a designated authority was to take action to compile and furnish the following lists in respect of reproductive material that could be collected there from and furnish to the coordinating agency, which would then prepare a national register to incorporate information obtained from the states and circulate the same to other states for their information and use.

1. List of seeds that are available in different seed zones (Source Identified Material).
2. List of Seed Production Areas.
3. List of Plus trees.
4. List of Seed Orchards.
5. Test results of progeny and provenance trials, as and when conducted and analyzed.

The Indo-Danish Project had prepared various proformae for recording and submitting the above information.

Minimum Requirements for Forest Reproductive Material.

The scheme sets out the minimum requirements for forest reproductive material separately for a) selected reproductive material, b) seed orchards and cultivars, and c) tested reproductive material. This is on the lines of the OECD and EU schemes.

Organization and Issuing Certificates

The seed certification will be implemented in the states by designated authority responsible to the government and will usually be the Conservator of Forests, Research/Development or equivalent officer.

The project also envisaged the constitution of working groups to prepare a detailed programme for future work on individual, or a group, of species occurring in the same region for example, chir, deodar, fir, spruce etc. in the Himalayan region.

Seed Testing

Material in this section is mainly drawn from Khullar *et al.*, 1992. Efficiency and success in raising plants in the nursery and in their subsequent establishment in forest plantations depend to a great extent on the quality of the seeds used. It follows that foresters, nurserymen, seed dealers and others need accurate estimates of the quality of the seeds in which they deal or which form the basis of their afforestation projects (Turnbull 1975).

For the users, the most important object of seed testing is the provision of an accurate estimate of the capacity of a given seed lot to produce healthy, vigorous plants suitable for field planting. In the present context, seed "quality" refers to the physiological vigour of the seed rather than its genetic quality.

The beginning of seed testing in forestry was made in the country in the year 1962, when a Seed Testing Laboratory was established at the Forest Research Institute, Dehra Dun, during the Fourth Five Yea Plan. The Forest Research Institute has also started provided technical assistance for the establishment of Seed Testing Laboratories in the states.

Procedures for the Development of Seed Testing Rules

A major obstacle in taking up testing and issuing seed analysis certificate for forestry seed so far, was the absence of formalized procedures for testing them. In developing rules for seed testing, well-defined objectives have served as guidelines (Justice, 1972). The objectives, which are sought to be incorporated in seed testing programmes, may be listed as follows: a) to provide methods by which the quality of seed samples can be determined accurately, b) to prescribe methods by which seed analysts, working in different laboratories within the country or in different countries, can obtain uniform results, c) to relate the laboratory results, as far as possible, to planting value, d) to complete the test within the shortest possible time.

For the development of rules for the testing of seeds of Indian tree and shrub species, the constitution of technical committees for such quality factors as purity, germination test, including the determination of hard seeds, examination of noxious weed seeds, viability of tree seeds by the tetrazolium procedure, seed health condition i.e., seed borne organisms, genuineness of species, and sub-species, cultivar, moisture content, provenance or locality of harvest, unit weight of seeds, and homogeneity of seed lots should be considered (Khullar et *al.*, 1992).

Essentials of seed testing

Seed tests are important on two occasions – first, immediately following extraction and cleaning; and second, prior to planting (Bonner, 1974). Seed Testing is a streamlined process, which begins with the drawing of a sample and ends with the interpretation and application of test results. In practice, only a small proportion of a seed lot is tested and results are applied to the whole seed lot. This is called "sampling"

In India, the ICFRE and particularly Forest Research Institute, Dehra Dun has developed seed testing rules for some important forestry species (Khullar et al., 1992). The authors have also given sample weights of seed for "submitted" and "working" samples for important Indian tree species. Details about maintaining seeds in storage for orthodox and recalcitrant seed have also been provided. More details are discussed in a subsequent chapter.

5

A Study on Gangetic Dolphins And Conservational Issues

HELINA JOLLY

Introduction

"As the godess Ganga descended from the locks of lord Shiva's hair to earth , she was beautiful and was also accompanied by the cows of heaven.She came to prithvi (earth) on her vahana ".This was the reference of the River Ganga in the ancient veda and the vahana and the cows described here are gangetic dolphins.Dolphins were part of human culture and cultural evolution.Dolphins are aquatic mammals that are closely related to whales and porpoises. There are almost forty species of dolphins and they evolved about ten million years ago, during the Miocene era. The name is originally from Ancient Greek äåëößò (delphís; "dolphin"), which was related to the Greek äåëöõò (delphys; "womb"). The animal's name can therefore be interpreted as meaning "a 'fish' with a womb". It is also said to be derived from the word "adelphis" meaning" brotherly love,showing how significant a role these mammals have played in the emotional evolution of human beings. Dolphins tend to be very social animals, swimming in social groupings called pods. These groups, however, are very flexible and fluid, not at all like the social unit we refer to as a family.Dolphins mainly have two families the Delphinidae (oceanic dolphins) and the plantanistoidea (freshwater dolphins).Most of the common dolphins are the oceanic dolphins and are seen in dolphinariums.There are about four obligate river dolphins in the world and these are highly threatened by the activities of human beings. River dolphins are now facing extinction due to habitat loss, hunting by humans, and naturally because of their low numbers.The fresh water river dolphins are blind and mostly prefer murky water and they are endemic to those regions .It is known that earlier they had a wider distribution. but now they are in need of our help and cooperation. Out of the four freshwater dolphins, three are described below:

Baiji. The Chinese River Dolphin (*Lipotes vexillifer*)

The Critically Endangered Yangtze river dolphin, or baiji,can live only in freshwater and has very poor eyesight. It once lived in the lower and middle reaches of the Yangtze River,Fuchun

River, and in the Dongting and Poyang Lakes, China. Today it is the world's most endangered cetacean. Fewer than 100were thought to survive in the middle reaches of the Yangtze. However, a 2006 survey failed to sight any individuals, raising fears that the species is one step close to extinction.

Boto : The Amazon River Dolphin (*Inia geoffrensis*)

The pink river dolphin or boto, the Amazon river dolphin, can live only in freshwater. It is found throughout much of the Amazon and Orinoco River Basins in Bolivia, Brazil, Colombia, Ecuador, Guyana, Peru, and Venezuela. It is the most abundant freshwater cetacean and probably numbers in the tens of thousands.

Bhutan : The Indus River Dolphin (*Platanista minor*)

Closely related to the Ganges river dolphin, the Endangered Indus river dolphin, or bhulan, is found in Pakistan's Indus River It lives in freshwater and is essentially blind. It once ranged throughout the Ganges-Brahmaputra-Meghna and Karnaphuli-Sangu river systems of Nepal, India, and Bangladesh, from the Himalayan foothills to the Bay of Bengal.

Origin of Dolphins

One theory suggests that cetaceans are descendants of the Mesonychid, an animal that lived about 55 million years ago. This terrestrial mammal ranged from the size of a small dog to the size of a large bear. Another theory, based on DNA, molecular, and genetic studies, suggests that cetaceans share a common heritage with that of the hippopotamus. This theory would place cetaceans in theArtiodctyl group composed of even toed ungulates (deer, sheep, cows, pigs, hippos, etc.) New skeletal discoveries seem to back this claim, giving possible morphological evidence as testimony. Scientists have recently discovered ancient whale skeletons in Pakistan with well-preserved anklebones displaying features similar to that of the Artiodactyl group.The fossil is named as pakicetus, which showed asimilarity with to whales. Whales and dolphins are close relatives and a bottlenose dolphin with legs (or 'remnants of legs') has been found by Japanese researchers. An embryo of a Spotted Dolphin in the fifth week of development showed that hind limbs were present as small bumps (hind limb buds) near the base of the tail.

Aquatic Ape Theory

The aquatic ape theory was first put forward by Alister Hardy, research scientist and professor of zoology ,Oxford University, back in 1960 when he suggested that many of the characteristics that make the humans different from apes can be supported by proposing an aquatic phase in human evolution. Our aquatic phase is theorized to have taken place between nine and three million years ago, at a time which corresponds with the emergence of the dolphin of today. In the works of the theorist Sir Alister Hardy, award-winning writer Elaine Morgan, and in the studies by Dr. Michel Odent - world famous surgeon and pioneer into human water births - dolphins, humans and apes are likely to have evolved from a common ancestor millions of years ago. In fact, researchers at the Texas A&M University have found that out of the 22 dolphin chromosomes 13 are the same as humans."If we can show that humans are similar to dolphins, and anything that endangers dolphins is an equal concern for humans, it may be easier to persuade governments to become serious about combating industrial pollution and keeping oceans clean," says Busbee., a famous scientist working on dolphins. Even I personally believe this. To make any human being a conservationist he or she should have a kind of kin feeling towards that animal. People oppose and support this statement, but all that finally matters, is whether these animals, which are result of the beautiful phenomenon known as evolution, are properly conserved.

The Gangetic River Dolphin (*Platanista gangetica*)

Habitat

The Ganges River dolphin, or susu, inhabits the Ganges-Brahmaputra-Meghna and Karnaphuli-Sangu river systems of Nepal, India, and Bangladesh

Features

A long thin snout, rounded belly, stocky body and large flippers are characteristics of the Ganges River dolphin. Its eye lacks a lens so the dolphin is blind. The species has a slit similar to a blowhole on the top of its head, which acts as a nostril. The dolphin has the peculiarity of swimming on one side so that its flipper trails the muddy bottom and this helps the dolphin to search for food Being a mammal, the Ganges River dolphin cannot breathe in the water and must surface every 30-120 seconds. Because of the sound it produces when breathing, the animal is popularly referred to as the 'Susu'

IUCN STATUS : endangered

NUMBER OF INDIVIDUALS : 1200 to 1800 (wwf India)

Why is Conservation A Big Issue ?

Have you always wondered why the conservation of dolphins is a big issuep. To understand that we really need to comprehend the beauty in diversity of our country. The Gangetic river dolphins are endemic to Ganga and its tributaries. The river Ganga is about 2000 km long with a very large flood plain, which is the rice bowl of the whole country. Intensive agricultural practices are being carried on the banks of the river .The river banks are also beaming with many industries such as coal, iron and jute .These industries and agriculture are supported by the banks of this majestic river and more than that it is where the 10% of the world population of fresh water dolphins are concentrated. The Ganga river is not just a river as it has got deep emotional links to the people of India. This is the holy river where people have got many emotional ties and where religion is tied to the strings of history , so any activity of conservation should be in such a way that it should not harm the feelings of the population and their emotional pillars.

Threats faced by the Gangetic Dolphins

- Habitat Degradation
- Population Segregation
- Prey Depletion and Fishing
- Pollution
- Poaching

HABITAT DEGRADATION

The construction of at least 50 dams and dams within the known range of the dolphin has (Smith et al. 2000) dramatically affected its habitat, abundance, and population structure Changes in the land use pattern and deforestation due to increased human pressure, have caused the river banks to slump and erode, reducing flows and eliminating the deep pools because of silting. Such an effect has considerably reduced the habitable stretch of the river and the dolphins are forced into those areas where human activities are extreme. This may be one of the reasons why the dolphin population has depleted during the recent years in tributaries. The locals admit that the riverbanks of smaller tributaries have constantly undergone erosion and the width of these rivers has drastically widened within the last two decades. Humans first cleared the land of the flood plain for agricultural activity and with agricultural activity came the industries and human settlement.Humans took control of even more area of the Gangetic plain and it also resulted in deforestation .Along with deforestation came problems like soil erosion , silting etc which depleted the quality of the river water and the quantity also, so this caused the degradation of the only habitat for these fresh water tigers, as Gangetic dolphins are called , and now the high conservational threats are faced by it.

Polulation Segregation

Dams and bridges and barrages constructed across the Ganges and its tributaries have caused one major problem for the Gangetic dolphins and that is population segregation.This led to the separation of the small population of Gangetic dolphins into meta populations .These populations of dolphins are facing problems like the genetic inbreeding depressions .It is only the small populations that are drastically affected by the environmental disasters and human

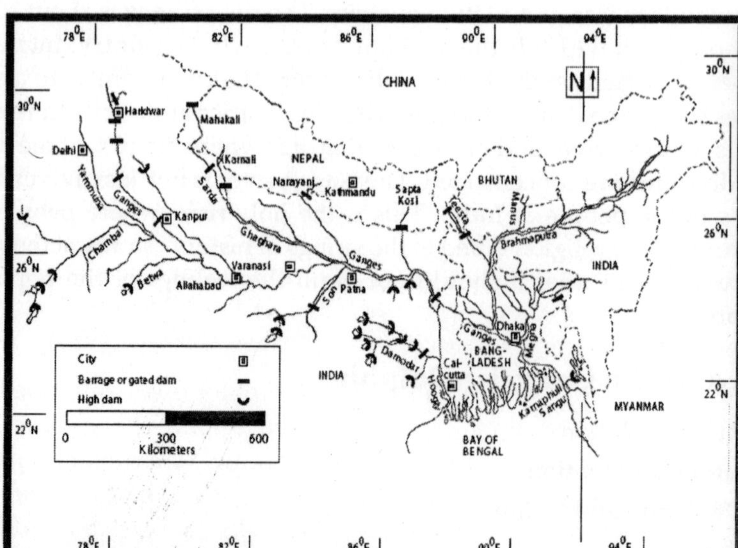

Figure 1. Distribution of the Ganges River Dolphin (*Platanista gangetica gangetica*)

exploitations In addition to fragmenting dolphin populations, dams and barrages degrade downstream habitat and create reservoirs (known as head ponds (or pondage in India) in the case of barrages) with high sedimentation and altered assemblages of fish and invertebrate species. The insufficiency of water released downstream by this barrage has eliminated the dry-season habitat for more than 300 km, or until the Ganges (Padma)-Brahmaputra confluence (Smith et al. 1998). It has also allowed salt water to intrude an additional 160 km into the Sundarbans Delta (Rahman 1986), further decreasing the amount of suitable habitat for this obligate freshwater dolphin (Reeves et al. 1993).

Some of the major threats faced by the gangetic dolphins due to these dams and barrages are summarized below

- fragmentation of the dolphin metapopulation,
- reduction or elimination of habitat simply in terms of the dry-season flow,
- "escapement" of dolphins into canals where they are unlikely to be able to get back into rivers and are therefore doomed,
- cascading effects from interrupted migrations of prey organisms, degradation of prey spawning habitat etc.,
- contaminant flux leading to significant changes in chronic and/or acute exposure to toxins,
- loss of complexity (channelization, sediment entrapment upstream of dams, etc.) making the rivers less habitable for dolphins, etc, downstream effects on the ecology of the delta (e.g., saline encroachment loss of sediments)

PREY DEPLETION AND FISHING

Most of the people living on the banks of the river have fishing as one of the important means of their livelihood.Though the commercial fishing of dolphins has been reduced in recent times, in some parts of the country it's still the fish or the saus mach which is highly prized .Due to extensive fishing using modern fishing gear, the accidental deaths of many dolphins has resulted. These blind dolphins are often victims of getting entangled in the nets and are often injured due to the propellors of the boats. The fishermen commonly use three types of fishing gear, cast nets, gill nets and hook lines. It has been stated that a floating gillnet used in a flooded river appears deadly since it is not easily detected visually and acoustically by the dolphins. The xets are set across the river during the monsoon (May-July) when dolphins lose their sonar ability due to the murky runoff and become easy victims (Shrestha,1995).

One of the other major problems faced by the dolphins is the excessive fishing which has resulted in the depletion in the prey, due to which the dolphins have started migrating to the lower streams and also to the canals for irrigation .Once they are reaching anywhere here, then they are doomed further leading to population segregation, as low water level has a more adverse effecton dolphin prey availability. Additionally, gill's nets are destructive as they entangle fish fauna of all sizes and thus pose a dire threat to the breeding of fish populations.

POLLUTION: The Ganga river is the area which has a lot of industries and these industries discharge pollutants into the river stream, polluting the river and making it unhabitable for these endemic animals. The agricultural activities include the use of various chemicals which enter the river system during floods and leaching activities take place. Dolphins, being the apex predators of the rivers they occupy are the highest level in the food chain and so they also face the problems of bio accumulation a nd in turn this leads to the toxicity and finally the death of these animals .Organochlorine and butyltin concentrations in samples from the tissues of Ganges dolphins were high enough to cause concern about effects (Kannan et al. 1993, 1994, 1997; Senthilkumar et al. 1999). Pollutant loads can be expected to increase with industrialization and the spread of intensive agricultural practices facilitated by water diversion. River dolphins may be particularly vulnerable to industrial pollution because their habitat in counter-current pools downstream of confluences and sharp meanders, often places them in close proximity to point sources in major urban areas (e.g., Allahabad, Varanasi, Patna, Calcutta, and Dhaka). Furthermore, the capacity of rivers to dilute pollutants (e.g., arsenic, DDT) and salts has been drastically reduced in many areas because of upstream water abstraction, diversion, and impoundment. Again, this problem is bound to worsen as more development takes place. In the intensification of agriculture, farmers are using higher yielding but lower pest resistant varieties of crops, along with the application of higher chemical fertilizers and pesticides. At the same time, the open border with India has resulted in local farmers acquiring a higher amount of chemical fertilizers such as Urea and Potash from India, at lower rates. These agrochemicals harm the dolphins both directly and indirectly through the food chain(Behera2005)

POACHING AND HUNTING : Gangetic dolphins are not hunted for a commercial purpose in India, but in certain countries, dolphin meat is considered to be a delicacy and so dolphins from India are hunted and exported to those countries where these are highly priced gourmet meats . Gangetic dolphins were caught and the blubber or their fat was used to form the emulsion for the boats and the oil was also used as the bait for attracting the fish.

The deliberate killing of river dolphins is believed to have declined in most areas but it still occurs at least occasionally in the middle Ganges near Patna, India (Smith and Reeves 2000, Sinha 2002), in the Kalni-Kushiyara River of Bangladesh (Smith et al. 1998), and in the upper reaches of the Brahmaputra River in Assam, India (Mohan et al. 1997). Even in India some of the tribal fishing communities kill dolphins in the upper Brahmaputra for their meat and fishermen kill dolphins in the middle reaches of the Ganges for their oil, which is used to attract fish. In addition, the animals are sometimes killed by fishermen in retaliation for'stealing' fish.

A lot of myth and mystery associated with these animals has gone on and so they are often the victims of human craziness .They have been killed for their meat which is supposed to have medicinal properties.These kinds of human activities again prove that even the level of literacy cannot shatter the strong walls of superstitions associated with human minds.

CONSERVATION OF GANGETIC DOLPHINS: Have you often wondered Why is this all happening ? It is all because one group of higher vertebrates (who call themselves highly civilized animals) have themselves chosen the task of protecting and helping other animals or living beings. We all know that with whoever else we share this third planet from the Sun is not just our home , but it is the home of all living organisms.When the world realized the conditions of these aquatic mammals they got together and thought of various ways to protect the river dolphins .Some of these/ ways are given below

1.APPENDIX I OF CITES

It is the convention providing the rules and regulations in the international trade of endangered animals and its products, convention on international trade of endangered species. Threatened organisms listed on three appendices depending on level of risk

Appendix 1 - all international trade banned

Appendix 2 - international trade monitored and regulated

Appendix 3 - trade bans by individual governments, others asked to assist

CITES administered by UN Environment Programme

2.THE YEAR 2007 AS THE YEAR OF THE DOLPHIN

The UN Convention on Migratory Species, together with its specialized agreements on dolphin conservation ACCOBAMS and ASCOBANS and the WDCS, the Whale and Dolphin Conservation Society, have declared 2007 to be the Year of the Dolphin.This was declared to bring about more awareness among the public for the importance of knowing about a policy for the conservation of the gangetic dolphins

3. Whale and Dolphin Conservation Society (WDCS

Wildlife charity and environmental organizations which help in the conservation of the dolphins and whales internationally .They help and assit various projects in countries related to dolphin conservation

4. Convention on the Conservation of Migratory Species of Wild Animals (also known as CMS or the Bonn Convention

This aims to conserve terrestrial,marine and avian migratory species throughout their range.Though dolphins are not obligate migratory creatures they they are included under this convention because of their wide distribution and highly fragile habitat

5. IUCN Red List of Threatened Species (IUCN Red List or Red Data List)

Created in 1963 it has the most important role in the conservation of plant and animal species.Gangetic dolphins are also considered to be one of the endangered group of animals requiring the need of conservation.

CONSERVATION AT THE NATIONAL LEVEL

The indian Government was also not far behind the conservation activities for the gangetic dolphins. Gangetic dolphins are included under the schedule 1 of the Indian Wildlife act 1972. whereby poaching of the animal is a criminal offence leading to imprisonment. India designated six new Ramsar sites at the 9th Conference Contracting Parties in Uganda India has declared six more Ramsar sites and one among them is the upper reaches of the Ganga , which is the major habitat of the dolphins. The Upper Ganga River (Brijghat to Narora Stretch). 08/11/05; Uttar Pradesh; 26,590 ha; 28°33'N 078°12'E. is a shallow river stretch of the great Ganges with intermittent small stretches of deep-water pools and reservoirs upstream from barrages. The river provides a habitat for IUCN Red listed Ganges River Dolphin, Gharial, Crocodile, 6 species of turtles, otters, 82 species of fish and more than hundred species of birds. This river stretch has high Hindu religious importance for thousands of pilgrims and is used for cremation and holy baths for spiritual purification. The major threats are sewage discharge, agricultural runoff, and intensive fishing. Conservation activities carried out are in the form of plantation to prevent bank erosion, training on organic farming, and lobbying to ban commercial fishing. Ramsar site no. 1574

Ganges River dolphin (Vikramshila Gangetic dolphin Sanctuary), India

The Vikramshila Gangetic Dolphin Sanctuary in Bihar, India, is the only designated protected area for this endangered dolphin in Asia but the sanctuary is suffering largely due to lack of funds and proper research in this field

NEED FOR CONSERVATION

Dolphins are the apex predator of the ecosystem and they help in maintaining the ecological balance by maintaining the prey / predator ratio.These animals survive only in those waters which have a lot of fish which is their main food source . So the higher the number of dolphins in the river, indicates the better quality of the water. More than that to conserve dolphins is our duty and responsibility towards the younger generation because these animals are our national treasure and are to be protected but the conservation has a lot of limitations .There is the lack of proper study done about the behavioural and breeding patterns of the dolphins .The migratory route of these animals is not known well.For the proper monitoring, the proper study is important .Some of the other reasons are the lack of exchange of data among the countries to have combined monitoring of these animals .It is also difficult to educate the fishermen about the significance of these animals in the rivers of the Ganga. For the fishermen, these are a prized catch required for sustaining their life.

Conclusion

Though conservation measures are going on actively, the population of the gangetic dolphins is still reducing .The Chinese dolphin is said to be extinct and next is said to be the susu ..

Gangetic dolphin .It took around 50 million years for the dolphins to evolve but it is sad to see that it will take only about 5 years for the anthropogenic activities to remove them from the face of earth. It is hard to accept but it's the truth the bipeds with the "highly developed grey and white matter" are responsible for this. As is always said" told the tale without a happy ending does not have an ending". So human beings have got time to improve.The change should be at the root level. Children who are the future adults should be educated properly with the concepts of conservation and it's need.People involved in the poaching and hunting of these animals should be given proper punishment. The conservation of Gangetic dolphins are in the hands of each of us . It is true that if each individual pledges money for the conservation of these silent, blind animals, then we can all make a great difference. A difference in attitude, a difference in the deed, and a difference in the words of ours can definitely bring about this big difference in the lives of the Gangetic dolphins

References

1. HIMAL SOUTH ASIAN October 2002 report on gangetic dolphin
2. WWF gangetic river dolphin
3. Copy of IUCN 2007 red list of endangered species
4. Dolphin Ape Human by Paula Peterson (Earth Code International
5. Evolution of river dolphins (The royal society, Healy Hamilton) Status, distribution and conservation threats of gangetic dolphin in karnali River, Nepal (May 2006)

6

Assessment of Diversity and Relationship among Cotton Germplasm for Drought Resistance and Fibre Quality Traits using Morphological, Physiological and Molecular Tools

A. GOPIKRISHNAN, N. JAGADEESH SELVAM, R. RAVIKESAVAN AND N. MANIKANDA BOOPATHI

Introduction

Cotton is playing a key role in the economic and social affairs of the global farmers. It is an important natural fibre crop and cultivated in more than 80 countries of the world; but ten countries *viz.*, India, China, USA, CIS, Brazil, Pakistan, Turkey, Mexico, Egypt and Sudan account for nearly 85 per cent of the total production (Van Esbroeck and Bowman, 1998). India ranks first in the cotton growing areas and > 65 per cent of the cotton crop in India is grown under rainfed conditions. Depending upon the climate and the crop growing period, cotton needs 700-1200 mm water to meet its maximum water recuirement. The water requirement is low during the first 60-70 days after sowing (DAS) and is highest during the flowering and boll development stages. Rainfed cotton yields are low owing to the erratic and uneven distribution of rainfall, which coincides with the flowering and boll development stages. The rainfed crop depends on residual soil moisture and several measures have been recommended to conserve rainwater in situ or harvest excess run-off and recycle it at critical phases. Cotton cultivation on ridges across the slopes conserves more water, reduces soil erosion and improves yield. As suggested in the cotton package of practices by the Directorate of Cotton Development, Mumbai could collect excess water into farm ponds and recycle it to provide one life saving irrigation at the critical early boll development phase and has may improve rainfed cotton yields. However, genetic improvement of cotton cultivars for drought resistance may be a simple solution to circumvent the above said problems.

Virtually all of the worldwide production of cotton is from two closely related allotetraploid species *Gossypium hirsutum* and *G. barbadense* and 90% is from upland cotton, *G. hirsutum.*

Intensive selection for agronomic traits has apparently resulted in a narrowed *G. hirsutum* gene pool so that there is little genetic polymorphism between existing accessions of cultivated cotton (Abadalla et al., 2001). This resulted in a poor resistance to biotic and abiotic stresses. The effective use of cotton germplasm in genetic improvement programs depends on the extent of genetic variation for desirable alleles and the accurate characterization of the variability within and among germplasm accessions in the collection. Thus, genetic diversity is desirable for long term crop improvement and reduction of vulnerability to important crop pests and diseases and abiotic stresses. However, many successful cotton cultivars have been developed from closely related parents, but limited yield gains in recent years have led some to advocate more extensive use of exotic germplasm (Meredith, 1991). The primitive cotton genotypes possess the potential for increasing the genetic diversity in cotton breeding programmes and possibly are a reservoir of novel alleles for important traits.

Considerable resources have gone into the collection, development, evaluation and maintenance of germplasm resources and a large amount of diverse cotton germplasm is being maintained at the TNAU. However, the genetic diversity has not yet done and such information may identify the diverse parents with superior quality that are likely to produce improved progeny. Limited pools of cotton germplasm have been previously characterized with morphological traits, isozymes and molecular markers such as random amplified polymorphic DNA (RAPDs) (Tatineni et al., 1996; Iqbal et al., 1997) and restriction fragment length polymorphism (RFLPs) (Wendel and Brubaker, 1993; Brubaker and Wendel, 1994; Meredith, 1995). However, the level of polymorphism detected by these techniques was generally low. These marker types are difficult to scale up for genotyping large germplasm collections efficiently. DNA marker systems for germplasm genotyping must be accurate, highly informative, cost effective and amendable to automation. The use of SSR markers can fulfill these requirements and fluorescence based SSR genotyping of cotton race stocks were done and shown that the SSR markers provide an accurate way of determining genetic diversity at the molecular level (Liu et al., 2000).

Publically available SSR markers which span all over the cotton genome are being used for genetic diversity analysis in this study. Useful allelic variations for important traits could be identified in these species and transferred to the cultivated species to create favourable new gene combinations. Therefore, the objective of this study was to evaluate the genetic diversity among the cotton germplasm maintained at the TNAU at the morphological, physiological, biochemical and molecular level. This information will be used to develop mapping populations which would be useful to achieve the long term goal of development of high yielding cotton cultivar under water limited environments through marker aided selection (MAS).

Materials and Methods

Plant materials

Totally eighty six *G. hirsutum* germplasm accessions which included 1) LRA5166, 2) Anjali, 3) Pusa9217, 4) Sumangala, 5) TCH1218, 6) TCH1705, 7) H109/BN, 8) TCH1589, 9) AC738, 10) F2036, 11) MCU5, 12) JK87682, 13) TCH1693, 14) HL34, 15) KC2, 16) TCH1696, 17) R16, 18) TSH288, 19) TCH1002, 20) PAR N1.20/352, 21) TCH1652, 22) HLS72, 23) GISV9716/2, 24)

CN3002, 25) TCH1649, 26) ALABAR51, 27) O41-6, 28) B-6-1350, 29) PABLA 400-2, 30) S-47, 31) REBA-TK, 32) SANZPONNA, 33) UA 31-62, 34) MCU1, 35) 170CO2M, 36) 5143, 37) Halden4, 38) CTI30-10DELTA, 39) EC3556, 40) Hancock, 41) Indore1, 42) K232, 43) Moore Special, 44) K3103, 45) DPL15, 46) PK863, 47) Acc. No. 10.1, 48) Acc. No. 9.68.5, 49) SH169, 50) SH169ND, 51) H492, 52) SH469(1.1), 53) SH131, 54) Big Boll Trump, 55) SH469, 56) RS271, 57) RS4001, 58) RS284, 59) LH33, 60) TCH1223, 61) TCH1233, 62) TCH1569, 63) DK820, 64) Acala44, 65) Acala glandless, 66) 43/85EL613, 67) 71/85-1842-1-1, 68) TCH1390, 69) Abadhita, 70) MCU11, 71) MCU12, 72) TCH1627, 73) TCH1452, 74) ARB9701(2028), 75) ARB8908(2024), 76) RAH III (2023), 77) NDLH1678(2019), 78) RAH100(2026), 79) CNH301(2021), 80) TCH1609, 81) SVPR2, 82) Surabi, 83) SVPR3, 84) MCU13, 85) MCU7 and 86) MCU9 were used in this study. The reason for selecting these accessions is that these lines have superior fibre quality traits and hence are routinely used in the breeding programme. Further, they have shown at least some moderate resistance to water stress during our previous experiments. Seeds were obtained from germplasm bank maintained at the TNAU, Coimbatore and plants were raised as described below.

Pot experiments

In pot experiment I, all the 86 accessions were grown in 40 X 30 cm pots filled with clay : red soil : sand in the ratio of 1:1:2, respectively. The pots with each accessions were grouped into two treatments *viz.*, irrigated control and water stress with two replications. Routine crop husbandry measures were followed regularly to all the pots except the irrigation to the water stressed pots. The water stress was imposed during the flowering stage (50 DAS). The duration of water stress was a continuous rain free 15 days and after that both control and water stressed pots were irrigated. At the last day of the water stress period, data on days to first flowering, plant drying (with 1-7 scoring where, 1= no drying, 2= started to wilting, 3= half of the plant wilted, 4= fully wilted, 5= started to dry, 6= half of the plant dried and 7= completely dried) and relative water content were recorded. Three days after relieving water stress, data on stress recovery (with 1-5 scoring where, 1= not recovered, 2= start to recover (from wilting to turgid), 3= half of the leaves of the plant recovered, 4= three fourth of the leaves of the plant recovered and 5= completely recovered), basal root thickness (cm), root length (cm), number of lateral roots, plant height (cm), shoot weight (g; fresh and dry weights) and root weight (g; fresh and dry weights) were recorded. From this data total plant weight (g) and root to shoot dry weight ratio were derived.

Based on the results of the experiment I, 15 entries were selected (described in results and discussion section) and again evaluated for drought resistance in pot experiment II with the same protocol described above but water stress was imposed both in the flowering and boll formation stages. All the data that has been recorded during pot experiment I was also documented in addition to the total chlorophyll content and proline as described below.

Chlorophyll Content Estimation

The Chlorophyll content of leaf samples was determined by the following protocol: Briefly, 250 mg of fresh leaf samples obtained from both irrigated control and water stressed pots was weighed and transferred to a mortar. The sample was macerated with 10 ml of 80% acetone and transferred to a centrifuge tube. The contents were centrifuged at 4,000 rpm for 10 minutes. After centrifugation the supernatant was collected in a volumetric flask and made up to 25 ml by acetone. The optical density (OD) of the unknown sample was measured at 480, 510, 645, 652 and 663 nm by a UV-VIS Spectrophotometer and 80% acetone was used as a blank to calculate the total and individual components of the chlorophyll content.

Proline Estimation

The proline content of leaf samples was determined by the method described by Bates et al., (1973). In short, 250 mg of fresh leaf samples were collected from irrigated controled and water stressed pots and were weighed and transferred to a mortar. The sample was macerated with 10 ml of 3% sulphosalicylic acid and transferred to a centrifuge tube. The contents were centrifuged at 4,000 rpm for 10 minutes. After centrifugation, 2 ml of the supernatant was collected in a test tube. To this, 2ml of acid ninhydrin, 2 ml of glacial acetic acid and 2 ml of 6M orthophosphoric acid were added. The test tubes were kept in a water bath for one hour and then cooled under tap water. The solution was transferred to a separating funnel and 4 ml of toluene was added and mixed for 30 seconds or until it formed two different layers. The colourless bottom layer was discarded and the upper pink layer was collected to record the OD reading at 520 nm. A standard proline solution was prepared and used to calculate the proline content of an unknown sample by simply plotting the OD value with the respective proline standard.

DNA extraction, PCR profile and gel electrophoresis

DNA was isolated by essentially following the method described by Callahan and Mehta, (1991) and the DNA quality and quantity was tested by some 0.8 % agarose gel electrophoresis. Cotton SSR primers were synthesized based on the published sequence information and the PCR reaction mix preparation and PCR conditions were performed as described by Lacape *et al.* (2007). The 4% Metaphor agarose gel electrophoresis was done to separate the alleles generated by SSR primers and the gel was photographed using gel documentation system (Alpha Imager, USA).

Genetic diversity estimation

The amplified fragments of each SSR marker were scored as "1" and "0", where "1" indicated the presence of a specific allele (band) and "0" indicated its absence. The genetic diversity estimate related analyses were done using NTSYSpc ver.2.02i by feeding the data using qualitative data option (and for morphological traits interval data option was used; Rohlf, 2000). Genetic similarities (GS) between pairs of accessions were measured by the DICE similarity coefficient. Genetic distances between pairs of lines were estimated as GD or D = 1 - GS. The clustering of accessions was done based on a similarity matrix using an unweighted pair group method with arithmetic average (UPGMA) algorithm following SAHN module. The clustering result was used to construct a dendrogram following TREE module (Ali *et al.*, 2008).

Results and Discussion

There may be sufficient allelic variation, mutation or recombination in the mating of closely related individuals to result in improved agronomic performance. The large number of successful cultivars developed as reselections indicates that there was sufficient recombination in matings of closely related parents to improve agronomic performance. May et al., (1995) noted that although regionally adapted lines were closely related, they contained a wide range of diversity for maturity and fibre properties. However, to get genetically improved genotypes for biotic and

Table 1. Variations in trait mean values among cotton germplasm under well watered and water stressed conditions in pot experiment I.

Name	DFF^C	DFF^S	PD^S	SR^S	BRT^C	BRT^S	RL^C	RL^S	NLR^C	NLR^S	PH^C	PH^S	RWC
Anjali	.	.	1	5	2.7	2.6	12.5	22	6	8.5	63	53	57.32
Pusa9217	.	.	4	5	3.6	3.5	16.8	22.3	11	12	65	56.5	75.28
Sumangala	66	63	3	5	4	3	24	21.75	12	9.5	65	60	74.03
TCH1218	.	65	1	5	3.4	3.1	26.3	22.75	3	13	63	58.75	85.56
TCH1705	66	60	1.5	5	4.4	2.8	28	29.65	20	10	64	52	83.89
H109/BN	64	62.5	4	5	4.7	3.35	33	30.5	7	11	69.5	64.75	39.82
TCH1589	66	.	3	5	4.5	2.45	30	27.4	15	18	70	65.5	73.11
AC738	66	64	1	5	4.4	3.25	32	24.5	15	5.5	71	67.5	86.2
F2036	65	61	4	5	4.2	3.95	30	26.25	12	14.5	59	54.5	74.33
MCU5	.	60.5	1	5	2.5	3.3	26.5	33	4	10	65.3	59.5	57.4
JK87682	.	.	1	5	3.3	2.6	30.5	23	6	5.5	65.3	62.5	62.09
TCH1693	.	.	1	5	3.5	2.5	30	29	10	5.5	70	68.75	71.99
MCU12	.	64	1	5	3.5	3.15	24.5	32	3	6	58	60.2	83.82
HL34	.	57	1.5	5	3	2.6	28.5	21.75	5	10.5	63.2	56.5	58.57
KC2	64	63	4	5	3.2	2.95	29.5	22	14	15	61	59.9	81.42
TCH1696	57	63	6.5	5	3.6	3.25	27	17.5	17	11.5	64.5	55.15	56.45
R16	62	58.5	4	5	4	3.3	27	28.7	20	15.5	58	52.15	83.61
TSH288	63	64.5	2	5	3.8	3	29	25.75	24	16.5	64	53.55	68.66
TCH1002	.	.	1	5	4	2.4	29.5	18.4	7	5	61	55	52.96
PAR N1.20/352	.	53.5	5	5	4.5	4.55	30.5	25.25	17	24.5	54	63.5	49.43
Surabi	63	63	3	5	4.7	4.4	29.9	25.4	11	17.5	49	61	59.11
TCH1652	64	60.5	2.5	5	4.6	4.75	30.5	20.55	12	17	72	57	48.99
HLS72	.	59.5	6	5	.	3.7	.	25.3	.	17.5	.	55.5	67.09
GISV9716/2	.	65	1	5	5.5	3.5	30.1	29	11	10	77	53	71.64
CN3002	.	57	4	5	4.4	4.1	30.3	30	9	11.5	62	66	53.53
TCH1649	57	63	5	5	4	5.15	30	29.75	18	13	65	62.5	74.36
ALABAR51	.	.	3	5	.	3.75	.	29.75	.	10	.	51	67.03
O41-6	.	.	1	.	.	.	.	.	.	.	.	.	.
B-6-1350	.	61.5	5	5	3.8	4.1	30	20.3	14	11.5	48	58	64.29
PABLA 400-2	.	.	3	5	5.2	3.5	29.9	20.6	13	9	55	56	61.04

Name	DFF^C	DFF^S	PD^S	SR^S	BRT^C	BRT^S	RL^C	RL^S	NLR^C	NLR^S	PH^C	PH^S	RWC
S-47	.	66	4	5	4.3	4	20.9	20.6	17	15	68	54.5	52.77
REBA-TK	.	62	3.5	5	.	3.6	.	25.5	.	16.5	.	59.5	60.96
SANZPONNA	55	66	3	5	3.2	4.45	30.1	25.2	16	12	59	57	55.88
UA 31-62	.	.	1	5	.	3.2	.	29.8	.	4	.	47	53.87
MCU1	.	60	5	5	4.5	4.75	20.9	27.75	21	10.5	70.5	55	66.23
170CO2M	66	64	4	5	5	4.75	20.1	20.2	20	8.5	70.4	60.3	70.63
5143	66	64	6	4	5.5	4.1	30.5	24.9	14	14	90.1	60.45	80.34
Halden4	.	63	4.5	5	4.3	3.9	20.2	20.4	9	9.5	60.6	44.5	65.93
CTI30-10DELTA	55	57.5	4	3	6	3.4	20.4	24.65	10	14	80.4	55.25	59.92
EC3556	64	55	6	3	6	2.8	20.5	25	9	15	70.5	70	60.7
Hancock	.	64	4	5	.	3.6	.	20	.	12	.	50	70.62
Indore1	50	.	4.5	5	5.3	4.35	30	30.5	16	12	80.9	55.1	62.88
K232	.	.	.	.	.	.	.	.	.	.	.	.	.
Big Boll Trump	64	64	1	5	5.5	3.45	27	24.5	14	11	60	45.45	81.96
Moore Special	.	55	5	5	5.8	5	24	28	9	11	60.7	40.5	64.1
K3103	.	.	1	5	5	2.75	29	25.5	9	4	50.2	30.2	30.04
DPL15	.	.	1	5	4	4	27	27	11	6	40.4	50	66.73
PK863	.	56	7	.	.	5	.	33	.	18	.	71	68.61
10.1	.	.	6	5	.	5	.	30	.	15	.	61	64.68
9.68.5	.	.	.	.	5	.	25	.	5	.	59	.	.
SH169	.	.	.	.	.	.	.	.	.	.	.	.	.
SH169ND	.	55	4.5	5	4	4.15	28	31	8	10	75	49.5	72
H492	63	.	3	5	5.6	3.3	26	23.5	8	11	57	38.5	60.95
SH469(1.1)	.	.	2	5	5.4	3.75	30	28	13	6	62	43.5	58.61
SH131	66	.	1	5	6.2	3.55	34	33.5	5	8	59.5	37	69.99
SH469	.	.	.	.	.	.	.	.	.	.	.	.	67.97
RS271	.	62	4.5	5	.	3.8	.	28	.	8	.	50.5	63.46
RS4001	.	.	1	5	4.5	1.5	20	14	5	3	42	18	64.8
RS284	.	.	5	5	5.4	3.3	25	20	6	8	58	53	64.12
LH33	62	50	5	4	6.3	4.5	30	27	8	9	93	54	70.65
SVPR2	.	62	2	5	.	4.8	.	31	.	15	.	53	81.13
TCH1223	.	.	1	5	5.8	4	39.5	38	11	10	68	56	46.85

Name	DFF^C	DFF^S	PD^S	SR^S	BRT^C	BRT^S	RL^C	RL^S	NLR^C	NLR^S	PH^C	PH^S	RWC
TCH1233	66	66	1	5	5.5	3.5	30	28	12	4.5	92	52	81.23
TCH1569	66	.	2	5	6.3	4.5	33	25.5	15	5	90	46.5	59.31
DK820	.	.	.	.	.	.	.	.	.	.	.	.	.
Acala44	.	.	5	5	.	4.5	.	25	.	10	.	46	71.89
Acala glandless	.	.	3	5	4.7	3.7	35	19	8	4	64	43	46.15
43/85EL613	65	.	2	5	6.5	4	26	29.5	10	7.5	74	42	67.32
71/85-1842-1-1	.	.	1	.	5.8	.	29	.	10	.	73	.	.
TCH1390	.	.	1	.	5.5	.	27	.	14	.	69	.	.
Abadhita	.	60.5	5.5	5	5.7	4.4	24	28	10	13	69	54.5	60.89
MCU11	.	.	.	.	.	.	.	.	.	.	.	.	.
MCU7	57	62	4.5	5	7	4.3	31	33	16	4	55	47.5	59.25
SVPR3	.	55	6	3	.	5	.	30	.	10	.	32.5	82.6
TCH1627	.	.	1	5	.	3.5	.	32	.	7	.	29	36.43
TCH1452	.	.	4	5	6.3	2.7	35	32	7	9	50	51	74.49
ARB9701(2028)	.	66	5	5	.	4	.	31	.	6	.	47	69.04
ARB8908(2024)	.	.	1	5	6.5	4	28	16	8	6	75	45.5	.
RAH III (2023)	66	66	2	5	4.3	4.05	31	24	6	5	70	44	83.87
NDLH1678(2019)	66	.	4	5	5.3	4	27	25.5	5	7	93	50.5	76.66
RAH100(2026)	.	.	.	.	.	.	.	.	.	.	.	.	.
CNH301(2021)	66	66	1	5	6.4	5.15	26	30	8	5.5	74	53.5	80.19
TCH1609	62	64	6	1	5.8	5.1	31	30	12	13.5	94	58	62.49
LRA5166	66	.	5	5	5.5	4.25	29	25.5	14	12.5	70	43.5	70.12
MCU13	62	.	2	5	.	7.3	.	30.5	.	11.5	.	71.5	59.06
MCU9	.	66	3	5	4.2	4.3	29	32	7	7	50	59.5	51.92
Mean	63	61	3.2	4.9	4.8	3.9	27.9	26.3	11	10.2	66.2	52.9	66.4
Range	50-66	50-66	1-7	1-5	2.5-7.0	1.5-7.3	12.5-39.5	14-38	3.0-24	3.0-24.5	40-94	18-71.5	30.04-
86.19													
S.D.	3.9	3.9	1.8	0.6	1	0.9	4.4	4.6	4.7	4.1	11.7	9.5	11.6

Where, DFF - days to 50% flowering, PD - plant drying, SR - stress recovery, BRT - basal root thickness, RL - root length, NLR - number of lateral roots, PH - plant height and RWC - relative water content; superscript C refers under control treatment (with regular irrigation) and S refers under water stress treatment. Dot (.) indicates data has not recorded.

Table 2. Genetic variation of representative cotton germplasm for morphological, physiological and biochemical traits under well watered and water stressed conditions in pot experiment II.

Traits	MCU5		MCU12		MCU13		TCH1218		TCH1002		KC3		SVPR2	
	Control	Stress	Control	Stress	Control	Stress	Control	Stress	Control	Stress	Control	Stress	Control	Stress
Root length	27.7±2.1	28.7±0.6	30.0±3.6	29.7±3.6	29.3±1.5	31.3±2.1	32.7±0.5	34.1±0.9	30.9±1.3	30.8±1.6	25.0±3.5	24.0±2.0	29.9±0.4	32.3±1.2
Number of lateral roots	11±3.6	12±2.6	11.7±2.1	10.0±2.1	17±2.6	15.3±1.5	17.7±6.8	14.0±3.5	7.0±1.0	6.0±2.0	18.0±5.2	10.7±1.2	16.3±0.6	18.7±4.7
Basal root thickness	3.6±0.9	3.3±0.3	4.6±0.6	3.1±0.6	3.9±0.2	3.8±0.3	4.5±0.5	3.8±0.3	3.9±0.1	3.3±0.6	4.1±0.1	3.8±0.3	4.2±0.4	4.0±0.5
Chlorophyll a (mg/g)	0.5±0.1	0.4±0.1	1.0±0.1	0.7±0.1	0.8±0.0	0.6±0.0	0.4±0.0	0.3±0.1	0.6±0.0	0.5±0.0	0.8±0.1	0.6±0.1	0.7±0.1	0.5±0.0
Chlorophyll b (mg/g)	0.3±0.0	0.2±0.0	0.4±0.0	0.3±0.0	0.4±0.0	0.3±0.0	0.3±0.0	0.2±0.0	0.3±0.0	0.3±0.0	0.4±0.1	0.3±0.0	0.3±0.0	0.2±0.0
Total chlorophyll (mg/g)	0.8±0.1	0.7±0.0	1.4±0.2	1.0±0.2	0.9±0.1	0.7±0.1	0.8±0.1	0.6±0.1	1.1±0.0	0.7±0.5	1.5±0.2	1.2±0.1	1.1±0.1	0.7±0.0
Proline (mg/g)	0.6±0.0	1.0±0.0	0.3±0.0	0.4±0.0	0.4±0.0	0.6±0.1	0.3±0.0	0.6±0.0	0.5±0.0	0.9±0.1	0.5±0.0	0.9±0.1	0.4±0.0	0.7±0.0
2.5% span length	34.0±0.0	31.0±1.0	31.2±1.0	30.1±1.0	32.3±0.2	31.0±0.5	32.1±0.2	31.1±0.6	27.9±0.1	26.7±0.1	28.3±0.2	27.4±0.4	26.4±0.1	25.3±0.2
Micronaire	3.8±0.1	3.5±0.2	4.1±0.1	3.7±0.1	3.9±0.1	3.6±0.1	3.4±0.2	3.3±0.1	4.1±0.1	3.4±0.1	5.4±0.1	5.1±0.1	4.8±0.1	4.5±0.1
Tenacity 3.2 mm (bundle strength) (g/tex)	24.0±1.0	22.0±1.0	19.3±0.2	19.0±0.2	23.8±0.3	22.2±0.7	19.0±0.1	19.0±0.1	19.5±0.3	19.4±0.3	18.5±0.2	18.3±0.2	16.8±0.1	15.9±0.1

abiotic stress resistances, it is necessary to identify genetically diverged parents that are adapted to target environment. A greater effort to introgress diverse germplasm into regionally adapted cultivars that do not carry a yield penalty may offer greater rewards in crop improvement and reduced vulnerability. In this context, the present study was designed to study the genetic variation of cotton germplasm maintained at the TNAU, which may help to understand the genetic variation among cotton accessions and create unique gene combinations necessary for improved cultivars.

Genetic variation in morphological and physiological traits among cotton germplasm

From the results of pot experiment I, it was noted that there were considerable genetic variations existing in cotton germplasm in expression of several morphological and physiological traits pertaining to the overall performance of cotton accessions under water limited environment (Table 1). Almost in all the accessions, flowering was induced early under water stressed conditions. The plant drying score has helped to identify the accessions that can withstand continuous water stress for 15 days and accessions that cannot. Interestingly, all the accessions were completely recovered after rewatering (Table 1).

Invariably, basal root thickness was more or less similar in all the accessions both under irrigated control and water stress conditions. However, there was a variation in root length and a number of lateral roots under irrigated control and water stress conditions. It was clear that some of the accessions have recorded reduced plant height, root length and number of lateral roots (Table 1) besides root shoot dry weight ratio (data not shown due to space restriction) under water stress. On the other hand, accessions such as Anjali, Pusa9217, MCU5, MCU12, MCU1, CTI30-10 Delta, EC3556, Moore Special, 43/85EL613 and Abadhita have shown better root systems under water stress. Nevertheless, TCH1693, BigBollTrump, DPL15, TCH1233, which have shown overall drought resistance have a similar root systems development to that of R16, CN3002, TCH1649, S-47, 170CO2M, Indore 1, SH169ND and MCU9, which have shown an overall susceptibility to drought both under irrigated control and water stress conditions. This indicated that mechanism(s) other than (and/or in addition to) the root system might be involved in drought tolerance during the flowering and boll formation phase water stress. Conversely, it should be noted that the root system development might be restricted although larger pots were used in this study. To rule out the experimental restriction to root system development, the selected cotton entries were evaluated under field conditions and the results were similar to those of the pot experiments.

Interestingly, a sizable number of cotton germplasm accessions such as TCH1218, TCH1705, AC738, MCU12, KC2, R16, 5143, BigBollTrump, SVPR2, TCH1233, SVPR3, RAH III (2023), NDLH1678 (2019) and CNH301 (2021) have retained >75% relative water content under water stress though they have shown a significant variation in their root system development. The above said 14 germplasm were selected besides KC3 (as local drought resistance check) and they have again been evaluated for drought resistance in the pot experiment II. During this experiment, water stress was imposed at two different phonological stages: flowering phase and boll formation phase. The results were almost similar to those of pot experiment I (Table 2). There was no significant difference among cotton germplasm for root traits even though they have again confirmed their overall drought resistant potential (Table 2). Water stress reduced

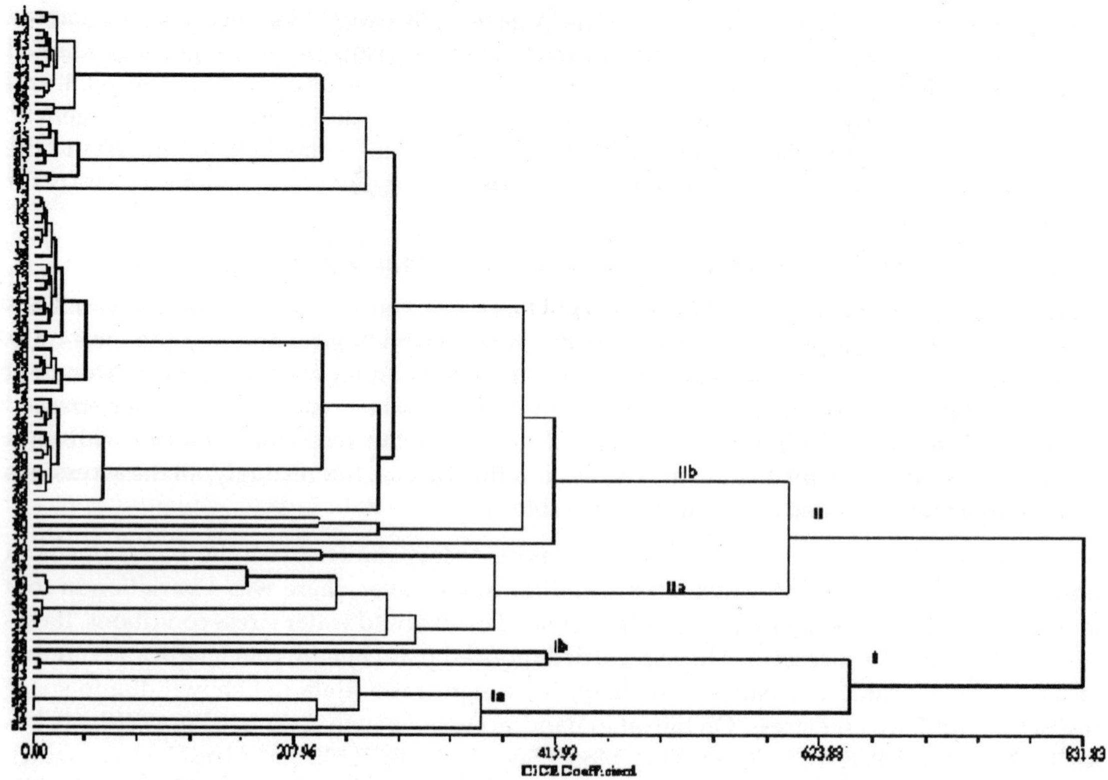

Figure 1. Dendrogram of 86 cotton germplasm revealed by cluster analysis of genetic similarity estimates generated by DICE coefficient based on morphological traits. Serial number of 1-86 represents the entries in the same order that are described in the materials and methods.

the total and individual components of the chlorophyll content and fibre quality traits such as microniare, 2.5% span length and tenacity 3.2 mm (bundle strength, g/tex). However, the total amount of proline has shown to be increased invariably in all the cotton germplasm screened under water stress conditions. This indicated that drought resistance in cotton may be greatly imparted through other mechanism(s) such as osmolyte synthesis other than (and/or in addition to) the root system.

Cluster analysis and genetic diversity

Since genotyping of the cotton germplasm is under progress, morphological traits were used to analyze the genetic relationship and diversity among cotton germplasm. Based on the data on morphological traits, the genetic similarity co-efficient was estimated for each pair of the 86 cotton accessions. The dendrogram (Figure 1) clearly discriminated all the accessions into two different major clusters. Cluster I comprised two subgroups: cluster Ia included TCH1569, BigBollTrump, RAHIII 2023, Abadhita, SH169, Indore1 and TCH1649. Interestingly, RAHIII 2023, Abadhita, SH169 and Indore1 grouped as single cluster and they have a DICE coefficient of 0.0. It is interesting to note that at least one of the parents of these accessions was common. Cluster Ib consisted Acc. No. 9.18.5, 43/85EL613 and Acc. No. 71/85-1842.1.1. On the other hand, cluster IIa contained PK863, MCU13, TCH1627, SH469, IC3556, PABLA900-2, Acc. No. 10.1, MCU11, Reba TK, CN3002, DK820 and PARN1:20/352 and rest of the accessions were grouped into cluster IIb.

Therefore, Figure 1 evidently indicated that grouping based on clustering analysis was in good agreement with the available pedigree and genetic background information. A number of diverse pairs of cotton accessions were identified which have shown significant difference for phenotype expression and thus grouped in different clusters. For example, SVPR2, MCU5, MCU12, TCH1218, TCH1002 and KC3 have grouped in different clusters. Interestingly, these six accessions have shown increased drought resistance compared with other accessions used in this study in both the experiments. Hence, these accessions were selected and they are now being used for mapping the population development. Although the choice of parents is often the most important decision in a breeding programme, little is known about the importance of parental genetic diversity to successful cotton cultivar development. However, there are conflicting views concerning the importance of genetically distant parents to cultivar improvement. While successful cultivars were developed from both distantly related and closely related cultivars the large number of cultivars developed as reselections or from the crossings of closely related parents, indicated that there was a sufficient variability or mechanism to create variability in closely related lines to make breeding progress. The greater than expected frequency of cultivars with closely related parents suggests that distantly-released germplasm was unadapted and matings were, for the most part, restricted to genetically similar regionally adapted lines (Van Esbroeck and Bowman, 1998). However, the few diverse germplasm lines that were introgressed into agronomically suitable germplasm became widely used. Unless progress is made in transferring useful allelic variation from diverse to adapted germplasm without negative agronomic effects, germplasm resources will probably remain underused and the trend towards increased genetic vulnerability will continue. Thus, this study has helped to select the cotton germplasm that has not only shown variation in drought resistance but is also genetically diverse. This will help to improve the drought resistance and yield in cotton under water-limited environments through MAS which is the ultimate goal of this study.

References

1. Abadalla, et al., 2001. Theor. Appl. Genet. 102: 222-229.
2. Ali, et al., 2008. Mol. Breeding 21 : 497-509.
3. Bates, et al., 1973. Plant and Soil, 39(1): 205-207.
4. Brubaker, C. L. and Wendel, J. E. 1994. Am. J. Bot. 81: 1309-1326.
5. Callahan, F. E. and Mehta, A. M. 1991. Plant Mol. Biol. Reporter 9(3) : 253-261.
6. Iqbal, et al., 1997. Theor. Appl. Genet. 94: 139-144.
7. Lacape, et al., 2007. Mol. Breeding 19 : 45 – 58.
8. Liu, et al., 2000. Crop Sci. 40: 1459-1469.
9. May, et al., 1995. Crop Sci. 35: 1570-1574.
10. Meredith, W. R. Jr. 1991. In. H. L. Shands and L. E. Weisner (ed.). Use of plant introductions in cultivar improvement. Part I. CSSA Spec. publ. 17. CSSA, Madison, WI.
11. Meredith, W. R. Jr. 1995. In: G. A. Constable and N. W. Forrester (ed). Proceedings of the World Cotton Research Conference I. Brisbane, Australia, 14-17 Feb, 1994. CSIRO, Brisbane, Australia.
12. Rohlf, F. J. 2000. NTSYS-pc: Numerical taxonomy and multivariate analysis system. Exeter Publishing, Setauket. pp119.
13. Tatineni, et al., 1996. Crop Sci. 36: 186-192.
14, Van Esbroeck, G. and D. T. Bowman. 1998. J Cotton Sci. 2: 121-129.
15. Wendal, J. F. and C. L. Brubaker. 1993. Am J Bot. 80(6): 71.

7

Fauna, Flora and Faith: Essence of Environement Alism Alism in Hindu Religious Thought

CHANDREYI BANERJEE AND GAUTAM BASU

Introduction

The man-environment relationship is as old as civilization itself. The environment includes our habitat and surroundings and encapsulates all living creatures, including man, within its purview. Over the years, the forms and modes of man's interaction with his environment have changed continually and inexorably. With the advancement of civilization man has also gradually evolved from being a "natural being" to an "economic being". Thus, while earlier human destiny was considered to be inextricably linked with the environment and conditioned by it, later humans emerged as prime modifiers of Nature themselves. This is propounded in the geographical school of thought wherein earlier the principle of environmental determinism, i.e. humanity as determined by Nature, was dominant but was subsequently replaced by the possibilistic viewpoint of humanity being the prime modifier of Nature.

In a world increasingly enamoured with technology, leading to exponential development in terms of material pleasure and prosperity, environmental problems such as pollution, deforestation and depletion of natural resources have affected almost every facet of human life at the individual, community, national and global level. With increased industrialization and materialism, environmental problems have only multiplied. Mankind has resorted to science and technology to find a solution to our environmental travails but the magnitude and viciousness of the problem defy any satisfactory permanent solution. Hence, over the last few decades the quest for an environment that can sustain man without degrading itself irreversibly has turned to the world's religious traditions based in antiquity. The effort is to develop a code of environmental ethics based on faith so that the environmental cause as a whole can garner greater and wider, broad-based support. Therefore, many stakeholders have sought to encourage the development of an environmental ethic within the Hindu religious tradition to practice and achieve holistic and sustainable development without irreparably damaging our fragile environment. Hinduism comprises the oldest continually observed corpus of religious traditions in the world and propagates the principle of "*dharanath dharma ucyate*": that which sustains all species of life and helps to an maintain harmonious relationship among them is dharma and that which disturbs the ecology is *adharma*.

Environmental elements in Hinduism

Hinduism offers a wealth of perceptions on the ways in which ecological principles are enmeshed in tradition and faith. Hinduism is often termed Sanatan Dharma, which means "the eternal essence of life". This essence of life is not only related to humans but unites all beings – human, animal and plant species – with the universe that surrounds them and ultimately with the original source of their existence, i.e. the Creator. The *Upanishads* say "*tat sristva ta devanu pravisat*": after creating the universe God entered into every object thus created to help it maintain its congenial interrelationship with all others. The divine is not external to creation but expresses itself as a part of it through natural phenomena. Thus, in the *Mundaka Upanishad* the Divine is described as:

> *"Fire is his head, his eyes are the moon and the sun;*
> *the regions of space are his ears, his voice the revealed Veda;*
> *the wind is his breath, his heart is the entire universe;*
> *the earth is his footstool, truly he is the inner soul of all."*

Sri Krishna in the *Bhagavad Gita* says, "I pervade the Universe; all objects in the universe rest on me as pearls on the thread of a garland". Hence in the Hindu belief there is no life that is inferior. All lives enjoy the same importance in the matrix of the universe and play fixed and ascertained roles in maintaining the ecological balance, but man being the most intelligent being and the only one with the power to reason has in modern times continually disturbed this balance wantonly and perniciously instead of sustaining and nurturing it. This is largely due to man's unrelenting and burgeoning needs coupled with his unbounded greed for material consumption and enjoyment.

The *Bhagavad Gita* conveys the message "conserve the ecology or perish". In the third chapter of the Gita it is said that a life led without tangibly contributing towards the preservation of the ecology is a life led in sin and a life futile and without use. The ecological cycle is explained in verses 3.14-16 of the *Gita*: "All living bodies subsist on foodgrain, which is produced from rain. The rains are produced by the performance of *yajnas* (sacrifice), and *yajna* is born out of prescribed duties. Regulated activities are prescribed in the *Vedas* which are directly manifested from the Supreme Personality of the Godhead. Consequently, the all-pervading Transcendence is eternally situated in acts of sacrifice. One who does not follow in human life the cycle of sacrifice established by the *Vedas* certainly lives a life of sin. Living only for the satisfaction of the senses, such a person lives in vain."

Several studies on the ecological aspects of Nature are recorded in the *Vedas*. The *Vedas* have held forth various rituals that extol the earth (*bhu*), the atmosphere (*bhuvah*) and the sky (*sva*). The *Atharva Veda* contains 63 verses known as "*Bhumisukta*" devoted to praise of Mother Earth (*prithvi*) and highlighting the dependence of man on Her. It says: "Earth is my mother. I am her son". In the same *Atharva Veda;* fire (*agni*) was thought to be a form of energy which exchanges its form immutably and indestructibly (verse 111:21:1). The pagan element in Hinduism advocated sustainable development by the deification of Nature as evinced in the hymns of the *Rig Veda* dedicated to *Surya* (sun), *Indra*, *Agni* (fire), Soma (moon) and *Usha* (dawn). Hymn IX of Book 10 of the *Rig Veda* recognizes the life-giving ability of *Ap* (water) not only physically but also spiritually. Its verse 6:48:17 exhorts man not to cut trees because trees are a deterrent to pollution. Verse 5:48 of the *Yajur Veda* calls on us not to disturb the sky and not to pollute the atmosphere. Verse 13:47 of the same text also urges all men not to kill an animal that is helpful to all; rather, one should attain happiness by serving (such animals).

Hinduism stresses on simple living and inner transcendental peace. This means that the search for material possessions, and the consumption of material goods and energy such possessions bring, should not be allowed to dominate life. Life's main purpose is to discover the spiritual nature and the peace and fulfillment that it heralds. Thus, there is a prayer in the *Sama Veda* called "*Shantipath*" which says: "May there be peace in the sky; may there be peace in the mid-region; may there be peace on earth; may there be peace in the waters; may the medicinal plants ... the forests be peaceful; may there be peace in gods; may Brahma be peaceful; may there be peace and peace only; may such peace come to us".

Two fundamental Hindu tenets are:

i. *R'ta* – which is the sense of fundamental order and balance in the universe – must be observed and sustained by obeying the ancient laws; and,

ii. Dharma – which emphasizes the need to act for the sake of good in the world ("*sarva bhutah hita*") and to behave and conduct ourselves in such a manner as to maintain natural harmony – must be a paramount and inalienable part of our lives.

Another important tenet of Hinduism is the Doctrine of Reincarnation, according to which the Supreme Being is reincarnated or reborn in the form of various species. Lord Krishna said: "this (his) form is the source and indestructible seed of multifarious incarnations within the Universe, and from every particle of this form are created the different living entities, including demigods, animals, plants, human beings and others (*Srimadbhagavad Gita Mahapurana 1:3:5*)". Among the various incarnations of God are *matsya* (fish), *kurma* (tortoise), *varaha* (boar), *narasimha* (combination of man and lion), etc. Moreover, in Hindu mythology, there is a close and inalienable relationship between the various deities – who are all different aspects of the same divine power – and their individual animal or bird mounts.

Furthermore, the belief in the Doctrine of Karma and the associated cycle of birth and rebirth wherein a person may be born again as an animal or a bird or a tree or any other creature depending on the *karma* (i.e., actions committed in the previous births) entails that Hindus are called upon to accord to other species not only respect but also reverence.

Hinduism is all-inclusive; it covers all facets of life, not only religious but also economical and political. The duty of governments and the ruling class in regard to the environment has been prescribed in the *Sukraniti* as: "wealth and life are preserved by men for enjoyment. But what avail a man to have wealth and life who has not protected the land?" Similarly, the *Charaka Samhita* says that destruction of forests is akin to the destruction of the state, and reforestation is an act of rebuilding the state and advancing its welfare. It also enjoins upon all Hindus to protect animals as a sacred duty. Therefore, for Hinduism, the concept of environmentalism is not a modern phenomenon but one of the challenges of modern Hinduism is to reclaim its firm roots in terra firma.

Hinduism, Indian philosophy and environmentalism

The *Vedas* are the oldest extant literary monument of the Hindu religious tradition. The origin of Indian philosophy may be traced to the Vedas. If we go by the fact of accepting the authority of the *Vedas*, the nine schools of Indian Philosophy – Charvaka (materialism), Jainism, Buddhism, Sankhya, Yoga, Vaisheshika, Nyaya, Mimamsa and Vedanta – may be categorised into the "*nastikas*" and the "*astikas*". While the former three schools denying the authority of the *Vedas* are called "*nastikas*", the latter six believing in it are called the "*astikas*".

To begin with the *astika* school, Sankhya is the oldest in this line of thought. The Hindu ideals of ancient Vedic times became formalized into this system that promoted care for the ecology as a part of intrinsic Hindu belief and attitude. Sankhya maintains a dualism between *Purusa* and *Prakriti*. *Prakriti* is viewed as the root cause of the material world or the world of objects. It is ever-active with unlimited power, and is also called *Shakti*. The extreme subtleness of *Prakriti* makes it imperceptible; we infer its existence through its products. Purusa is the self, the subject, the knower and the foundation of all knowledge. This is epitomised in Nature. The entire world with all animate and inanimate objects around us, or the environment surrounding us, is called *Prakriti*. It, too, is subtle and its existence may be inferred through the various resources it gives us. *Purusa* comprises all humanity possessing the ultimate knowledge of these resources since man is the only rational being. So the role of *Purusa* or humanity is then to protect *Prakriti* and help all lives and other objects in the universe to play their roles predestined in the natural environment effectively. But *Purusa* (humanity), it seems, is not interested in protecting Prakriti, but has instead exploited her indiscriminately. Therefore, to defend herself, Prakriti sometimes sheds her subtle form and we see her infuriated and powerful external manifestation through the occurrence of droughts, floods, earthquakes and tsunamis.

The Yoga system of philosophy advocates control over the body, the senses and the mind. It recommends perfection of the being, renounces sensual attachment and passions, and lays down certain ethical commandments popularly known as "*Ashtanga Yoga*" (Eight-fold Path of Discipline). They are *yama, niyama, asana, pranayama, pratyahara, dharana, dhyana* and *samadhi*. Of these, *yama* and *niyama* may be considered important from the environmental point of view guiding human behaviour in relation to other humans, living creatures and non-living resources. *Yama* means abstention and includes:

i. Non-violence (*ahimsa*) towards all animate and inanimate creations;

ii. Truth (*satya*), shunning the use of resources obtained by illegitimate means and avoiding;

iii. Destruction and vandalism (*asteya*);

iv. Celibacy (*brahmacharya*) because humans need to keep their numbers in check for fear of irreversibly disturbing the population-resource nexus; and,

v. Not coveting resources and wealth beyond requirement (*aparigraha*).

These are to be realized through various *niyamas* or self-culture and self-discipline, which include:

i. *Shaucha* relating to: (a) cleanliness of one's mind, body and surroundings; and, (b) ridding oneself of undue lust (*kama*), anger (*krodha*), greed (*lobha*), attachment (*moha*) and conceit and vanity (*ahankara*);

ii. Santosh (contentment);

iii. Tapas (austerity);

iv. Swadhaya (introspection of the self); and,

v. Ishwar pranidhan (prayer and meditation).

The deification of Nature as is prevalent in Hinduism has been formalized into the *Vaisheshika* denotation of the five great elements (*mahabhuta*), namely, earth (*prithvi*), water (*ap*), fire (*tejas*), air (*vayu*) and ether (*akasha*). The meditative and ritual processes of Hinduism entail awareness of these constituents of materiality (*dravya*). Daily worship (*puja*) employs and evokes these five powers. According to the *Vaisheshika* philosophy, the five sensory organs (*jnanendriyas*) with the functions of sight, smell, taste, touch and sound are derived from these five gross physical elements of fire, earth, water, air and ether respectively.

These gross physical elements are believed to be woven together to form the universe where each is dependent on the other. If a disturbance is made in one part of this web, its balance will be upset which will cause disharmony somewhere else. This disharmony may not merely be in the outside world but also in the internal constitution of our body.

The Vaisheshika School deals mainly with seven "*padarthas*" or categories, namely: (i) *dravya* (substance); (ii) *guna* (quality); (iii) *karma* (action); (iv) *samanya* (generality); (v) *vishesha* (particularity); (vi) *samavaya* (inherence); and (vii) *abhava* (non-being). This may be interpreted as: The *dravyas* or substances make up different composite wholes in Nature each with distinct *gunas* or qualities. *Guna* inheres in a whole and is static and permanent. Each of these composite wholes performs some action or *karma* which is dynamic and transient but like *guna* it also inheres in a substance. Though each of the composite wholes is exclusive and different, they have particularity (*vishesha*) and yet they have also generality (*samanya*) in them owing to their common characteristics of being constituted of the same *dravyas*. Therefore, there should be *samavaya* or inseparable eternal relation among them, which when absent may lead to *abhava* or non-existence of the entire world.

The Nyaya system is closely allied to the Vaisheshika system. While the former develops logic and epistemology, the latter develops metaphysics and ontology. Both agree that the bondage of the soul is due to ignorance of reality and that liberation is due to the right knowledge of reality. Similarly, the different environmental problems we are facing today are essentially due to our ignorance of the proper use of the earth resources and our exploitative attitude. It can be solved by proper knowledge and understanding of the use of resources and their optimum and sustainable harvest. Hence, in the first stanza of the *Isha Upanishad* it is said that "resources are given to mankind for their living; knowledge (*isha*) of using them is necessary".

The Doctrine of Karma finds its manifestation in the Nyaya Theory of Causation. The Nyaya, like the Vaisheshika, believes in teleological creation. It advocates theism and spiritualism and refers to God as the Creator, Preserver and Destroyer of the world and introduces the elements of love and devotion. The world is seen as an effect and hence it must have an efficient cause, which is God. The elements forming the world are inactive and cannot lead to any formation unless God gives them motion. God is also the moral governor of all beings who decides the fruits of our actions. We reap the fruits of our own actions. Merit and demerit accrue from our actions. This Doctrine of Karma, so dominant in Hindu tradition, is very important from the standpoint of environmental ethics. It points out that we must carefully decide our actions and keep sustainable development in mind or our unbridled exploitation of Nature may hurtle us nearer to "Doomsday".

The word "*mimamsa*" is applied to the interpretation of the Vedic rituals. Mimamsa and Vedanta are treated as allied systems of thought as both are based on and both try to interpret the *Vedas*. While *Mimamsa* deals with the earlier portion of the Vedas, i.e. the *Mantra* and the *Brahmana* dealing with rituals and sacrifices; the *Vedanta* deals with the latter portion, i.e. the *Upanishads* disseminating knowledge of the reality. *Mimamsa* deals with Dharma and *Vedanta* deals with Brahma

Dharma is the subject of inquiry in *Mimamsa*. It is a command or injunction which impels men to action. Of the four human values, only dharma and *moksha* (liberation) deal with true spirituality and are revealed by the Vedas, while *artha* and *kama* deal with common morality and are learnt by worldly intercourse. Dharma and *adharma* deal with happiness and pain to be enjoyed or suffered in the life beyond. According to Hinduism, the ultimate goal of human life is liberation described by the *mimamsikas* as "*apavarga*" or the attainment of heaven (*svarga*). Dharma, translated as religion, is the source by which liberation may be fully realized. Liberation is not the stillness of death but a dynamic harmony among all diverse facets of life. Humanity, as part of the natural world, can contribute through dharma to this natural harmony.

Finally, coming to the Vedanta School dealing with the knowledge of reality, its most popular viewpoints have been put forward by Shankaracharya as "*Advaita Vedanta*". It asserts that the ultimate reality is Brahma, which is pure consciousness and involves a vision of oneness – "*advaita*". The entire world created by the Lord (*ishvara*) is a manifest of this oneness. However, *jiva* (individual self) is ignorant of this essential unity due to *avidya* (ignorance), believes that only diversity is true and wrongly regards himself as the sole beneficiary. *Avidya* vanishes at the dawn of knowledge. At the same time, it dismisses the material world by referring it to a creation of *maya* (illusion). So in other words, it tries to renounce the materialistic pleasure that we crave for and that has become the hallmark of our lives in modern times.

Of the schools which renounce the Vedas, i.e. the *nastikas*, even Charvaka — which regarded sensual and material pleasure as the summum bonum of life — recognized the importance of the four elements of *vayu, prithvi, ap* and *agni*. It is of the view that bodies, senses and objects are the results of the different combination of elements.

The "*yamas*" of the Yoga School are contained in the *anuvratas* (minor vows) of Jain philosophy which are required to be fulfilled during the lifetime of man. They are the same as *ahimsa, asteya, satya, brahmacharya* and *aparigraha* of the Yoga School of philosophy.

The Hindu concept of "dharma" was adopted by Buddhism as the Doctrine of Pratityasamutpada or the Theory of Dependent Origination. Indian philosophy is deeply rooted in the *Vedas* which are the edifice of Hinduism. Simultaneously, the *Vedas* are also the true manifest records of environmentalism in Hindu religious thoughts. So, Hindu environmentalism and the ideas enshrined in the *astika* school of Indian philosophy may be said to be deeply interwoven. Not only this, even the *nastika* school – from which have emerged two great religions of the world, Jainism and Buddhism – have several ideas drawn from Hinduism which in turn have been inculcated into their spirit of environmentalism.

Modern ideas of environmentalism inspired by Hinduism

Religious interests are not the only ones that individuals are attempting to harmonize with environmental concerns. Many have turned to the feminist cause with the belief that its principles have much in common with those of environmentalism. It is supposed that if the two causes collaborate they might bolster each other's efforts and bring about positive changes more effectively. Environmentalism and feminism have emerged from the same theoretical structure of oppression, exploitation and abuse, of the natural world in the former case and women in the latter. A new movement, "ecofeminism", has emerged from the threshold of both.

The ecofeminist movement traces its origin in the 1970s when feminist Francois d'Eaubonne coined the expression "*ecological feminiane*". Later, Maria Mies and Vandana Shiva, who have been active in the feminist and environmental movements respectively, co-authored a book titled *Ecofeminism*. Some consider ecofeminism to have a significant place in Hinduism. Though the Vedas appear to promote an environment-friendly attitude, they do not seem to take a keen interest in the status of women. However, "Shaktism" is an important branch within the Hindu tradition that lays a primary emphasis on goddess worship and on the concept of *Shakti* that may be simply translated as "power". In *Shakta Tantra* women are being perceived as the goddess in human form. By the same rationale, earth (*prithvi*) has been portrayed as feminine.

Shaktism, especially its Tantric aspects, believes that the most effective way to halt and reverse the abusive treatment of women and Nature is to go to the source, which means deconstructing the destructive principles that have allowed such treatment to persist. Not only does the Shakta tradition provide a prominent place for the feminine principle, *Shakta Tantra* reinterprets the traditional hierarchical system which exists in Hinduism in favour of the non-dualist concept of "*ardhanarishwar*".

The new environmentalism in the form of "deep ecology" has gained considerable popularity in the West in the recent past. Many environmental activists now claim that deep ecology is their guiding principle. The term was coined by Arne Naess. Naess pointed out that a shallow but influential ecological movement and a deep but less influential one compete for our attention. He characterized the former as one that fights environmental degradation to preserve human health and affluence while the "deep" ecological movement operates out of a deep-seated respect and even veneration for ways and forms of life, and accords them an "equal right to live and blossom".

Naess noted the identification of the "self" with "Self" in the sense that is used in the *Bhagavad Gita* (i.e. as the unity which is one) as the source of deep ecological attitudes. In other words, he linked the tenets of his approach to ecology with what may be termed "self-realization". This connection may be summarized as:

i. Self-realization presupposes a search for truth;
ii. All living beings are one;
iii. Violence against oneself makes complete self-realization impossible;
iv. Violence against a living being is violence against oneself; and,
v. Violence against a living being makes self-realization impossible.

Gandhi, a true 'Hindu environmentalist'

Mahatma Gandhi, the 'Father of the Nation' of modern India, was a practitioner of sustainable development in the real sense of the word. His life and work is an environmental legacy for all humanity. A true Gaian, he understood the primordial man-Nature relationship and his theory and philosophy of life, society and politics are in consonance with it. Thus, he remarked, "the earth provides enough for everyone's need but not for everyone's greed".

Years before India became independent, when asked if he would like to see Free India to be as developed as her colonial ruler, Gandhi replied in the negative with the question that if it took Britain the pillage of half the resources of the world to reach its contemporary levels of development, then how many worlds would India need to do the same? Gandhiji was against the "oil-coal-metal economies" and formulated a pattern of life for man's peaceful, purposeful and happy existence based on three cardinal principles:

i. Simplicity, i.e. one of contentment based on the possession of only the bare necessities of life;
ii. Slowness, for life to be smooth and free from tension and from the din and bustle of everyday urban life; and,
iii. Smallness, as a good life can be lived only in a small community.

Gandhiji's thoughts were profoundly inspired by Hindu religious thought and philosophy. He too believed in the divinity in all life forms and a fundamental unity in diversity. He took up the cause of women and other oppressed sections of society and was a practising "*yogi*" who observed all *yamas* and *niyamas*. He encouraged the three 'S's of:

i. Indigenous capability and local self-reliance (*swadeshi*);
ii. Self-rule and local self-governance (*swaraj*); and,
iii. Welfare of all (*sarvodaya*).

Sarvodaya is derived from the Sanskrit words "*sarva*" meaning 'all' and "*udaya*" meaning 'awakening', or the awakening of all in every respect integrating everything pertaining to man, society and the environment. It can be realized through various stages:

i. *Purushodaya* (personal awakening);

ii. *Kutumbodaya* (family awakening);

iii. *Gramodaya* (village community awakening);

iv. *Nagarodaya* (urban community awakening);

v. *Deshodaya* (national awakening); and,

vi. *Sarvodaya* (universal awakening).

This is true because the enemy of the Nature is latent within each one of us and only when we control our personal greed can we think of the welfare of the world order.

Conclusion

Hinduism offers unique resources for the creation of an earth ethic. If instead of considering it as merely a religion we treat it as an all-inclusive philosophy then Hinduism too may provide the essence of the ideals that comprise and sustain "ecosophy". The famous Hindu dictum of *vasudhaiva kutumbakam* (universal brotherhood) might lead to the enhancement of the core man-environment relations. The variegated theologies of Hinduism suggest that:

i. The earth can be seen as manifestation of the goddess and must be treated with respect;

ii. The five elements hold great power;

iii. Simple living might serve as a model for the development of sustainable economies; and,

iv. The concept of 'dharma' can be re-interpreted from an earth-friendly perspective.

This Vedic view of the earth is based on a spiritual foundation which may provide a firm and effective platform for a pan-global ecological movement today. Gandhi as a world leader had already shown the way a century ago.

References

1. Brown, E.H. (1970): *Man Shapes the Earth*, Geographical Journal, Vol. 136.

2. Dikshit, R.D. (1998): *Geographical Thought: A Contextual History of Ideas*, Prentice Hall of India Pvt. Ltd., New Delhi , India.

3. Hick, John H. (2003): *Philosophy of Religion*, Fourth Edition, Prentice Hall of India Pvt. Ltd., New Delhi , India.

4. Hiriyanna M. (1993): *Outlines of Indian Philosophy*, First Indian Edition, Motilal Banarsidass Publishers Pvt. Ltd., Delhi, India.

5. Jha Sreekrishna (1997): *Mahatma Gandhi–An Environmentalist with a Difference*, Gandhian Institute of Studies, Varanasi.

8

Biodegradation of Aminobenzenesulfonates by a Bacterial Co-culture

SURJEET SINGH, S.K.AWASTHI, PRATIMA BAJPAI, POONAM SINGH AND LEELA IYENGAR

Introduction

Aminobenzenesulfonates (ABS) are important constituents of many azo dyes, pesticides and pharmaceuticals (Lidner, 1985). The presence of the sulfonic acid group on the aromatic ring confers the Xenobiotic and polar characters to these compounds, as these compounds are seldom synthesized in nature (Laskin and Lechevelier, 1984). Further, the sulfonic acid group is completely dissociated at the physiological pH balance level making these compounds highly polar, thus limiting their entry into the cell in the absence of specific transport proteins. Nevertheless, a few mixed cultures and pure bacterial strains, which could utilize ABS as the sole carbon and energy source, have been isolated. The studies on 2-ABS degradation, with *Alcaligenes* sp. strain O-1, have shown that this strain can utilize 2-ABS, 4-toluenesulphonate and benzenesulfonate as growth substrates (Kneimeyer et al., 1990). On the other hand, 4-ABS degrading pure cultures exhibit a very narrow specificity and can utilize this specific isomer. One approach is to develop separate enrichment or pure cultures for the degradation of each of these isomers and explore their suitability for the degradation of mixed substrates. Few ABS degrading cultures have been recently isolated in our laboratory. The aim of this study is focused on the specificity to culture towards each ABS isomer as well as on the degradation of mixtures of these isomers using a combination of these cultures.

Materials and Methods

Bacterial cultures

4-ABS degrading bacterial strain PNS-1 were isolated from the enrichment culture developed using activated sludge from an aerobic biological unit treating domestic wastewater in Kanpur City. Bacterial consortium (BC), which can utilize 2-ABS as the sole carbon and energy source, was derived from the sludge taken from an effluent treatment plant of a large chemical manufacturing industry located at Rasayani, India.

Medium and culture conditions

The growth medium (MM) used, in the present study, consisted of the following constituents: 12.0 g Na_2HPO_4, 2.0 g K_2HPO_4, 0.5 g NH_4Cl, 0.1 g $MgCl_2.6H_2O$ and 0.05 g $CaCl_2.2H_2O$ per liter of distilled water. 1 ml filter sterilized trace element solution (Kneimeyer et al., 1990) and required volumes of ABS from a stock solution (5 gL^{-1} neutralized to pH 7.0 using 1N NaOH) were added to the growth medium after sterilization. Final pH of the medium was 7.0 ±0.2. Cultures were grown in 100 ml liquid medium taken in 250 ml Erlenmeyer flasks, which were kept in a rotary shaker incubator (120 rev min^{-1}) at 35±2 °C.

ABS degradation and bacterial growth

ABS degradation was monitored at an initial concentration of 400 mg/l. Strain PNS-1 or BC, grown up to a late exponential phase and was used as the inoculum (10% v/v) for studies on the degradation of individual isomers. At the Initial biomass optical density at 555 nm was generally below 0.1. Aliquots were withdrawn periodically. Bacterial growth was determined by measuring the turbidity at 555nm. Samples were then centrifuged at 1100×g (3400 rpm) and ABS was estimated in the supernatant. Uninoculated controls with the organic carbon source were always included in experiments.

The degradation of ABS mixtures was studied at an individual ABS isomer concentration of 400 mgL^{-1} and cultures of strain PNS-1 and BC were used as the inocula. Kinetic studies on ABS removal were carried out, after growing the mixed culture for three cycles on mixed ABS substrates. Chemical oxygen demand (COD) was determined with 0.45 ì membrane filtered culture samples taken out just after inoculation and at the end of the exponential growth phase.

Analytical Procedures

Biomass growth was monitored at 555 nm against distilled water in a UV-Visible Spectrophotometer. An optical density of 1.0 represented 340 mg cell dry weight per liter. ABS isomers were estimated by measuring the absorbance at their $ë_{max}$ in the UV-Visible Spectrophotometer. When used in combination, 2 and 4-ABS were quantified by measuring the absorbance at 244 nm for total ABS removal and at 290nm for 2-ABS removal, as 4-ABS did not exhibit the absorbance at 290 nm. Chemical oxygen demand (COD) was determined by a close reflux method as per the procedure given in the standard methods (Greenberg et al., 1992).

Results

ABS degrading bacterial cultures

Enrichment cultures for 2-ABS and 4-ABS degradation were developed in batch cultures under aerobic conditions with the specific isomer as the growth substrate. Strain PNS-1 was isolated from 4-ABS degrading enrichment cultures. The Sequencing of the PCR-amplified 16S rRNA gene and its comparison with available information in the database showed that the strain PNS-1 clustered with á subclass of proteobacteria and was in close similarity with the genus, *Agrobacterium* (Singh et al., 2006). The sequence has been deposited in the NCBI GenBank Pubmed database (htpp://www.ncbi.nlm.nih.gov/) with accession number AY 00762361.

Cultivation of an enrichment culture for more than a year on 2-ABS, as the sole carbon and energy source, yielded a stable bacterial consortium (BC), which consisted of two bacterial

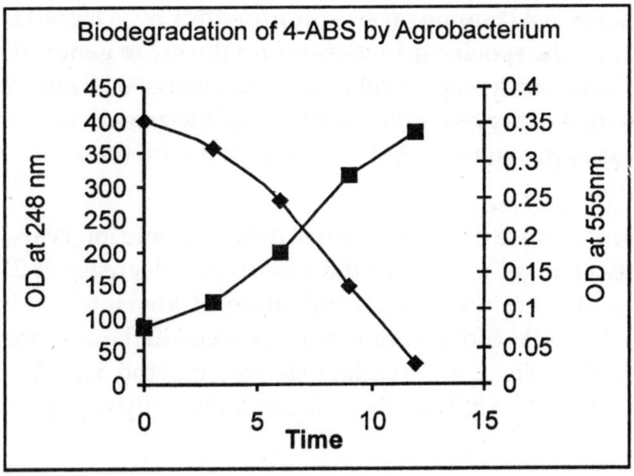

Figure 1. Degradation of ABS by Agrobacterium sp. strain PNS-1(●) 4-ABS and (■) Growth.

strains. Present investigations were carried out with 4-ABS degrading *Agrobacterium* sp. strain PNS-1 and 2-ABS degrading BC.

Degradation of ABS isomers by Agrobacterium sp. strain PNS-1 –

Kinetics of 4-ABS degradation by strain PNS-1 is shown in Fig.1. 4-ABS were rapidly utilized and >95% degradation was observed in 12 h. The specific growth rate (ì) and mean generation time were calculated to be 0.19 h^{-1} and 3.6 h respectively. Biomass yield was 0.34 mg/mg 4-ABS degraded. Mineralization of the substrate was ascertained by, COD analysis of the culture

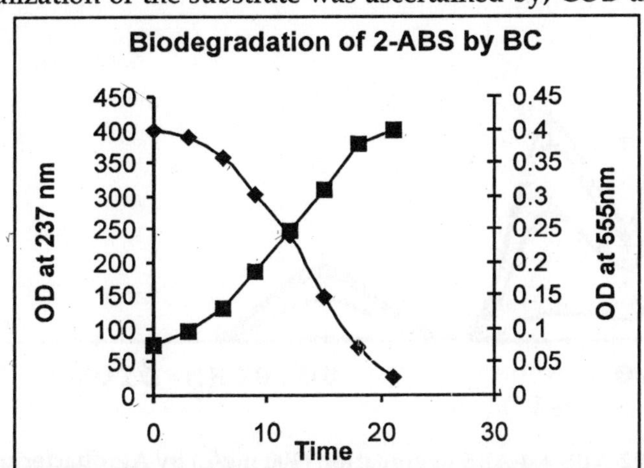

Figure 2. Degradation of 2-ABS by Bacterial Consortium (●) 2-ABS and (■) Growth.

filtrate prior to and after the growth of strain PNS-1 on 4-ABS up to a late exponential phase. Theoretical COD for ABS is 1.4 mg/mg ABS.

Degradation of ABS by BC

Fig. 2 presents the kinetics of 2-ABS removal and the growth of BC. Degradation of 400 mg L^{-1} 2-ABS required around 21 h. The specific growth rate and the mean generation time for BC was calculated to be 0.104 h^{-1} and 6.65 h respectively. Biomass yield of 0.38 mg/mg 2-ABS degraded was marginally lower to that observed with 4-ABS. Thus, the growth rate of BC with 2-ABS as the growth substrate was approximately half as compared with the strain PNS-1 on 4-ABS at equimolar concentrations.

UV-Visible spectra of culture filtrates drawn at different time intervals, during the growth phase, did not show any changes except for the absorbance decrease at 237 nm (spectra not shown). This showed that there was no accumulation of any intermediate during 2-ABS degradation. No change in 2-ABS concentration was observed in the absence of the culture or in the presence of heat killed BC. These observations showed that the decrease in absorbance was due to biodegradation of 2-ABS. COD analysis indicated extensive mineralization of 2-ABS **(Table 1).**

Batch degradation studies were conducted, using either 2- and 4-ABS or 2-, 3- and 4-ABS at an initial substrate concentration of 400 mgL^{-1} of each isomer, using the co-culture of strain PNS-1 and BC. When 2- and 4-ABS were used together as growth substrates, 4-ABS was undetected beyond 12 h and >90% 2-ABS degradation was observed in 21 h **(Fig. 3).** Further neither

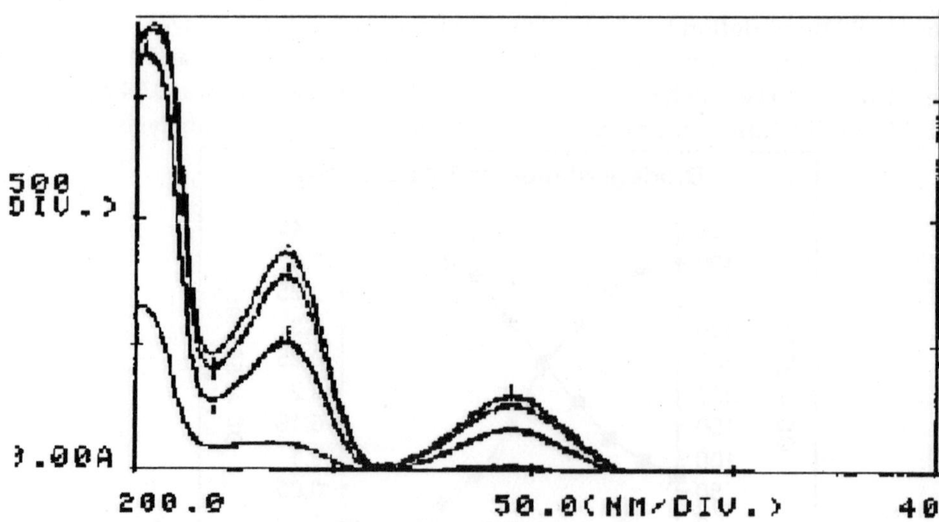

Figure 3. UV spectra of 2-ABS + 4-ABS degradation (400 mg/L) by Agrobacterium sp. strain PNS-1 and two membered bacterial consortiums.

2- or 4-ABS was preferentially utilized. UV-Visible spectrum as well as percent COD removal **(Table 1)** after the growth of the co-culture indicated mineralization of both these isomers.

Table 1. COD removal for 2-ABS (400 mg/l) +4-ABS (400 mg/l)

Influent COD (mg/l)	Effluent COD (mg/l)	% of COD Removal
1180	180	83.63
1100	80	92,72
1120	100	91.07

CONCLUSION AND DISCUSSION

There are few reports on mixed and pure bacterial cultures, which can utilize specific ABS isomers as the sole carbon and energy source. Studies on the mineralization of a combination of these isomers by a co-culture are reported in this communication.

2- and 4-ABS degrading cultures were developed in the laboratory using a batch enrichment technique. It was observed that 4-ABS degrading enrichments could be developed with several inocula. *Agrobacterium* sp. strain PNS-1 was isolated from one such enrichment. On the other hand, 2-ABS degrading bacterial consortium could be derived only from one source inoculum. Both strain PNS-1 and BC were highly specific and could utilize only 4-ABS and 2-ABS respectively.

A co-culture of Strain PNS-1 and BC could utilize 2- and 4-ABS. Complete Biodegraded of 2- and 4-ABS by the co-culture was shown by COD analysis. 3-ABS could not be utilized by the co-culture.

Bacterial genes encoding enzymes required for the biodegradation of aromatic pollutants are often regulated in response to the availability of the respective substrate. However, if a rapidly metabolizing carbon source is additionally present (which is often the case in wastewaters), then the synthesis of peripheral enzymes required for the pollutant degradation, can be affected. It will be interesting to determine whether the gene for the degrading enzymes is presented on bacterial plasmid or on chromosomal.

References

1. Coughlin, M.F., Kinkle, B.K., Bishop, P.L. High performance degradation of azo dye acid orange 7 and sulphanilic acid in a laboratory scale reactor after seeding with cultured bacterial strains. Wat. Res. 2003; 37:2757-2763.
2. Feigal, B.J., Knackmuss, H.J. Syntrophic interactions during degradation of 4-aminobenzenesulfonic acid by a two species bacterial culture. Arch. Microbiol. 1993; 159: 124-130.
3. Greenberg, A.E., Clrsceri, L.S., Eaton, A.D. APHA Standard Methods for Examination of Water and Wastewater. 18 ed, Washington, D.C. APHA; 1992. ISBN: 0-87553207.

4. Junker, F., Leisinger, T., Cook, A.M. 3 Sulphocatechol 2,3 dioxygenase and other dioxygenases (E.CI.13.II.2 and ECI.14.12.) in the degradative pathway of 2-aminobenzenesulfonic acid and 4-toluenesulphonic acids in Alcaligenes sp. Strain O-1. Microbiol. 1994; 140: 1713-1722.

5. Kneimeyer, O., Probian, C., Rosello-Mora, R., Harder, J. Anaerobic mineralisation of quarternary carbon atoms: isolation of denitryifying bacteria on dimethylmalonate. Appl Environ. Microbiol. 1999; 65: 3319-3324.

6. Laskin, A.I., Lechevalier, H.A. Microbial products, CRC Handbook of Microbiology, (2nd edn.) CRC press; 1984. Boca Raton, Florida, 5, 111-127, 576.

7. Lidner, O. Benzenesulfonic acid and their derivaties, In: Ullmann's Encylopedia of Industrial Chemistry, 5th edn., ed. Gerhartz W.A3, Weinheim Germany, VCH Verlagsgesellschaft; 1985. ISBN 3-527-20103-3.

8. Perei, K., Rakhely, G., Kiss, U.I., Polyak, B., Kovcas, K.L. Biodegradation of sulfanilic acid by *Pseudomonas paucimobilis*. Appl. Microbiol and Biotechnol. 2000; 55: 101-107.

9. Singh, P., Birkeland, N.K., Iyengar, L., Gurunath, R. Mineralization of 4-aminobenzenesulphonate (4-ABS) by Agrobacterium sp. strain PNS-1. Biodegradation 2006; 17: 495-502.

10. Tan, N.G.C., van Leeuwen, A., van Voorthuizen, E.M., Slenders, P., Prenafeta-Boldu, F.X., Temmink, H., Lettinga, G., Field, J.A. Fate and biodegradability of sulfonated aromatic amines. Biodegradation. 2005; 16: 527-537.

11. Thurnheer, T., Kohler, T., Cook, A.M., Leisinger, T. Orthanilic acid and analogues as carbon sources for bacteria: Growth physiology and enzymic desulphonation. J.Gen. Microbiol. 1986; 132: 1215-1220.

12. Thurnheer, T., Cook, .A.M., Leisinger, T. Co-culture of defined bacteria to degrade seven sulphonated aromatic compounds: efficiency rates and phenotypic variations. Appl. Microbiol. & Biotech. 1988; 29: 605-609.

13. Thurnheer, T., Zurrer, D., Hoglinger, O., Leisinger, . T., Cook, A.M. Initial steps in the degradation of benzenesulfonic acid, 4-toluene sulfonic acid and orthanilic acid in Alcaligenes sp. Strain O-1. Biodegradation 1990; 1:55-64.

9

Biodiversity and Role of Microbiology and Biotechnology in Environment Biopesticide

IYER VARSHA

Biodiversity

Biodiversity is the shortened form of two words-'biological' and 'diversity'. It refers to all the varieties of life that can be found on earth as well as to the communities that they form and the habitats in which they live.

The Convention of Biological Diversity gives a formal definition of biodiversity in its Article 2...'Biological diversity means the variability among living organisms from all sources including inter alias, terrestrial,marine and other aquatic ecosystems and the ecological complexes of which they are part; this includes diversity within species, between species and of ecosystems.'

Biodiversity is described by two parameters: "(i) point or alpha diversity represented by the number of species in a specified area and (ii) beta diversity represented by the turnover of species across space. The diversity increases with the total number of individuals encompassed and thus with the increase in the area sampled and the productivity per unit area . Diversity depends on how large are speci' ranges and following two scenarios may be found: (i) if the range is large, a-diversity is independent of the area sampled, (ii) if diversity is low with the species' ranges being small and non-overlapping, many parks will be needed to protect diversity.

Importance of Biodiversity

Biological diversity has direct consumption value in food; agriculture, medicine and industry .It also has aesthetic and recreational value. Biodiversity maintains the ecological balance and the continuous evolutionary processes. The indirect ecosystem services provided through biodiversity are photosynthesis, pollination, transportation, chemical cycling, soil maintenance, climate regulation, air and water system management, and waste treatment and pest control.

Current Levels of Biodiversity

Due to the destruction of natural habitats by human interference, biodiversity is being lost at a fast rate, particularly in the tropical regions, where not only biodiversity is high, but also the human interference is high due to population pressure. It is estimated that almost 40% of the net primary productivity (NPP-which makes food supply of all animals) on land is directly used or lost due to the activities of a single species - Homo sapiens. Further the human population is projected to become 10 billion in the next 50 years, and this is supposed to be accompanied with 5-10-fold increase in economical activity. If anything close to this level of increase in population and economical activity be witnessed in the future, then the World's biodiversity is destined to be completely wiped out from the surface of the Earth. This demonstrates the crisis of the loss of biodiversity and the need to conserve it.

The ecosystem services are provided by biodiversity at such a grand scale, that there is no possible substitute for these services even with the advances in knowledge made in recent decades. Recent attempts for finding inorganic substitutes for these services have met with little success. These attempts include the following (i) synthetic pesticides, (ii) inorganic fertilizers, (iii) chlorination for purification of natural waters, (iv) dams for flood and drought control, (v) air-conditioning for high temperatures etc. All these substitutes destroy the environment, and often require a large energy subsidy, thus adding to the environmental problems. Therefore, these are not satisfactory, thus forcing us to think now of Biopesticides, Biofertilizers, afforestation, etc.to deal with these problems .In other words, we need to restore biodiversity, which is being destroyed due to our use of synthetic devices as substitutes for ecosystem services.

Nature of Problems

A Pesticide is a pest killing agent (the Latin word-cida means to cut or kill) .The term usually refers to one or more materials developed and used to destroy a broad range of specific pests. Pesticides may be defined as 'any substance used for controlling, preventing

destroying, repelling or mitigating any pest'.Pesticides are a key input in meeting world food needs and a factor overcoming the continuing periods of serious food shortages which have occurred since the Second World War. Modern technological innovation in agriculture has helped to boost the production of food grains by making judicious use of agrochemicals including pesticides. Each year pest species consume or destroy an estimated 48% of the world's food supply. An estimated 5 billion pounds of toxic chemicals are used annually worldwide to control plant and animal pests. The developing countries use 20% of pesticides; whereas, the U.S.A alone use about 20% of the total pesticides! According to the World Health Organization, technical pesticides data contains a total of 759 chemical and biological pest control agents. Of the seven hundred plus chemical pesticides considered to be in current use, 33 have been classified as extremely hazardous to human health and have a high environmental impact.

For example

Inorganic Herbicides

Ammonium sulphate

Sulphuric acid

Sodium borate

Sodium chlorate

Organic Herbicides

Disodium methanearsonate

Monosodium methanearsonate

Carbonates

Thiocarbamates

Aliphatic acids

Phenol derivatives, Diphenylethers

Fungicides

A fungicide refers to chemicals capable of preventing or eradicating diseases caused by fungi and bacteria .The pathogens of bacterial plant diseases may sometimes be controlled with bactericides or antibiotics. Fungicides are generally not effective against viral diseases.

However, because insects, mites, or nematodes transmit many viral diseases, their control is sometimes possible by using insecticides, matricides and nematicides to destroy these vectors.

For examples

Inorganic Fungicides

a) Sulfur fungicides (elemental form)

b) Copper fungicides

 i) Cupric carbonates

 ii) Copper hydroxide

 iii) Copper oxy chloride

 iv) Copper sulphate

 v) Cuprous oxide

Organic Fungicides

a) Dithiocarbamates

b) Organometallic compounds

c) Systematic Fungicides

d) Other Organic Compounds

Fumigants:

Fumigants are substances or a mixture of them, which produce gas, vapour, fumes or smoke intended to kill insects, nematodes, bacteria or rodents. Usually, fumigants are volatile liquids or solids, or gaseous substances already consisting of small molecules, which often contain halogen radicals. They are used to disinfect buildings, stored products or soils.

Examples:

Chloropicrin	Methyl bromide
Ethylene dibromide	Naphthalene
Formaldehyde	Phosphine

·Pesticides are generally applied either as a preventative treatment to protect crops or stored products against diseases or infestations by animal pests, or as a curative treatment to destroy or limit population development of noxious organisms. The application of chemical pesticides may be targeted to field crops, plantation crops, protected crops, seeds and plant materials, stored products, buildings, surface water, right-of-way- areas, etc.

Chemical pesticides will continue to play a vital role in increasing agricultural production and in securing the supply of food and fibres needed. However, it is essential that the reasons for using pesticides and the consequences be carefully analyzed in order to obtain maximum benefits from their application, while at the same time preventing and remedying their possible hazardous effects on non-target organisms and the environment.

Advantages of Chemical Pesticides

Pesticides are an effective and reliable means for controlling pests and diseases and preventing losses in the field and in storage. Pesticides offer protection against many diseases, except for those inflicted by particular bacteria, viruses and mycoplasms.

- Pesticides are used when there is no time left for limitation by a build up of natural enemies and the ripening produce is being threatened. Serious infections and outbreaks can be limited or possibly even controlled due to the fast action of chemical pesticides.
- Chemical pesticides are effective under very diverse ecological conditions and they are less dependent on the scale of the operation than the various forms of cultural and biological control.
- The use of chemical pesticides is only the option in case of intensive pest problems in agriculture, horticulture, household pest and rodent control and storage protection.
- In modern plant protection, the proper preventive use of pesticides, such as the disinfections of soils and seeds, and the protective spraying of fungicides are indispensable methods.
- Pesticides are readily available, in developing countries, at prices affordable even to poor farmers.
- Pesticides consumption is often stimulated through liberal subsidizing by National Government and through credit schemes, which include the supply of fertilizers and pesticides.

Disadvantages of Chemical Pesticides

A) Chemical pesticides, either of a specific type or a chemical group of pesticides, sometimes cause the development of resistance in the population of organisms, many species of insects, mite's nematodes, microorganisms and weeds.

B) When the chemical pesticides become gradually less effective, which necessitates the application of higher dosages, higher application frequencies for the mixture and rapid replacement of those chemicals become new products. This is so called "Pesticides trademill".

C) Improper application of pesticides may cause unwanted or hazardous effects and be a grave source of environmental contamination

D) Pesticides selectivity and dosage need to be taken into account with regard to the specific sensitivity of the target organisms and natural enemies .The application technology must correspond with their behaviour and environmental condition.

E) A profound knowledge and skill is needed for the responsible use of chemical pesticides; qualities which are often scarcely available among farmers and plant protection technicians in developing countries.

F) The cost of chemical pesticides is rising steadily .The increasing trend of the rising price of chemical pesticides will adversely affect farmers as farm crop prices have generally remained static at levels below the world market values.

G) The increasing complexity of newer pesticides requiring a higher investment in research and manufacturing results in a higher retail price for the user.

H) The long-term persistence of chemical pesticides in the environment and its inherent toxicity is of grave concern for non-target organisms and mankind.

FOLLOWING IS A PRESENTATION OF ONE OF THE CHEMICAL PESTICIDE i.e. DDT TO SHOW ITS EFFECT ON THE ENVIRONMWNT.

Properties and chemistry of DDT

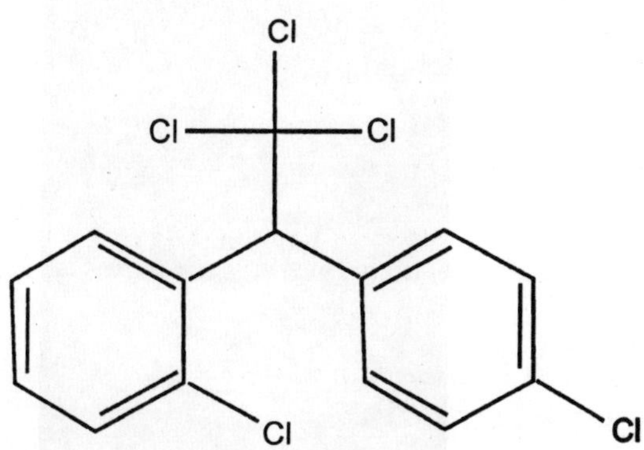

***o,p*-DDT, a minor component in commercial DDT.**

DDT is an organochlorine insecticide, similar in structure to the pesticides dicofol and methoxychlor. It is a highly hydrophobic, colorless, crystalline solid with a weak, chemical odor. It is nearly insoluble in water but has a good solubility in most organic solvents, fats, and oils. DDT does not occur naturally, but is produced by the reaction of chloral (CCl_3CHO) with chlorobenzene (C_6H_5Cl) in the presence of sulfuric acid, which acts as a catalyst. Trade names that DDT has been marketed under include Anofex, Cesarex, Chlorophenothane, Dedelo, Dicophane, Dinocide, Didimac, Digmar, ENT 1506, Genitox, Guesapon, Guesarol, Gexarex, Gyron, Hildit, Ixodex, Kopsol, Kybosh, Neocid, OMS 16, Pentachlorin, Pennsalt, Puritan, Rukseam, R50, and Zerdane.

Commercial DDT is actually a mixture of several closely related compounds. The major component (77%) is the *p,p* isomer which is pictured at the top of this article. The *o,p* isomer (pictured to the right) is also present in significant amounts (15%). Dichlorodiphenyldichloroethylene (DDE) and dichlorodiphenyldichloroethane (DDD) make up the balance. DDE and DDD are also the major metabolites and breakdown products of DDT in the environment. The term "**total DDT**" is often used to refer to the sum of all DDT related compounds (*p, p*-DDT, *o, p*-DDT, DDE, and DDD) in a sample.

Production and use statistics

From 1950 to 1980, when DDT was extensively used in agriculture, more than 40,000 tonnes were used each year worldwide, and it has been estimated that a total of 1.8 million tonnes of DDT have been produced globally since the 1940s. In the U.S., where it was manufactured by Ciba, Montrose Chemical Company and Velsicol Chemical Corporation, production peaked in 1963 at 82,000 tonnes per year. More than 600,000 tonnes (1.35 billion lbs) were applied in the U.S. before the 1972 ban, with usage peaking in 1959 with about 36,000 tonnes applied that year.

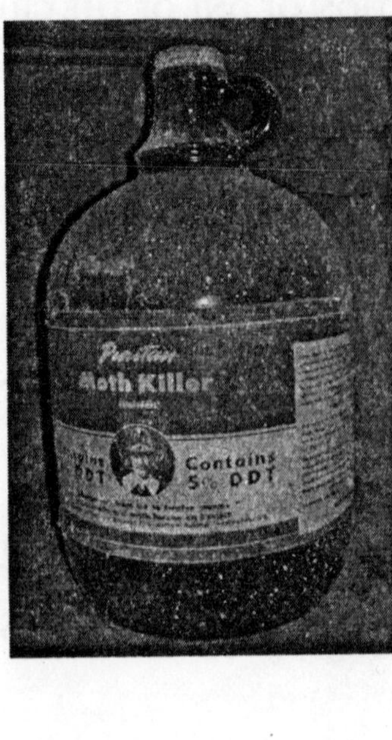

Currently about 1,000 tonnes of DDT are used annually worldwide in vector control operations.India and China are the only countries still producing and exporting the chemical.

Mechanism of action

DDT is moderately toxic, with a rat LD_{50} of 113 mg/kg, and has potent insecticidal properties; it kills by opening sodium ion channels in insect neurons, causing the neuron to fire spontaneously. This leads to spasms and eventual death. Insects with certain mutations in their sodium channel gene may be resistant to DDT and other similar insecticides. DDT resistance is also conferred by up-regulation of genes expressing cytochrome P450 in some insect species.

History

First synthesized in 1874 by Othmar Zeidler,DDT's insecticidal properties were not discovered until 1939 by the Swiss scientist Paul Hermann Müller, who was awarded the 1948 Nobel Prize in Physiology and Medicine for his efforts.

Use in the 1940s and 1950s

DDT is the best-known of a number of chlorine-containing pesticides used in the 1940s and 1950s. It was used extensively during World War II by Allied troops in Europe and the Pacific as well as certain civilian populations to control the insect vectors for typhus and malaria (nearly eliminating typhus as a result). Entire cities in Italy were dusted to control the typhus carried by lice. DDT also sharply reduced the incidence of biting midges in Great Britain, and was used extensively as an agricultural insecticide after 1945.

In 1955, the World Health Organization commenced a programme to eradicate malaria worldwide, relying largely on DDT. The programme was initially highly successful, eliminating the disease in "Taiwan, much of the Caribbean, the Balkans, parts of northern Africa, the northern region of Australia, and a large swathe of the South Pacific" and dramatically reducing mortality in Sri Lanka and India. However resistance soon emerged in many insect populations as a consequence of the widespread agricultural use of DDT. In many areas, early victories against malaria were partially or completely reversed, and in some cases rates of transmission even increased.The programme was successful in eliminating malaria only in areas with "high socio-economic status, well-organized healthcare systems, and relatively less intensive or seasonal malaria transmission".

DDT was less effective in tropical regions due to the continuous life cycle of mosquitoes and poor infrastructure. It was not pursued at all in sub-Saharan Africa due to these perceived difficulties, with the result that mortality rates in the area were never reduced to the same dramatic extent, and now constitute the bulk of malarial deaths worldwide, especially following the resurgence of the disease as a result of microbe resistance to drug treatments and the spread of the deadly malarial variant caused by *Plasmodium falciparum*. The goal of eradication was abandoned in 1969, and attention was focused on controlling and treating the disease. Spraying programmes(especially using DDT) were curtailed due to concerns over safety and environmental effects, as well as problems in administrative, managerial and financial implementation, but mostly because mosquitoes were developing resistance to DDT. Efforts were shifted from spraying to the use of bednets impregnated with insecticides and other interventions.

Silent Spring and the U.S. ban

As early as the 1940s, scientists had begun expressing concern over possible hazards associated with DDT, and in the 1950s the government began tightening some of the regulations governing its use.However, these early events received little attention, and it was not until 1957 when the *New York Times* reported an unsuccessful struggle to restrict DDT use in Nassau County, New York that the issue came to the attention of the popular naturalist-author, Rachel Carson. William Shawn, editor of *The New Yorker*, urged her to write a piece on the subject, which developed into her famous book *Silent Spring*, published in 1962. The book argued that pesticides, including DDT, were poisoning both wildlife and the environment and were also endangering human health.

Silent Spring was a best seller, and public reaction to it launched the modern environmental movement in the United States.DDT became a prime target of the growing anti-chemical and anti-pesticide movements, and in 1967 a group of scientists and lawyers founded the Environmental Defense Fund (EDF) with the specific goal of winning a ban on DDT. Victor Yannacone, Charles Wurster, Art Cooley and others associated with the inception of EDF had all witnessed bird kills or declines in bird populations and suspected that DDT was the cause. In their campaign against the chemical, EDF petitioned the government for a ban and filed a series of lawsuits. Around this time, toxicologist David Peakall was measuring DDT levels in the eggs of peregrine falcons and California condors and finding that increased levels corresponded with thinner shells.

In response to an EDF suit, the U.S. District Court of Appeals in 1971 ordered the EPA to begin the de-registration procedure for DDT. After an initial six-month review process, William Ruckelshaus, the Agency's first Administrator rejected an immediate suspension of DDT's registration, citing studies from the EPA's internal staff stating that DDT was not an imminent danger to human health and wildlife.However, the findings of these staff members were criticized, as they were performed mostly by economic entomologists inherited from the United States Department of Agriculture, whom many environmentalists felt were biased towards agribusiness and tended to minimize concerns about human health and wildlife. The decision not to ban thus created public controversy.

The EPA then held seven months of hearings in 1971-1972, with scientists giving evidence both for and against the use of DDT. In the summer of 1972, Ruckelshaus announced the cancellation of most uses of DDT—an exemption allowed for public health uses under some conditions.

The U.S. DDT ban took place amid a climate of growing public mistrust of industry, with the Surgeon General issuing a report on smoking in 1964, the Cuyahoga River catching fire in 1969, the fiasco surrounding the use of diethylstilbestrol (DES), and the well-publicized decline in the bald eagle population.

Restrictions on usage

In the 1970s and 1980s, the agricultural use of DDT was banned in most developed countries. DDT was first banned in Hungary in 1968 then in Norway and Sweden in 1970 and the US in 1972, but was not banned in the United Kingdom until 1984. The use of DDT in vector control has not been banned, but it has been largely replaced by less persistent, and more expensive, alternative insecticides.

The Stockholm Convention, ratified in 2001 and effective as of 17 May 2004, outlawed several persistent organic pollutants, and restricted the use of DDT to vector control. The Convention was signed by 98 countries and is endorsed by most environmental groups. Recognizing that a total elimination of DDT use in many malaria-prone countries is currently unfeasible because there are few affordable or effective alternatives for controlling malaria, the public health use of DDT was exempted from the ban until such alternatives are developed. Malaria Foundation International states:

The outcome of the treaty is arguably better than the status quo going into the negotiations over two years ago. For the first time, there is now an insecticide which is restricted to vector control only, meaning that the selection of resistant mosquitoes will be slower than before.

About 1,000 tonnes of DDT per year is still used today in countries where mosquito-borne malaria is a serious health problem.Use of DDT in public health to control mosquitoes is primarily done inside buildings and through inclusion in household products and selective spraying; this greatly reduces environmental damage compared to the earlier widespread use of DDT in agriculture. It also reduces the risk of resistance to DDT.This use only requires a small fraction of that previously used in agriculture; for example, to spray 1,700 homes, the required amount of DDT is estimated to be roughly equal to the amount that might have been used on 0.4 km^2 (100 acres) of cotton during a typical growing season in the U.S.Despite the worldwide ban on agricultural use of DDT, some farmers in India are known to still use it in crop production.

Environmental impact

DDT is a persistent organic pollutant with a half life of 2-15 years, and is immobile in most soils. Its half life is 56 days in lake water and approximately 28 days in river water. Routes of loss and degradation include runoff, volatilization, photolysis and biodegradation (aerobic and anaerobic). These processes generally occur slowly. Breakdown products in the soil environment are DDE (1,1-dichloro-2,2-bis(p-dichlorodiphenyl)ethylene) and DDD (1,1-dichloro-2,2-bis(p-chlorophenyl)ethane), which are also highly persistent and have similar chemical and physical properties. These products together are known as "total DDT". DDT and its breakdown products are transported from warmer regions of the world to the Arctic by the phenomenon of global distillation, where they then accumulate in the region's food web.

DDT and its metabolic products DDE and DDD magnify through the food chain, with apex predators such as raptors which have a higher concentration of the chemicals, stored mainly in body fat, than other animals sharing the same environment.This is called as **BIOMAGNIFICATION**. In the United States, human blood and fat tissue samples collected in the early 1970s showed detectable levels in all samples. A later study of blood samples collected in the latter half of the 1970s (after the U.S. DDT ban) showed that blood levels were declining further, but DDT or metabolites were still seen in a very high proportion of the samples. Biomonitoring conducted by the CDC as recently as 2002 shows that more than half of subjects

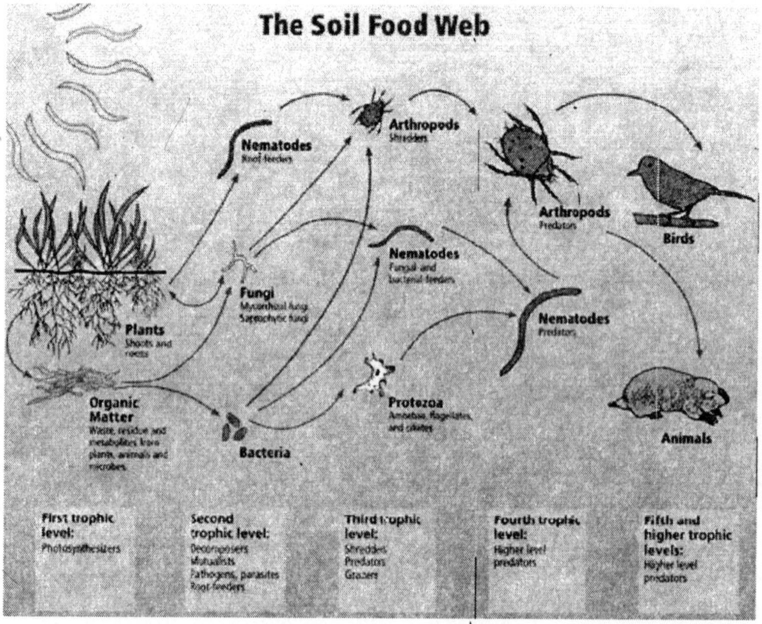

tested had detectable levels of DDT or metabolites in their blood, and of the 700+ milk samples tested by the USDA in 2005, 85% had detectable levels of DDE.

DDT is a toxicant across a certain range of phyla. In particular, DDT is a major reason for the decline of the bald eagle in the 1950s and 1960s as well as the brown pelicanand the peregrine falcon. DDT and its breakdown products are toxic to embryos and can disrupt calcium absorption, thereby impairing eggshell quality. Studies in the 1960s and 1970s failed to find a mechanism for the hypothesized thinning. However, more recent studies in the 1990s and 2000s have laid the blame at the feet of DDT. Some studies have shown that although DDT levels have fallen dramatically, eggshell thickness remains 10–12 percent thinner than before DDT was first used. DDT is also highly toxic to aquatic life, including crayfish, daphnids, sea shrimp and many species of fish. DDT may be moderately toxic to some amphibian species, especially in the larval stages. In addition to acute toxic effects, DDT may bioaccumulate significantly in fish and other aquatic species, leading to long-term exposure to high concentrations.

Effects on human health

The effects of DDT on human health are disputed since studies have yielded conflicting results.

Toxicity

Acute

DDT is classified as "moderately toxic" by the US National Toxicological Programme and "moderately hazardous" by the WHO, based on the rat oral LD_{50} of 113 mg/kg. It is not considered to be acutely toxic, and in fact it has been applied directly to clothes or used in soap.

Chronic

- Occupational exposure to DDT was associated with reduced verbal attention, visuomotor speed, sequencing, and with increased neuropsychological and psychiatric symptoms in a dose-response pattern (ie, per year of DDT application) in retired workers aged 55–70 years in Costa Rica. DDT or DDE concentrations were not determined in this study.
- Farmers exposed to DDT occupationally have an increased incidence of non-allergic asthma.
- Organochlorine compounds in general and DDE specifically have been linked to diabetes. A study of Native Americans exposed to DDE primarily from eating contaminated fish found that elevated blood DDE levels were associated with an increased incidence of diabetes. These results are consistent with previous studies on diabetes incidence and organochlorine exposure. A recent study of Mexican Americans yielded similar results.
- urate poisoning.

Cancer

- In 1987 the EPA classified DDT as a class B2 *probable* human carcinogen based on "Observation of tumors (generally of the liver) in seven studies in various mouse strains and three studies in rats. DDT is structurally similar to other probable carcinogens, such as DDD and DDE." Regarding the human carcinogenicity data, they stated "The existing epidemiological data are inadequate. Autopsy studies relating tissue levels of DDT to cancer incidence have yielded conflicting results."
- A study of malaria workers who handled DDT occupationally found an elevated risk of cancers of the liver and biliary tract. Another study has found a correlation between DDE and liver cancer in white men, but not for women or black men. An association between DDT exposure and pancreatic cancer has been demonstrated in a few studies, but other studies have found no association. Several studies have looked for associations between DDT and multiple myeloma, and testicular, prostate, endometrial, and colorectal cancers, but none conclusively demonstrated any association.
- A Canadian study from 2007 found a positive association between DDE and non-Hodgkin Lymphoma.
- A recent study in the Journal of the National Cancer Institute concluded that DDE exposure to may be associated with testicular cancer. The incidence of seminoma in men with the highest blood levels DDE was almost double that of men with the lowest levels of DDE.

DDT use against malaria

The World Health Organization estimates that there are between 300 million and 500 million cases of malaria every year, resulting in more than 1 million deaths, with about 90% of these deaths occurring in Africa, mostly to children under the age of 5.

Most prior use of DDT was in agriculture, but the controlled use of DDT continues to this day for the purposes of public health. Current use for disease control requires only a small fraction of the amounts previously used in agriculture, and at these levels the pesticide is much less likely to cause environmental problems. Residual house spraying involves the treatment of all interior walls and ceilings with insecticide, and is particularly effective against mosquitoes, which favour indoor resting before or after feeding. Advocated as the mainstay of malaria eradication programmes in the late 1950s and 1960s, DDT remains a major component of control programmes in southern African states, though many countries have abandoned or curtailed their spraying activities. South Africa, Swaziland, Mozambique and Ecuador are examples of countries that have very successfully reduced malaria infestations with DDT.

Today there is debate among professionals working on malaria control concerning the appropriate role of DDT. The range of disagreement is relatively narrow: Few believe either that large scale spraying should be resumed or that the use of DDT should be abandoned altogether. The debate focuses on the relative merits of DDT and alternative pesticides as well as the complementary use of interior wall spraying, insecticide-treated bed-nets, and other mosquito control techniques.

The growth of resistance to DDT and the fear that DDT may be harmful both to humans and the environment led donor countries and various national governments to restrict or curtail

Ladybird larva eating wooly apple aphids

Lacewings are available from biocontrol dealers.

the use of DDT in vector control. At the same time, the use of DDT as an agricultural insecticide was often unrestricted, and restrictions were often evaded, especially in developing countries where malaria is rife, so that resistance continued to grow.

Since the appointment of Arata Kochi as head of its anti-malaria division, the WHO has shifted its position in this controversy, from primary reliance on bed-nets to a policy more favorable to DDT. Until an announcement made on 16 September 2006, the policy had recommended the indoor spraying of insecticides in areas of seasonal or episodic transmission of malaria, but a new policy also advocates it where continuous, intense transmission of the disease causes the most deaths. In 2007, the WHO clarified its position, saying it is "very much concerned with health consequences from use of DDT" and reaffirmed its commitment to phasing out the use of DDT.

Natural Predators

Examples of predators

Ladybugs, and in particular their larvae which are active between May and July in the northern hemisphere, are voracious predators of aphids such as greenfly and blackfly, and will also consume mites, scale insects and small caterpillars. The ladybug is a very familiar beetle with various colored markings, whilst its larvae are initially small and spidery, growing up to 17 mm long. The larvae have a tapering segmented grey/black body with orange/yellow markings and ferocious mouthparts. They can be encouraged by cultivating a patch of nettles in the garden and by leaving hollow stems and some plant debris over winter so that they can hibernate.

Hoverflies resemble slightly darker bees or wasps and they have characteristic hovering, darting flight patterns. There are over 100 species of hoverfly whose larvae principally feed upon greenfly, one larva devouring to fifty a day, or 1000 in its lifetime. They also eat fruit tree spider mites and small caterpillars. Adults feed on nectar and pollen, which they require for egg production.

Dragonflies are important predators of mosquitoes, both in the water, where the dragonfly naiads eat mosquito larvae, and in the air, where adult dragonflies capture and eat adult mosquitoes. Community-wide mosquito control programs that spray adult mosquitoes also kill dragonflies, thus removing an important biocontrol agent, and can actually increase mosquito populations in long term.

Parasitoid insects

Most insect parasitoids are wasps or flies. Parasitiods comprise a diverse range of insects that lay their egg on or in the body of an insect host, which is then used as a food for developing larvae. Parasitic wasps take much longer than predators to consume their victims, for if the larvae were to eat too fast they would run out of food before they became adults. Such parasites are very useful in the organic garden, for they are very efficient hunters, always at work searching for pest invaders. As adults they require high energy fuel as they fly from place to place, and feed upon nectar, pollen and sap, therefore planting plenty of flowering plants, particularly buckwheat, umbellifers, and composites will encourage their presence.

Biological control with micro-organisms

Various microbial insect diseases occur naturally, but may also be used as biological pesticides. When naturally occurring, these outbreaks are density dependent in that they generally only occur as insect populations become denser.

Biopesticides

Biological pests control agents are naturally occurring or genetically modified agents that are distinguished from conventional pesticides by their unique modes of action,low use volume and target species-specificity.

Microbial Pest Control Agents include formulations either of naturally occurring infective microorganisms ,such as bacteria,fungi,viruses,protozoa and mycoplasmas or genetically modified microorganisms that are used to achieve natural control of pests.Such agents are called Biopesticides.Microorganisms are playing an increasingly important role in agriculture as biopesticides to increase crop yield.

Following is a **case study** on use of Bacillus thuringiensis (Biopesticide)on Cotton plant.

Cotton is in many ways an ideal candidate for introduction to cotton-growing countries as the pilot and model GM crop. Its principal use as a fiber crop, rather than a food/feed-crop, facilitates its regulation and acceptance by the public at large. From a biosafety viewpoint it is a self-pollinating tetraploid that will not outcross with native diploid cottons and the movement of the large pollen, which is not dispersed by wind, is limited to a few meters. Cotton is not found as a weed in the global production areas and **Bacillus thuringiensis**(Bt) is unlikely to confer an advantage that would result in Bt cotton establishing as a weed. Thus, the potential biosafety consequences are negligible due to the limited movement of pollen, natural genetic barriers that preclude outcrossing with native cotton, with no known compatibility with any wild relatives. The safety of the Cry1Ac protein is well documented and the Cry1Ac gene is very unlikely to confer any competitive advantage. With the adoption of any technology, there is always a risk that unintended or unforeseen effects could present new challenges. However, with the significant and substantial proven benefits that Bt cotton offers developing countries, the greatest risk is not to explore the technology, and thus be certain to suffer the consequences

of inferior technology that will disadvantage farmers in developing countries who have to compete in international markets.

The six countries that have the potential for significant benefits from Bt cotton have either already adopted the technology, (China, India, USA and Australia) or are exploring its development (Pakistan and Brazil). The challenge is to provide the same opportunity for the potential beneficiary countries, with small to modest areas of cotton, in the developing world where several factors preclude access to Bt cotton. It is important that these smaller cotton-growing countries with resource-poor cotton farmers are offered the option of commercial access to Bt cotton so that they are not disadvantaged by being denied the significant benefits that accrue to adopters of the technology. There are 30 such developing countries, 21 in Africa, five in Asia and four in Latin America that grow small to modest areas of cotton that are potential beneficiaries of commercial Bt cotton but because of various constraints do not have the option to explore the potential benefits that Bt cotton offers in their own countries.

Other case studies on use of biopesticide on Cotton plant are as follows:

Insect Resistant Cotton (1)

Bt cotton varieties were introduced in 1996, providing control of three major cotton insect pests: tobacco budworm, cotton bollworm and pink bollworm. These varieties offer an alternative to conventional insect spray programs. Tobacco buds were particularly heavy in 1995, causing severe yield loss in some areas. Alabama growers, who on average experienced a 29% yield loss due to bollworm/ budworm despite seven insecticide applications, sustained the worst damage. These losses were attributed to the ineffectiveness of pyrethroid insecticides against budworm, due to the development of resistant populations in some states.

The adoption of Bt varieties was extremely rapid in states that experienced resistance problems (Arizona, Alabama, Georgia, Florida).Growers in BWEP areas are advised to plant Bt cotton due to the effects of the weevil sprays on predators of bollworms/budworms The impacts of the adoption of Bt cotton varieties include a reduction in yield losses due to Bt target pests, reductions in insecticide use, and cost savings.

In recent years, the bollworm/budworm has become significantly less troublesome in the Southeast (Georgia, Alabama, Florida) narrowing the economic difference between Bt and non-Bt acreage.

Insect Resistant Cotton (2)

The fall armyworm, soybean loopier and the beet armyworm are destructive migratory pests of many crops in the southeastern US. Damage caused by fall armyworms on cotton is from their feeding on the fruit. Once loppers begin feeding on the outer canopy, they can completely defoliate the plant in 36 to 48 hours. Young beet armyworm larvae feed together and gradually disperse as they grow. They skelotenize leaves.

Transgenic Bt cotton has been commercially available in the United States since 1996. Bt cotton has demonstrated remarkable control of some lepidopteran pests, particularly the tobacco budworm and the pink bollworm. Since its release into commercial markets, Bt cotton seldom, if ever, has required supplemental insecticide control for these two pests. Control of the bollworm has been less dependable. Common lepidopteran pests such as fall armyworms, beet armyworms and soybean loppers are even more tolerant than bollworms. Supplemental foliar insecticide applications have been used in many Bt cotton fields to control economically damaging populations of fall armyworms, beet armyworms, soybean loppers and especially bollworms. Approximately 36% of current Bt cotton acreage is treated for bollworms (1.9 million acres) with 527,700 pounds of chemical active ingredients.

Approximately 65000 bales valued at $19 million were lost to bollworms on Bt cotton acreage in 2000. For beet armyworm/fall armyworm/soybean looper control, approximately 21% of the current Bt cotton acreage is treated with 458,955 pounds of chemical active ingredients. Approximately 12,000 bales valued at $3.6 million were lost to loopers/armyworms on Bt cotton acreage and in an unacceptable control of bollworms and other lepidopteran pests such as beet armyworms, fall armyworms and soybean loopers, prompted the development of a new genetically modified cotton that contains two separate crystalline proteins.

Bt cotton I will likely be phased out and completely replaced with Bt cotton II; a process that will take several years. It is estimated that Bt cotton II will be adopted on the same acreage that is currently planted with Bt cotton I at an increased cost of $2/A. The major impact of BT cotton II would be an elimination of current losses and spraying costs due to bollworms/loopers/armyworms on BT cotton acreage.

Conservation

Biodiversity conservation involves all the ecological processes that have maintained the area's biodiversity such as predation, pollination, parasitism, seed dispersal and herbinary and complex interactions between several species of plants and animals.

Protection requires specialized knowledge or solutions and an understanding of ecological processes. The conservation of biological diversity is essential for the survival of the human race. Biodiversity conservation outside the areas where species naturally occur is known as ex-situ conservation, while conservation in their natural habitat is termed in-situ conservation.

References

1. Bains, William (1993): Biotechnology from A to Z. *New Scientist 138: 41.*

2. Bull, A.T. (1991): Biotechnology and Biodiversity. In: *The Biodiversity of* Microorganisms and Invertebrates: Its *Role in Sustainable* Agriculture. ed. D. L. Hawksworth. Wallingford, U.K.: CAB International Pp. 203-19.

3. Bull, A.T.; M. Goodfellow and J.H. Slater (1992): Biodiversity as a Source of Innovation in Biotechnology, *Annual Review of* Microbiology 46:219-52.

4. Chivian, E.; M. McCally; H. Hu and A. Haines (1993). Critical *Condition, Human Health and* the *Environment, Cambridge,* Massachusetts: MIT Press.

5. Cohen, M.L. (1994): Emerging Problems of *Antimicrobial* Resistance, *Annals of Emergency Medicine* 24(no. 3): 454-56.

6. Colborn. T. (1994): The Wildlife/Human Connection: Modernizing Risk Decisions. *Environmental Health Perspectives* 102 (sup. 12): 55-59,

7. Epstein. P.R. (1995): Emerging Diseases and Ecosystem Instability: New Threats to Public Health. American *Journal of Public Health* 85(no. 2): 168-72.

10

The Contribution of the Peduncle and Penultimate Internode for the Mobilization of Stem Reserves in Wheat under Terminal Drought

Reza Shahryari

Introduction

There is a need to increase wheat productivity world wide, in particular in developing countries and a further is to increase the wheat yield potential genetically so it is important for us to understand the physiological and genetic basis of yield (Yang et al., 2006). Most wheat-producing regions of the world are subject to water deficits during some part of the growing season and grain filling is maintained by a high contribution from the assimilation before and immediately after anthesis and remobilization of vegetative reserves during kernel growth (Bidinger et al., 1977, Moustafa et al., 1996, Royo et al., 1999 and Shahryari et al., 2008).

There are two component traits involved in the extent of the contribution of stored reserves to grain yield in wheat (Ehdaie and Waines, 1996). The first component is the ability to store assimilates in the stem and the second component is the efficiency with which the stored reserves are mobilized and translocated to grain. The second component is a function of the genotype's sink strength, which is dependent on the number of grains per spike and grain weight. (Ehdaie et al., 2006a and 2006b).

In all grain crops the supply of assimilates to the developing grain originates both from current assimilation transferred directly to kernels and from the remobilization of assimilates stored temporarily in vegetative plant parts (Gebbing et al., 1999). The reserves deposited in vegetative plant parts before anthesis may buffer grain yield when conditions become adverse to photosynthesis and mineral uptake during grain filling (Austin et al., 1977; Tahir and Nakata, 2005). Grain growth and development in wheat depend on three C sources: (i) carbohydrate produced after anthesis and translocated directly to the grains, (ii) carbohydrate produced after anthesis but stored temporarily in the stem before being remobilized to the grains, and (iii) carbohydrate produced before anthesis is stored mainly in the stem and remobilized to grains during grain filling (Gallager et al., 1975; Daniels et al., 1982; Kobata et al., 1992 and Ehdaie et al., 2006a).

The major condition for stem reserves for grain filling is sufficient carbohydrate storage before the grain filling. This may be partially linked to plant traits that promote high yield potential, at least during the pre-flowering growth stages (Blum, 2005). Blum et al. (1991) and Wardlaw and Willenbrink (2000) found that the response to drought was the earlier mobilization of non-structural reserve carbohydrates from the stem and leaf sheaths, which is responsible for the formation of most of the grain dry mass at maturity.

Flag leaf photosynthesis alone cannot support both respiration and grain growth under terminal stresses (Rawson et al., 1983). Therefore, a substantial amount of the carbohydrates used during grain filling in wheat must come from reserves assimilated before anthesis (Gent, 1994).

The amounts of accumulated and mobilized stem reserves are either estimated indirectly by monitoring the changes in internode dry weight (Hunt, 1978; Pheloung and Siddique, 1991; Borrell et al., 1993; Shakiba et al., 1996; Cruz-Aguado et al., 2000), or directly by measuring internode WSC content during the grain-filling period (Davidson and Chevalier, 1992; Kiniry, 1993; Blum et al., 1994, Shakiba et al., 1996). The remobilization of dry matter may help to maintain grain growth when post-anthesis photosynthesis is reduced (Solfied et al., 1977; Papakosta and Gagianas, 1991 and Fathi, 2006). Wheat crops grown in dry land areas may depend more on stem reserves for grain filling than crops grown under well-watered conditions. Under terminal drought, there is a rapid decline of photosynthesis after anthesis that limits the contribution of current assimilates to the grain (Johnson et al., 1981). The wheat canopy respires rapidly during grain filling (Gent and Kiyomoto, 1985; McCullough and Hunt, 1989).

The estimated contribution of stored assimilates to grain yield in wheat depends on the genotype, experimental conditions, and the method of measuring stored carbohydrates. Stored reserves and their contribution to grain can be estimated by measuring postanthesis changes in internode dry matter (Hunt, 1979; Pheloung and Siddique, 1991; Borrell et al., 1993; Shakiba et al., 1996; Cruz-Agado et al., 2000), and/or changes in the internode water-soluble carbohydrate content during the grain filling period (Davidson and Chevalier, 1992; Kiniry, 1993; Blum et al., 1994; Shakiba et al., 1996), or estimated by the difference in the canopy dry weight at anthesis and at maturity excluding the grains (Takami et al., 1990; Flood et al., 1995; Ehdaie and Waines, 1996).

Longer stems and greater specific weight increases with storage. (Blum et al., 1997). The Rht1 and Rht2 dwarfing genes of wheat were found to reduce reserve storage by 35 and 39%, respectively, as a consequence of a 21% reduction in stem length (Borrell et al., 1993). The Rht1 allele in 'Maringa' wheat reduced the amount of stored reserves contributed by the peduncle and penultimate internode to grain by 56 and 40%, averaged over both well-watered and droughted treatments, as a result of an 18 and 16% reduction in the internode length, respectively (Shakiba et al., 1996). The Rht-B1b and Rht- D1b (previously known as Rht1 and Rht2, respectively) dwarfing genes of wheat were found to reduce reserve storage by 35 and 39%, respectively, the as a consequence of a 21% reduction in stem length (Borrell et al., 1993 and

Ehdaie et al., 2006b). Developmentally, the potential stem reserve accumulation and subsequent mobilization in wheat depends on stem length and stem specific weight (stem weight/stem length) (Ehdaie et al., 2006b). However, the mobilization efficiency, defined as the percentage of maximum stem weight mobilized, was lower in tall than in dwarf genotypes under favorable conditions (Ehdaie et al., 2006a).

In wheat, particularly, special attention should be given to the stems, since competition exists between the growing upper internodes and reproductive organs in the weeks before anthesis (Wardlaw, 1968; Bingham, 1972; Patrick, 1972; Brooking and Kirby, 1981; Siddique, Kirby and Perry, 1989), the outcome of which depends on both genotype and environment. Besides, there is good evidence that temporary storage is very important under stress conditions (Bidinger, Musgrave and Fischer, 1977; Blum et al., 1994; Wardlaw and Willenbrink, 1994; Seidel, 1996). Recently, genotypic differences in the patterns of allocation and mobilization of dry matter in the stems of wheat have been demonstrated (Blum et al., 1994).

The researchers demonstrated substantial genotypic variation for stem length and stem weight in different internodes of the main stem (Ehdaie et al., 2006a). Stem internodes including the peduncle, penultimate, and the lower internodes are components of the stem in wheat. The partitioning of stem length into different internodes had a bearing on the reserve accumulation before and after anthesis (Ehdaie et al., 2006b). The only major exception that constitutes a form of an effective dehydration tolerance mechanism in crop plants is stem reserve utilization for grain filling under drought stress (Blum, 1998 and Blum, 2005). This is a harmonised whole-plant process that allows effective grain filling when whole-plant photosynthesis is inhibited by stress during grain filling. It is a tolerance mechanism that allows grain filling in dehydrated or over-heated cereal plants, which can account for up to 90% of total grain weight under stress (Blum et al. 1994; Asseng and van Herwaarden 2003). Stem reserve utilization has been found to be an effective yield-supporting mechanism under drought stress (Hossain et al. 1990; Pheloung and Siddique 1991; Gavuzzi et al. 1997; Yang et al. 2002; Asseng and van Herwaarden 2003; Plaut et al. 2004). Although some stem reserve mobilization may support grain filling under non-stress conditions, reserve mobilization is noticeably induced by drought stress during grain filling (Blum et al. 1994; Palta et al. 1994; Yang et al. 2001; Plaut et al. 2004 and Blum, 2005).

An exceptional case for a very effective desiccation tolerance mechanism in crop plants is the sustained or even up-regulated stem reserve mobilization under stress, which is the only option for grain filling in a standing dehydrated crop. This capacity is mutually exclusive with non-senescence (Blum, 2005).

The area of farming in Iran is more than 12 million ha, which are mostly devoted to wheat production (6.2 million ha) (Fathi, 2006). Bread wheat (Triticum aestivum L.) production in the Ardabil region, Iran is subjected to water deficit after anthesis. Improving the genetic adaptation of wheat to drought stress represents one of the main objectives of regional breeding programmes (Shahryari et al., 2008).

The aims of this study were to examine the role of upper internodes after anthesis as a source of material for grain growth and to evaluate the possible factors limiting grain filling in wheat in Ardabil conditions. We evaluated the hypothesis that peduncle and penultimate internode length, weight, and specific weight of wheat genotypes affects the accumulation and mobilization of these internode reserves. This estimates the magnitude of mobilized dry matter and the mobilization efficiency in upper internodes of the main stem.

Materials and Methods

Forty two bread wheats were examined for postanthesis changes in internode dry weight under well-watered and drought field conditions in 2006-2007 to estimate the accumulation and mobilization of internode reserves. The experiment was in a twice replicated simple rectangular lattice design, conducted under irrigated and terminal stress conditions. The main tillers were harvested from the soil surface at maturity stage. The samples were dried in a warm house for 72 hours. Then, each main tiller was divided into spike and stem; then leaf blades and leaf sheaths were removed from the stem. Each stem was divided into two segments, namely peduncle (first internode below the spike), penultimate internode (the internode below the peduncle). The length and weight of each segment was measured, and then its specific weight (linear density) was calculated as the ratio of its weight to its length. The magnitude of mobilized dry matter in each internode segment was estimated as the difference between postanthesis maximum and minimum weight. Mobilization efficiency of dry matter in each internode segment was estimated by the proportion (%) of mobilized dry matter relative to postanthesis maximum weight of that segment. Analysis of variance (ANOVA) was performed for each character measured. The mean of the samples was used in the statistical analysis. Means were compared using the LSD test.

Results and Discussions

Little is known about the extent of genotypic variation for stem storage capacity and of the efficiency with which stored reserves are mobilized and transported to grain in wheat. Knowledge of the relationship between stem reserves and the utilization with plant morphological characteristics, grain growth, grain yield, and its components could be used to develop wheats more adaptable to harsh environments (Ehdaie et al., 2006a). However, studies using post- anthesis desiccation to simulate drought have demonstrated a genotypic variation in the reduction in kernel weight under stress in wheat (Blum et al., 1983; Turner and Nicolas, 1987; Hossain et al., 1990), which suggested a genotypic difference in the ability to remobilized dry matter in order to maintain grain growth (Fathi2006). Dry matter accumulation and mobilization varied along the stem in well-watered and drought field conditions (Ehdaie et al., 2006b). So, results of this investigation showed that there were significant differences between measured characters for environmental conditions and genotypes. Internode length, weight and specific weight were reduced under drought (Table 1). The relative importance of current assimilation and remobilization changes among genotypes is strongly related to environmental conditions.

Table 1. Means of length of main stem peduncle and penultimate internodes for environmental conditions in wheat genotypes

Environmental condition	Length (cm)		Weight (mg)		Specific weight (mg/cm)	
	Peduncle	Penultimate internode	Peduncle	Penultimate internode	Peduncle	Penultimate internode
Moisture treatment					10.6	17.3
Well-watered	27.2 a	17.6 a	327 a	341 a		
Droughted	24.8 b	16.8 b	229 b	243 b		

In temperate cereals, the importance of stem reserve utilization for grain filling under terminal drought and heat is derived mainly from the preanthesis stem-storage capacity (Bonnett and Incoll, 1992; Borrell et al., 1993 and Ehdaie et al., 2006a). Relationships among internode characteristics and accumulation and mobilization of peduncle and penultimate internodes were determined. Furthermore, our data (Table 2) demonstrated that mobilized dry matter was less in drought than in well-watered conditions for peduncle (76vs. 96mg) and for penultimate internode (81 vs. 93mg). Drought conditions increased the mobilization efficiency in the peduncle and penultimate internode by 8.3 and 13.6 %, respectively. Thus, the contribution of the penultimate internode was higher than the peduncle in the mobilization of stem reserves under the condition of this experiment.

Ehdaie et al.(2006 b) found that the contribution of the lower internodes as a segment to stem mobilized dry matter, on average, was 55%, that of the penultimate was 27% and that of the peduncle was 18%. So, it was reported that changes in stem mass of wheat after anthesis are related to changes in stem carbohydrate content (Blum et al., 1994; Wardlaw and Willenbrink, 1994; Stone et al., 1995); internodes reach their maximum length a few days after anthesis (Evans, Wardlaw and Fischer, 1975); and the amount of assimilates mobilized from the stems is unrelated to the culm length (Rawson and Evans, 1971). However, storage increases with longer stems and greater specific weight (Blum et al., 1994 and Ehdaie et al., 2006b).When photosynthetic activity is depressed by drought or heat after anthesis, grain filling becomes more dependent on mobilized stem reserves, which may represent 22 to 60% of the dry matter that accumulates in the grain (Bidinger et al., 1977; Bell and Incoll, 1990; Davidson and Chevalier, 1992; Blum et al., 1994; Ehdaie et al., 2006b).Van Herwaarden et al. (1998) found that the apparent contribution of stored assimilates to grain yield was 37–39% under high rainfall conditions during grain filling, but rose to 75–100% under dry field conditions.

Table 2. Postanthesis maximum and minimum means for peduncle and penultimate internode weight, the estimate of peduncle and penultimate internode mobilized dry matter (MDM), and the mobilization efficiency (ME) in wheat under well-watered and drought field conditions

	Well-watered				Droughted			
Mean	Max	Min.	MDM	ME	Max.	Min.	MDM	ME
		mg		%		mg		%
Peduncle	361	268	96	24	268	191	76	26
Penultimate internode	388	295	93	22	283	202	81	25

The difference in the plant height range was from 57.2 to 100 cm. The total mean of 42 genotypes for this trait was 74.8 cm. There were not significant differences (at a probability level of 1%) between genotypes for plant height in comparison of two environmental conditions. This means that all the of wheat reached a maximum length in height before terminal drought, but under the study, genotypes showed significant differences (at probability level of 1%) for a grain yield. Total mean of yield in this research was 4.47 ton/ha. Total mean of yield in a well-watered condition was 5.6 ton/ha and 3.3 ton/ha under drought conditions. Terminal drought decreased 2.3 ton/ha grain yield of wheat. In this study, we did not find a linear correlation between grain yield and plant height in both environmental conditions.

Interaction between genotypes x environmental conditions was significant (at probability level of 1%) for grain yield. Drought conditions decreased the grain yield of all genotypes.

Acknowledgements

The instructive advice of my Professors E.Gurbanov (Baku State University, Faculty of Biology, and Azerbaijan), A.Gadimov (Azerbaijan National Academy of Sciences, Institute of Botany) and M. Valizade (Tabriz University, Agriculture Faculty) are greatly acknowledged.

The author also thanks D.Hassanpanah (Ardabil Agricultural and Natural Resources Research Centre) for assistance in statistical analysis.

Author thanks F. Peyghami in charge of the technical part of the study conducted at the Agricultural Research Station of Islamic Azad University, Ardabil branch in Iran. He also thanks V. Ebrahimi Mollabashi for his contribution to the experiment.

References

1. Asseng, S and van Herwaarden A.F .2003. Analysis of the benefits to wheat yield from assimilates stored prior to grain filling in a range of environments. Plant and Soil 256: 217–229.
2. Austin, R.B., Edrich, J.A., Ford, M.A., Blackwell, R.D., 1977. The fate of dry matter, carbohydrates and 14C lost from the leaves and stems of wheat during grain filling. Ann. Bot. 41: 1309–1321.

3. Bidinger, F.R., R.B. Musgrave and R.A. Fischer, 1977. Contribution of stored pre-anthesis assimilate to grain yield in wheat and barley. Nature (London), 270:431–433.

4. Bingham J. 1972. Physiological objectives in breeding for grain yield in wheat. Proceedings of the 6th Eucarpia Congress, Cambridge, 15-29.

5. Borrell, A., L.D. Incoll, and M.J. Dalling. 1993. The influence of the Rht1 and Rht2 alleles on the deposition and use of stem reserve in wheat. Ann. Bot. (Lond.) 71:317–326.

6. Blum, A. 2005. Drought resistance, water-use efficiency, and yield potential—are they compatible, dissonant, or mutually exclusive? Australian Journal of Agricultural Research. 56: 1159–1168.

7. Blum, A., B. Sinmena, J. Mayer, G. Golan, and L. Shpiler. 1994. Stem reserve mobilization supports wheat-grain filling under heat stress. Aust. J. Plant Physiol. 21:771–781.

8. Blum A, Sinmena B, Mayer J, Golan G, Shpiler L. 1994. Stem reserve mobilization supports wheat-grain filling under heat stress. Australian Journal of Plant Physiology 21: 771-781.

9. Blum A, Shpiler G, Gonal G, Mayer J, Sinmena B .1991. Mass selection of wheat for grain filling without transient photosynthesis. Euphytica 54:111-116.

10. Borrell, A., L.D. Incoll, and M.J. Dalling. 1993. The influence of the Rht1 and Rht2 alleles on the deposition and use of stem reserve in wheat. Ann. Bot. (London) 71:317–326.

11. Brooking IR, Kirby EJM. 1981. Interrelationships between stem and ear development in winter wheat: the effect of norin 10 dwarfing gene, Gai} Rht2. Journal of Agricultural Science (Cambridge) 97: 373-381.

12. Cruz-Aguado, J.A., R. Rode´ s, I.P. Pe´ rez, and M. Dorado. 2000. Morphological characteristics and yield components associated with accumulation and loss of dry matter in internodes of wheat. Field Crops Res. 66:129–139.

13. Cruz-Aguado, J. A., F. Reyes, R. Rodes., I. Perez and M. Dorado . 1999. Effect of Source-to-sink Ratio on Partitioning of Dry Matter and 14C photoassimilates in Wheat during Grain Filling. Annals of Botany 83: 655-665.

14. Daniels, R.W., W.B. Alcock, and D.H. Scarisbrick. 1982. A reappraisal of stem reserve contribution to grain yield in spring barley (Hordeum vulgare L.) J. Agric. Sci. (Cambridge) 98:347–355.

15. Davidson, D.J., and P.M. Chevalier. 1992. Storage and remobilization of water-soluble carbohydrates in stem of spring wheat. Crop Sci. 32:186–190.

16. Ehdaie, B., G. A. Alloush, M. A. Madore, and J. G. Waines. 2006 a.Genotypic Variation for Stem Reserves and Mobilization in Wheat: I. Postanthesis Changes in Internode Dry Matter. CROP SCIENCE, VOL. 46:734-746.

17. Ehdaie, B., G. A. Alloush, M. A. Madore, and J. G. Waines. 2006 b. Genotypic Variation for Stem Reserves and Mobilization in Wheat: II. Postanthesis Changes in Internode Water-Soluble Carbohydrates. CROP SCIENCE, VOL. 46: 2093- 2103.

18. Ehdaie, B., and J.G.Waines. 1996. Genetic variation for contribution of preanthesis assimilates to grain yield in spring wheat. J. Genet. Breed. 50:47–56.

19. Fathi, G. 2006. Post-anthesis water stress and nitrogen rate effects on dry matter and nitrogen remobilization in wheat cultivars. J. Agron. 5(2): 267-277.

20. Flood, R.G., P.J. Martin, and W.K. Gardener. 1995. Dry matter accumulation and partitioning and its relationships to grain yield in wheat. Aust. J. Exp. Agric 35:495–502.

21. Gallagher, J.N., P.V. Biscoe, and R.K. Scott. 1975. Barley and its environment.V. Stability and grain weight. J. App. Ecol. 12:319–336.

22. Gavuzzi P, Rizza F, Palumbo M, Campanile RG, Ricciardi GL, Borghi B .1997. Evaluation of field and laboratory predictors of drought and heat tolerance in winter cereals. Canadian Journal of Plant Science 77: 523–531.

23. Gebbing, T., Schnyder, H., K¨uhbauch, W., 1999. The utilization of preanthesis reserves in grain filling of wheat. Assessment by steady-state 13CO2/12CO2 labelling. Plant Cell Environ. 22, 851–858.

24. Gent, M.P.N. 1994. Photosynthate reserves during grain filling in winter wheat. Agron. J. 86:159–167.

25. Gent, M.P.N., and R.K. Kiyomoto. 1985. Comparison of canopy and flag leaf net carbon dioxide exchange of 1920 and 1977 New York winter wheats. Crop Sci. 25:81–86.

26. Hossain ABS, Sears RG, Cox TS, Paulsen GM (1990) Desiccation tolerance and its relationship to assimilate partitioning in winter wheat. Crop Science 30: 622–627.

27. Hunt, L.A. 1978. Stem weight changes during grain filling in wheat from diverse sources. In S. Ramanujam (ed.) Proc. 5th Int. Wheat Genetics Symp. New Delhi, India. 23–28 Feb. 1978. Indian Soc. Of Genet. and Plant Breed., Indian Agric. Res. Inst.

28. Johnson, R.C., R.E. Witters, and A.J. Ciha. 1981. Daily patterns of apparent photosynthesis and evapotranspiration in a developing winter wheat crop. Agron. J. 73:414–418.

29. Kiniry, J.R. 1993. Nonstructural carbohydrate utilization by wheat shaded during grain growth. Agron. J. 85:844–849.

30. Kobata, T., J.A. Palta, and N.C. Turner. 1992. Rate of development of postanthesis water deficits and grain filling of spring wheat. Crop Sci. 32:1238–1242.

31. Mariotti, M., Arduini, I., Lulli, L., 2003. Traslocazione della biomassa nel frumento duro durante il grain-filling. In: Mori, M., Fagnano, M. (Eds.), Proceedings of the XXXV Congress of the Italian Society of Agronomy. Napoli, 16–18 September. Imago Media, Caserta, Italy, pp. 339–340.

32. McCullough, D.E., and L.A. Hunt. 1989. Respiration and dry matter accumulation around the time of anthesis in field stands of winter wheat (Triticum aestivum). Ann. Bot. (London) 63:321–329.

33. Moustafa, M.A., L. Boersma and W.E. Kronstad, 1996. Response of four spring wheat cultivars to drought stress. Crop Sci., 36(4):982–986.

34. Papakosta, K. D and A. A. Gagianas. 1991. Nitrogen and dry matter accumulation, remobilization and losses for Mediterranean what during grain filling. Agron. J. 83: 864- 870.

11

Ecophysiological Analysis of Vernacular Plant Species Used for Ecosystem Recovery of Coalmine Overburden Dumping Site

H. P. DEKA BORUAH, N. SAIKIA, J. GOGOI, N. PATHAK AND A. K. HANDIQUE

Introduction

Economic growth and demands for energy by society in order to meet their daily needs, accelerates the mining of coal and other mineral wealth. In India too as a raw material supplier, the mining industry expanded rapidly to meet the demands of an ever increasing economic a growth of the country. Approximately 95% of a non-renewable energy sources, 80% of industrial raw materials and over 75% of agricultural means of production are depending mainly on mining industries in developing countries (Zhu, 2000; Li, 2006). Mining activity leads to the transformation of fertile land into infertile land, forest land to de-forestation and generates huge wastelands (Akala, 1995; Cherfas, 1997; and Gogoi *et al.*, 2007). Besides this opencast coalmines generate huge overburden (OB) dumping sites where no ecological succession at the initial stage can take place. In general, the establishment of an appropriate plant species takes a minimum period of 100 years and to reach the climax state of plant succession nearly 1000 years (Dobson *et al.*, 1997).

Boulder, cobble, pebbles, mine reject materials with no true soil characteristics, elevated concentrations of trace and heavy metals, highly acidic (pH >2.0 to 2.5) or in basic(pH >8.0) nature are the geochemical and biochemical properties of mine OB (Over burden), (Deka Boruah, 2006; Maiti, 2007). These mine OB's are of an ecological concern because of the artificial bare land created in the dumping site and also inability of plant species to grow (Li, 2006; Maiti, 2007; Ghose, 2004). The trace and heavy metal content in mine OB's are above threshold limits (Ghose, 2004; Deka Boruah, 2006; Chaoji, 2002). Similar types of coalmine OB dumping sites are also available in and around opencast collieries of the North Eastern Coalfield, Coal India Limited (NECF, CIL) Margherita. The damage caused to the environment due to opencast coal mining is reported to be very high because of the generation of mine tailings at the ratio of 1:14

to coal compared with other collieries around the country (Chaoji, 2002). The mine OB's are highly acidic and contain 12% sulfur (Akala, 1995; Miao and Mars, 2000; Raju and Hassan, 2003).

Phytoremediation's considered to be one of the best methods available to re-clothe such ecologically denuded mine OB wasteland, but plant species grown in such adverse conditions, showed ecological changes at the morphophysiological, biochemical and in the molecular level (Larcher, 2003, Gogoi *et al.*, 2007; Gardner *et al.*, 1987). The plant behaviour shown by the plant species in different ecological niches at the physiological level are termed as the eco-physiology of the plant. Since the beginning of Charles Darwin (1880) there is an enumerable number of reports on plant adaptation and modification towards different ecological niche (Karban, 2008, Deka Boruah *et al.*, 2008). This behaviour is a form of phenotypic plasticity of a plant species in response to a stimulus (Silvertown and Gordon, 1989). This phenotypic behaviour ultimately has a physiological basis and is mediated by chemical reactions that are relatively rapid for the ecological adaptations of the plant. Therefore, plant response to different environmental conditions has been behind the intense research interest in several reviews (Novoplansky, 2002). Many studies on the eco restoration of mine OB dumping sites have been reported (Mendez *et al.*, 2007). A few studies on the response of plants to a hetergenity environment have been made (Darwin 1880, Braam, 2005; Smith, 2000; Bradshaw, 1965; Freeman et al, 1980), but so far no reports are available on eco-physiological responses of a plant species grown in a mine OB environment of a North-Eastern colliery of India. The plant potentiality to behaviour in *in situ* remediation at mine OB is an indicator of early recovery of mine OB in denuded land. In this paper an eco-physiological response of a plant species introduced into an ecosystem recovery of mine OB denuded land of Tirap colliery OB site, is described.

Materials and Methods

The site selected for the investigation was described in our earlier report (Gogoi *et al.*, 2007). Herbs (*Cynodon dectylon, Auxonopas sps, Cyperus sps, Oxalis sps,* etc.) shrubs (*Cassia streata, Mimosa pigra, Sesbania rostrata, Scoparia sps, Aspergus sps., Mimosa strigilosa,* etc.) and tree (*Tectona grandis, Dalbergia sisso, Gmelina arborea* etc.) were screened for recovery of mine OB damaged land. For comparision of plant survival i.e., natality mortality rate methods described by Misra (1992) were adapted.

The Population growth rate was calculated by the following formula [21].

Population growth rate = $\Delta N / \Delta t$

Where Δ(delta) means change in N= $(N_2 - N_1)$ the number of organisms with change in time $t(t_2\text{-}t_1)$

Natality (new seedling record) were assessed by = $\Delta Nn / N\Delta t$

Where ΔNn is production of new individuals in the population.

Natality rate per unit of population= $\Delta Nn / N\Delta t$ and specific mortality rate were expressed as % of initial population dying within a given time (Misra, 1992). Protocorm development and leaf numbers of the perennial plant species was done by counting the number of each individual bud developed within a respected time. The size of the leaves developed within different time intervals was measured by multiplying the universal constant.

Leaf size= L x B x 0.66

Where B= girth of a leaf, B= breadth of a leaves; and 0.66 is the universal constant

Root behaviour of annual plant species *S. rostrata, C. streata* was recorded according to Deka Boruah *et al.,* (2003). For this plant species supposed to be evaluated for root behaviour, the plot was lightly irrigated so that the substrate became loose. The plant species was than carefully uprooted without disturbing the roots and the root nature was determined.

Results

Except for *S. spontarum* no screened plant species survived > 70 percent. The vernacular plant species Bambosa, *T. grandis* showed 100% mortality after first year of plantation (Table 1). Approximately a 12% to 80% mortality rate was recorded for all the survived plant species in the subsequent year. The annual plant *C. winterianus, C flexosus, S. spontanum* all are monocot herbaceous species, *M. pigra, M. strigilosa, M. streata,* lianus species, and *C. streata* were screened for natality. The natality rate for the two consecutive years was found to be > 0.25% to > 0.62% while for leaves and shrubs species it was > 20% to > 60%.

The survival rate for the tree species was recorded over the year of observation (Table 2). This was observed in order to find out the tree species needed to achieve tertiary sere succession for ecorestoration of the mine OB dumping sites.

Table 1: Mortality and natality rate of the screened plant species grown in the mine overburden dumping site of Tirap colliery over the period of growth.

Name of plant	Mortality rate (%)		Natality rate (%)	
	1st year	2nd year	1st year	2nd year
Cymbopogon winterianus	50	30	0.58 (0.22)	0.42 (0.22)
C. flexosus	50	50	0.41 (0.21)	0.45 (0.20)
Saccharum spontaneum	80	70	0.25 (0.12)	0.62 (0.21)
Bambosa jati	0	0	0	
Mimosa pigra	80	0	62.50 (3.7)	0.65 (0.32)
M. streata	80	0	60.12 (3.8)	62.50 (3.3)
M. strigilosa	70	0	62.38 (5.3)	60.12 (3.8)
Sesbania rostrata	75	0	45.80 (3.5)	62.38 (3.7)
Cassia streata	75	30	20.83 (3.3)	45.80 (2.1)
C. coronns	50	10		
Tectona grandis	0	0		
Gmelina arborea	50	10		
Dalbergia sisso	80	50		
Olive	0	0		
Black berry	7	5		
Citrus	0	0		

Only few species which showed a minimum level of survival in first year are recorded. Data in parenthesis are standard deviation of observed values.

Table 2: Survival rate of tree species grown over the period of growth grown in mine overburden dumping sites of Tirap collieries

Plant species	% of survival rate over the period of observation				
	> 1 month	> 2 month	> 3 month	> 6 month	>12 month
Tectona grandis	>70	>70	0	0	0
Gmelina arborea	>90	>90	>50	>30	>20
Dalbergia sisso	>90	>90	>85	>85	>85
Olive	>100	> 100	0	0	0
Black berry	>80	>80	>10	>00	>5
Citrus	>60	>60	0	0	0

It was found that *T. grandis* and Olives show a survival rate of up to two months of plantation (Table 2). Subsequent to this all plant species died off. This survival for a month may be due to the stored metabolic activity of the planted sapling/stumps. On the other hand except *D. sisso* all the plant species showed a 6% to 22% mortality rate at the end of one year while *T. grandis,* olive and Citrus showed 100% mortality.

A protocorm development which was considered as the key indicator of the normal growth of a perennial plant species was recorded since the development (Table 3). The initiation of protocorm was found to be normal for *T. grandis* but it was delayed for a period of one week for *G. arborea* and *D. sisso.* A uniformity of protocorm development leaf size and leaf numbers was found for *G. arborea* but the shedding of leaves and premature falling of protocorm was found for *T. grandis* and Olive trees.

Table 3: Growth and development survived tree species grown over the period of growth in mine overburden dumping sites of Tirap collieries

Name of plant	Time of protocrome development (in days)	Average No of protocrome	Average No of leaf emergence	Leaf size (cm^2)
Tectona grandis	15	3	5.5	6.5
Gmelina arborea	22	2	2.35	10.45
Dalbergia sisso	25	2	2.5	3.5
Olive	27	3	5.35	4.5
Black berry	19	1	2.5	5.3

As per the usual agronomic practice bare minimum spacings were given to all the plant species (Table 4). Accordingly the canopy size and height of the plant species was recorded. A significantly reduced height and canopy size was found for the plant species grown in mine OB conditions.

Table 4: Canopy size and height of the survived tree species grown in mine overburden dumping sites of Tirap collieries

Name of plant	Height (cm)	Canopy size (sq. cm)
Cymbopogon flexosus	133.30 (1.15)	200 (2.15)
C. winterianus	132.25 (2.80)	202 (2.81)
Saccharup spontaneum	302 (21.25)	201.11 (3.10)
Mimosa pigra	–	540.00 (3.33)
M. strigilossa	–	552 .23 (4.23)
M. streata	–	543.15 (3.23)
Sesbania rostrata	100.25 (8.77)	96.2 (7.81)
Cassia streata	259.49 (4.90)	34.53 (3.32)
Gmelina arborea	132.21 (2.12)	203.29 (12.00)
Dalbergia sisso	348.77 (5.28)	245.91 (1.41)
Black berry	90.12 (3.61)	32.75 (1.63)

Table 5: Root behaviour of the plant species grown in mine overburden dumping sites of Tirap colliery.

Name of plant	Root length (cm)	No. of primary root	No. of secondary root	No of tertiary root
Mimosa pigra	38	31	54	35
Sesbania rostrata	52	18	21	35
Cassia streata	25	73	56	32

Root enumeration for the lianas species *M. pigra* shrubs *S. rostrata* (annual), *C. streata* (perennial) showed reduced growth (Table 5). The root length, number of primary roots, number of secondary roots and tertiary roots are about 3 fold to 4 fold less. Further the roots encompass only the shallow surface of the mine OB

Discussion

The contamination of the environment with the overburden of coalmining has become a world wide problem, affecting the ecosystem. This ecological degradation due to the high sulphur containing coal mining in the North Eastern part of India is also not except on. The species selection depends upon the future land use of the area and the soil condition of the area. The plant species were selected for the coal mine ecosystem recovery plan which was based on their adaptability under the coal mine stress conditions. Plants commonly employed for ecosystem recovery combine a high biomass plant with rapid growth, and a high hyper accumulation.

The selection of a plant species also varie with the nature of the mine OB spoils. The selected plant species were tree species Gomari (*Gmelina arborea*), Sisso (*Dalburgia sisso*), Titachapa (*Michelia champaka*) cover plant Cassia (*Cassia streata*) Dhaincha (*Sesbania rostrata*) Mimosa (*Mimosa pigra*) herbs Auxonopus, *Saccharumn spontaneum, Cymbopogon winterianus Jowitt,* and *C. flexosus.*

In the present investigation the vernicular plant species *Bambosa, Tectona grandis,* showed a 100% mortality rate after one year of planting. This was due to the adverse environmental condition of the, mine OB low pH balance level and the (1.5- 3.0) above threshold limit of trace and heavy metal which bore no true soil characters. On the other hand a 12-80% mortality rate was recorded the plants which and survived. The survival of the plant species *M. strigilosa, S. rostrata, S spontarum* were screened as tolerant plant species which can thrive under such a stressful environment. It was observed that the tree species are suitable to achieve tertiary succession. Such tree species (*Gmelina arborea, Tectona grandis, Dalburgia sisso* etc) may increase the soil fertility and this increase means that the increase of the C, N, content up to 6 times in surface soil in an 8 years old plantation (Garge and Jain, 1992; Gurge, 1998) will be beneficial.

The contribution of individual plant species to the plant community of the ecosystem was analyzed by measuring the canopy size. It was estimated by the ratio of the perimeter of the plant species to the total perimeter of the plant cover. Significantly, the highest (552 sq.cm) of the canopy size was recorded in the annual leguminous plant species *M. strigilosa* and the lowest canopy size was recorded in the tree species *D. sisso.*

Observation of the plant root morphology shows that the plant itself confirms that the plant root of Dhaincha (*S. rostrata*) shows the highest growth. Fine roots contribute sustainability in improving the soil structure, the soil pH, balance resource and the acquisition and water permeability. These results were supported by the work of Deka Boruah *et al.*, 2008. Soil nutrient influences the growth and development of plants. The growth and protocorm development of the plants were found to be 100 to 200 times lower that the plants grown in un-mined soil conditions. Several studies have suggested that C, N, P and moisture content may influence the quick succession and community structure (Tilman, 1986). Mine OB soils have no soil property only their skeletal material so for the restoration of such condition many plants should be tried so that the tolerant plant species can recover the denuded land for long term sustainability.

Conclusion

It can be concluded that the stunted growth and the lack of development of the adapted tree species *G. arborea, D. sisso,* is the characteristic feature of the plant grown under mine OB conditions. Premature falling of protocorm and reduced leaf size are found for the adapted plant species. This may be an adapted accommodation of a survived plant species. Further investigation of a step wise succession of plant species to achieve tertiary sere succession in an ecosystem recovery of mine OB wasteland is important.

12

Emerging Trends in Enhancement of Cotton Fiber Productivity and Quality Using Functional Genomics Tools

S. B. NIMBALKAR, DR. P. H. SAWANT AND DR. R. A. HEGDE

Introduction

Cotton (*Gossypium* spp.,), a highly valued agricultural commodity for more than 8000 years, has long been recognized as a vital component of the global economy[1]. Cotton production provides an income for approximately 100 million families, and approximately 150 countries are involved in cotton import and export[1]. Globally, cotton is grown on >32 million hectares (mha) with approximately 71% of the production in developing countries[2]. India has a larger area of cotton (9 mha) than any other country in the world[5] and produces almost three million tonnes of all qualities and staple lengths of cotton per year. Cotton provides a livelihood to more than 60 million people in India by way of support in agriculture, processing, and the use of cotton in textiles, and it also contributes ~30% of the Indian agricultural gross domestic product and thus cotton is a very important cash crop for Indian farmers[3]. Albeit India's cotton area represents now 25% of the global area of cotton, in the past it produced only 12% of world production. Yields of cotton in India are low, with an average yield of 300 kg/ha compared with the world average of 580 kg/ha[2]. The major limiting factors to both cotton production and quality in India are biotic and abiotic stresses. As with many cotton growing areas of the world, the major damage is due to insect pests, especially the bollworm complex, sucking pests and viruses. The productivity is still worsened by abiotic stresses such as drought and heat. It is worth mentioning here that most of the cotton in India is grown under rainfed conditions, and about a third is grown under irrigation[4], which also experiences water stress during certain growth periods. To cope with the growing demand on cotton fibre and its by products, the genetic enhancement of cotton is indispensable which will ensure competitiveness in the market of this natural-renewable product with petroleum-derived synthetic fibres, given the projected future decline in petroleum reserves. Moreover, modifications to expand the use of cotton seed derivatives for food and feed could profoundly benefit the diets and livelihoods of millions of people in food-challenged economies.

The genus *Gossypium* L. has long been a focus of genetic, systematic and breeding research. *Gossypium* spp. consists of at least 45 diploid and 5 allotetraploid species. The allotetraploid cotton species, which include two commercially important cultivated species, *G. hirsutum* L., and *G. barbadense* L., were generated by A- and D-compound genomes[5]. The best living models of the ancestral A- and D-genome parents are *G. herbaceum* and *G. raimondii*, respectively. These four genome groups have received special attention, since they have been domesticated for their abundant seed trichomes. The diploid donor of the allopolyploid A_T genome [where the T subscript indicates that the A genome in the tetraploid (AD) nucleus], was a species much like modern *G. arboreum* or *G. herbaceum*, whereas the allopolyploid D_T genome is derived from a progenitor similar to the modern *G. raimondii*. These well-established relationships provide a phylogenetic framework to investigate the evolution of gene expression both in terms of domesticated fibre production and polyploidy. Understanding the contribution of the A and D sub genomes to gene expression in the allotetraploids may facilitate improvement of fibre traits[1].

Interestingly, the A genome species produces spinnable fibre and is cultivated on a limited scale, whereas the D genome species does not. More than 95% of the annual cotton crop worldwide is G. *hirsutum* (Upland or American cotton) and *G. barbadense* (the extra-long staple or Pima cotton) accounts for less than 2%. Yield and fibre quality of upland cotton varieties, have declined over the last decade - a downward trend that has been attributed to erosion in the genetic diversity of breeding stocks and an increased vulnerability to environmental stresses[6]. The level of genetic diversity is low in *G. hirsutum*, especially among agriculturally elite types, as revealed by all means of assessment[6].

Increasing diversity is therefore essential to genetic improvement efforts. Each of the three major approaches to increasing genetic diversity – mutagenesis, germplasm introgression, and transformation - has advantages and disadvantages. Interspecific germplasm introgression is particularly attractive in that it utilizes a broad germplasm base, can be targeted to one or more specific traits/genes or modulated to include thousands of genes/even entire genomes, and is readily coupled to marker-assisted genome analysis and selection. Though, quantitative trait loci (QTL) mapping and marker aided selection have potential application in genetic improvement of the cotton for higher productivity, its application is not yet documented in cotton breeding programmes due to poor knowledge of the physiological and genetic nature of the fibre quality and productivity traits, low and complex heritability of investigated traits and the genotype X environment interactions etc., Although introgression of genes across species boundaries is difficult, it is quite desirable because the gene pools of cultivated species do not contain all of the desired alleles. Alternatively, mutagenesis and transgenic technology has been proposed. However, currently they have limited applications due to several technical reasons such as non availability of novel genes, lack of efficient method to alter/transfer large genetic element etc., Thus, overall the paucity of information about genes that control important traits impedes the genetic improvement of cotton in any one of the above said approaches.

Fibre represents over 90% of the total value of the cotton crop and the genetic improvement of the fibre properties is certainly a major target of trait besides the biotic and abiotic stress resistance types since these stresses also ultimately affect the final productivity and quality of fibre. The cotton fibre is a complex biological system that is the net result of the intricate interplay of elaborately developed regulated pathways consisting of literally thousands of genes and gene products, and as such has been a difficult subject to tackle using conventional approaches, since cotton does not lend itself easily to genetic analysis[7]. The high value per hectare of cotton and the global textile market demands an increased fibre uniformity, strength, extensibility, and quality to justify clearly the importance of new and innovative approaches towards evaluating and understanding genetic mechanisms of fibre qualities. Gene discovery and strategies to produce the desired pattern of expression and phenotype, are major goals that lay ahead for biotechnology programmes. In this context, the use of genomic tools and resources to facilitate breeding using molecular approaches and characterization of the cotton fibre transcriptome[1] are considered as key strategies for the genetic improvement of cotton. Besides its economic importance, cotton fibre is an outstanding model for the study of plant cell elongation, cell wall and cellulose biosynthesis. The fibre is composed of nearly pure cellulose, the largest component of plant biomass. Compared with lignin, cellulose is easily convertible to biofuels. Translational genomics of cotton fibre and cellulose may lead to the improvement of diverse biomass crops. Thus, functional genomics of cotton fibre development have several folds of applications.

Scope of functional genomics in cotton fibre improvement

It has been in the last decade only that researchers have begun to primarily focus on studying the underlying developmental mechanisms that control fibre properties as the basis for manipulating biological and cellular processes to improve fibre characteristics. The genetic complexity of the cotton fibre transcriptome is currently estimated to consist of approximately 18000 genes in the genome of cultivated diploid species. The cotton fibre transcriptome in the allotetraploid species is similarly estimated at approximately 36000 genes and to include homoeologous loci from both the A_T and D_T genomes[1]. The high genetic complexity of the fibre transcriptome is both in the diploid and tetraploid species and accounts for a significant proportion (45–50%) of all the genes in the cotton genome. However, despite approximately 1.5 million years of evolution following the polyploidization event, polyploidy has not been accompanied by rapid genome change, as the genome organization and gene sequences of orthologous loci from the A and D genomes of diploid and tetraploid cotton species are highly conserved. Moreover, spatial and temporal expression patterns have been evolutionarily conserved[8]. Thus, fibre gene function is highly conserved in the genomes of wild and cultivated species, as well as diploid and tetraploid species, despite millions of years of evolutionary history. The phenotypic variation in fibre properties therefore is more likely one of quantitative differences in gene expression as opposed to differences in the genotype at the DNA level[20]. Further studies, are required to understand the genes, their copy number and specific function in fibre development. The direct or indirect aims of all functional genomics programmes focussing on the fibre development are to define the function of the genes involved and to find candidate genes to improve fibre productivity and quality. This helps to sieve a smaller subset of genes which can be employed to draw up a final list of candidate genes (or master regulators) that could be used in future innovative breeding programmes for superior fibre productivity and quality.

As for methods of cloning fibre-related genes, differential screening of the fibre cDNA library (DD-RT PCR) was most popular in the earlier days. At the same time, sequencing randomly selected cDNA clones from the fibre library, PCR amplification using gene-specific probes, and suppression subtractive hybridization were shown to be useful in isolating cotton-fibre-related genes. Recently, expressed sequence tag (EST) projects and microarray technology have gained recognition as a means fox discovering novel genes. The results were confirmed by (reverse) northern blotting and/or quantitative real time PCR techniques. Thus techniques to monitor global gene expression rely either on hybridisation (microarrays) or on PCR (real-time PCR, cDNA-AFLP, Differential Display) or, more recently, on massively parallel signature sequencing (http://mpss.udel.edu) based methods. However, it is noteworthy to mention that a combination of different profiling methods is likely to be most informative, since no method is perfect and each one of them has its own pros and cons. At the time of this writing, microarrays are considered as a powerful method to simultaneously measure relative expression levels for thousands of genes. Two main types of microarrays have been developed in cotton, namely cDNA-based and oligonucleotide-based arrays. Based on their technical specifications cDNA and oligonucleotide microarrays may differ in sensitivity and dynamic range for detecting variation in the mRNA abundance as well as in the power to discriminate between related target sequences[8]. Limitations and constraints of the different microarray platforms or in comparison with PCR-based techniques have been emphasized in several instances. The principal limitation for the microarray technique in fibre transcriptome analysis, irrespective of its types, is that only a fraction of genes, those for which DNA sequences are available, can be investigated.

As of July 01, 2008, 370729 *Gossypium* sequences were in GenBank, including 39547 ESTs from *G. arboreum* (A), 63577 from *G. raimondii* (D), 266335 from *G. hirsutum* (AD tetraploid), 1023 from *Gossypium barbadense* and 247 from *G. herbaceum* var. *africanum* (http://www.ncbi.nlm.nih.gov/dbEST/dbEST_summary.html). Non-redundant ESTs have been used both to develop sequence- specific markers for genetic mapping and to construct microarrays for the identification of candidate genes involved in fibre cell initiation and elongation[1]. For example, Udall and team[9] created a long oligonucleotide microarray for cotton because of its low manufacturing cost, flexibility in design, homogeneous melting temperatures (Tm), and relative ease of adding probes. A small EST assembly (~45,000 ESTs) was previously used to generate oligonucleotide probes for cotton fibre[1]. A larger scale EST assembly (> 150,000 ESTs) was recently produced as a community-wide effort by cotton researchers[10]. Subsequent additions of cotton ESTs to GenBank (> 210,000 ESTs) have been compiled into a large EST assembly. These two assemblies constitute nearly all of the known genic sequence from cotton. An increasing number of sequence resources (bacterial artificial chromosomes (BACs) and ESTs) in *Gossypium* have been used to design fibre cDNA microarrays for functional studies by another group of scientists[11]. Recent efforts from the cotton community lead to the public release of 2 cotton microarrays (essentially from fibre ESTs), including a 24 K GeneChipÒ Cotton Genome Array from Affymetrix (http://www.affymetrix.com/products/arrays/specific/cotton.affx); and a 23 K oligonucleotide (60-70mer) microarray. It is worth mentioning that the sequences and annotations of all the probes are publicly available via a web-based query (http://cottonevolution.info) or by request.

Many potential sources of error can have a large impact on microarray experiments, such as inconsistency among multiple RNA extractions, reverse transcription, RNA amplification, and labelling, as well as different levels of background noise for each microarray. Many of these sources of error can be resolved by appropriate experimental design[12] and careful laboratory techniques; however, the quality of the microarray must often be assumed and is not under the control of the investigator. Hence, care has to be taken while doing microarray analysis in such a way that it provides a robust and reproducible platform for transcription profiling in cotton.

Though systematic transcriptomic approaches have been combined with genetic analyses, these studies do not address the occurrence of alternative splicing or the post-translational modifications of the proteins. In addition, proteins can move in and out of other macromolecular complexes, and thus modify their functionality. This level of complexity cannot be tackled using transcriptomics alone[13]. Thus, the study of proteomics and metabolomics are considered as a new and promising science area in cotton fibre development in addition to transcriptomics.

Molecular, cellular and developmental aspects of fibre development

Cotton fibres are seed hairs and they originate from the epidermal cells of the ovular surface. It is well known that fibre development is composed of four overlapping stages: fibre cell initiation and enlargement from -3 to 1 day postanthesis (DPA), fibre elongation after anthesis until 25 DPA, secondary cell wall cellulose deposition from 15 DPA to 50 DPA and fibre cell dehydration and maturation after 45 DPA[14]. Thus, the near-synchronous growth of 500000 terminally differentiated single-type fibre cells per ovule is characterized by four major discrete developmental stages - differentiation/initiation, expansion/elongation, primary cell wall (PCW) synthesis, secondary cell wall (SCW) synthesis and maturity. Among these developmental stages the productivity and quality of cotton depends mainly on two processes: fibre initiation, which determines the number of fibres present on each ovule, and fibre elongation, which determines the final length and strength of each fibre. As all plant cells undergo cell expansion to some degree during growth and development, rapidly elongating cotton fibres offer a unique single-celled model to study; i) the molecular and cellular mechanisms that regulate the rate and duration of cell expansion, and hence, govern cell size and shape, and in the case of cotton, important agronomic traits as well ii) cell wall development and iii) cellulose biosynthesis[8].

The type of growth mechanism (diffuse versus tip) associated with the exaggerated growth rate of rapidly elongating cotton fibres has been debated for decades. However, structural and physiological data provide compelling evidence for diffuse growth[15]. Interestingly, no genes known to be specific to tip-growing cell types (e.g. pollen and root hairs) have been identified in developing cotton fibres, further supporting a diffuse-growth mechanism during rapid polar elongation[8]. Polar elongation of developing cotton fibres via diffuse growth is controlled at the cellular level by the transverse orientation of microtubules, the sites of cell wall loosening, the cell wall deposition in the extracellular matrix, and polar vesicular trafficking[15].

Fibre length is dictated by the rate and duration of cell expansion, which is in turn, governed by developmental programmes that co-ordinately regulate cell turgor, the driving force of cell expansion, and cell wall loosening[16]. The transition from PCW to SCW synthesis, which occurs during the latter stage of expansion between ~16 and 21 DPA, is distinguished by the re-orientation of microtubules and cellulose microfibrils to steeply pitched helical arrays[17] in anticipation of SCW synthesis to produce a thick cell wall consisting of >94% cellulose. Coincident with the termination of the fibre elongation at ~21 DPA is an increase in fibre strength[18], presumably due to cross-linking of cellulosic and non-cellulosic matrices[19], that is also accompanied by a major loss (~ 36%) of high molecular weight noncellulosic polymers in the PCW[49]. Developmental programmes regulate the temporal synthesis of fibre PCW and SCW, which differ significantly in structure and composition. While the thin PCW (0.2–0.4 µm) deposited during fibre elongation contains <30% cellulose, the thick SCW (8–10 µm) is composed of >94% cellulose. In addition, the degree of the polymerization of the cellulose microfibrils also varies, being <5000 in PCW and ~14,000 in SCW[20]. In some domesticated varieties, cotton fibres may attain a final length of 6 cm or about one-third the height of an entire *Arabidopsis* plant[8].

Thus, rapid and simultaneous elongation occurs in millions of fibre cells in the cotton boll without concurrence of cell division and multicellular development. On the day of anthesis (flower opening), approximately one in four epidermal ovular cells has already been destined to become a cotton fibre, initially appearing as a spherical protrusion. Only about a third of all the epidermal cells become fibres, although the exact proportion varies between genotypes and in response to hormone levels. A greater understanding of the molecular processes that regulate which cells become fibres with molecular and physiological mechanisms of fibre development, could increase our ability to either breed for, or engineer, cotton plants with a higher density of fibres and, hence, a higher yield.

Functional genomics and fibre development

The cotton fibre transcriptome has attracted a lot of attention in recent years[21]. Many fibre-specific genes involved in fibre cell initiation, fibre elongation or cell wall biogenesis have been identified from the comparisons of normal (wild-type) versus fibre mutants of *G. hirsutum* species. Few reports have also investigated the mechanisms and genes underlying the important developmental differences between *G. hirsutum* and *G. barbadense*[22].

Most of the cotton fibre transcriptome was identified by a single gene discovery project[1] and numerous studies targeted the rapidly elongating cotton fibres for a number of the following compelling reasons: i) The rate and duration of fibre elongation governs important agronomic properties, such as yield and fibre length, ii) The exaggerated growth of elongating fibres, underpinned by high levels of metabolic activity was expected to be especially gene rich from a gene discovery perspective, and iii) Isolation of fibres free of contamination from surrounding complex tissues permit an unambiguous look at the transcriptome of a single cell within a biologically relevant framework[8].

A judicious selection of a cultivated diploid species (*G. arboreum* L.) as a model for fibre development proved especially rewarding, as the rate of gene discovery was enhanced at least two-fold simply by avoiding the redundancy due to polyploidy[1,8]. It is also important to identify the genes that are preferentially expressed in fibre tissues of the *G. hirsutum* L., which is highly adapted to the present environment and the most widely cultivated species. Further, similar biochemical pathways may have diverged during evolution and thus can create a possible drawback for the identification of the *G. hirsutum* fibre specific candidate genes when we translate the knowledge from *G. arboreum* EST projects. Hence, ESTs were developed from the fast elongating fibre of *G. hirsutum*[23]. Results from these studies were useful to conclude that the genes are very much essential for fibre development in the commercial cultivars.

At the time of this writing, the total mechanism of fibre development, cellulose biosynthesis and maturation has been unclear, even though the expression patterns and putative function of several arrays of genes have been analysed in detail as shown above. Thus fibre cells, though devoid of any critical functions in the cotton plant except seed dispersal, contain large numbers of active genes common to other cell types along with fewer active genes unique to fibre. More than half of all genes were up-regulated during at least one stage of fibre development. Genes implicated in vesicle coating and trafficking were found to be over expressed throughout all stages of fibre development, indicating their important role in maintaining rapid growth of this unique plant cell[24]. A second point that became clear during transcriptome analysis was that no major changes occur in the mRNA population during PCW and SCW synthesis stages and no subset of genes that are exclusively expressed during a given developmental stage was detected. Thus, it seems likely that most of the genes in fibre are active throughout the development of the fibre. Alternatively, gene transcription ceases early in the fibre development, but differential mRNA utilization occurs during growth and hence the protein content usually changes during later fibre developmental stages.

Databases

Comprehensive study of any genome depends on the availability of a saturated and fully integrated genetic and physical map of cotton. Hence, all the information should be collected and made publically available. There are several genome databases that are exclusively developed to serve the cotton research community. They are: The International Cotton Genome Initiative (ICGI; http://icgi.tamu.edu/), The Cotton Genome Database (CottonDB; http://www.cottondb.org;), The Cotton Microsatellite Database (CMD; (http://www.cottonssr.org), Comparative Evolutionary Genomics of Cotton (http://cottonevolution.info/), TropGENE Database (http://tropgenedb.cirad.fr/en/cotton.html), Cotton Genome Centre (http://cottongenomecenter.ucdavis.edu), The Cotton Diversity Database (http://cotton.agtec.nga.edu), the Cotton Portal (http://gossypium.info), National Center for Biotechnology Information (http://ncbi.nlm.nih.gov) for EST resources and BACMan resources at Plant Genome Mapping Laboratory (www.plantgenome.uga.edu). Cotton oligo-gene microarrays consisting of approximately 23,000 70-mer oligos designed from 250,000 ESTs can be found at http://cottonevolution.info/microarray. There is a great need to expand bioinformatic infrastructure for managing, curating, and annotating the cotton genomic sequences that will be generated in the near future. The cotton sequence database of the future should be able to host and manage cotton information resources in cotton using community accepted genome annotation, nomenclature, and gene ontology. Some existing databases may be upgraded to effectively handle a large amount of data flow and community requests, but additional resources will be sought to support key bioinformatic needs.

Future perspectives

Advances in technologies for harvesting specific cell types and in amplifying mRNA pools for expression profiling have stimulated studies of the transcriptome at the cellular level in plants. However, these experiments have a common obstacle of sample preparation and single cell-type isolation that could impact interpretations of gene expression. Measuring gene expression in a single, abundant cell type will not have much experimental induced error. The problem is more aggravated when different developmental stages of fibre tissues are studied. Thus, progress in the systematic survey of the genes crucial for fibre development is hampered by several factors including lack of suitable technique to separate and/or isolate fibre initials and other developmental stage specific tissues, non availability of complete cotton genome, common method to assess microarray data quality, control measures to avoid false positives, poor performance in evolving common strategy to analyze the high throughput gene expression profiles *etc.*, Careful experimental design and exploitation of suitable experimental samples along with the following considerations may fuel future research on cotton fibre development at molecular level.

Most of the microarray studies were conducted in controlled green house experiments (of course, in field conditions in some cases) and several experimental controls such as internal, positive and negative controls, calibration spike-in controls, transgene and vector controls, ratio spike in controls, blank and buffer controls interspersed among the cotton oligoNTs and biological and technical replications were included to avoid the experimental error. Despite the care that has been taken to ensure the genes that are expressed preferentially in fibre tissues, the impact of biotic and abiotic factors in developmental process of the fibre, which has a demonstrated influence on fibre quality, is not discussed in the available literature and remains to be tested. Given that the total number of genes in the cotton genome may be approximately 36000[1], the available cotton DNA chips represents ~40% of the cotton genome. Analysing all the cotton genes in a single experiment may give a different picture. Further, other issues such as gene redundancy in cultivated allotetraploid cotton, recalcitrant nature of matured fibres which does not lend them for the mRNA isolation *etc.*, may also hinder the progress. Hence, a proper method has to be identified to capture the novel genes involved in matured fibre development.

Though proteomics help to get comprehensive understanding on fibre development, it is hard to grasp fibre proteome due to several reasons such as the presence of interfering material such as cell walls, phenolic compounds and other secondary metabolites that will severely disturb the protein separation and analysis[25], loss of labile proteins during the preparation of experimental samples, poor resolution of basic glycoproteins by classical two-dimensional gel *etc.*, The fundamental challenge in transcriptomic and proteomic studies is the precise prediction of the structure and function of genes. The databases used for annotation are not complete though they have robust data on metabolic pathway[7]. Pathways associated with fibre development may not be identified fully due to poor representation of the annotated genes and their pathway on the microarray. It was also concluded that sequence similarity is insufficient evidence to predict accurately how the proteins work in plant cells and thus supporting the need for biochemical assay and other related studies[8]. In addition to that, quantitative data on proteome is still in its infant stage and protein-protein interactions and protein with other metabolites remains to be revealed. All this data combined with genetics, biochemistry and molecular biology should lead to a better understanding of the roles of genes/proteins in future fibre development.

Besides cDNA and oligonucleotide microarrays, tiling-path arrays have been used to study gene expression in plants[26]. The advantage of tiling-path arrays over conventional microarrays is that they are not stuck-up with the gene structure, and hence provide unbiased and more accurate information about the transcriptome. Further, they provide information about transcriptional control at the chromosomal level. The use of tiling-path arrays could help to provide novel information about the fibre transcriptome at the genome-wide level[26]. Further, the quickly expanding knowledge on gene function and the availability of whole genome sequences of model plants is expected to offer new perspectives to solve the problems encountered in genetically and physiologically complex traits in commercial crops - which is referred to as "plant translational genomics".

Jansen and Nap[27] proposed the merger of genomics and genetics into the novel concept of genetical genomics: the expression levels of genes or cluster of genes are analysed within a segregating population. In genetical genomics, gene expression profiles are quantitatively assessed within a segregating population, and expression quantitative trait loci (eQTL) can be mapped like classical QTLs[28]. The approach provides a novel way of discovering, at a genetic level, the regulators of gene expression acting either in *cis* or in *trans* relative to the target gene. The eQTL position may coincide with the gene itself displaying *cis* regulation or be different, thus revealing *trans*-acting factors controlling expression. A common feature of eQTL studies is the detection of 'hotspots' of *trans*-acting eQTL, interpreted as regions rich in regulatory genes that co-regulate many downstream targets. However, it should be noted that QTL regions often appear quite complex and approximate and may contain hundreds of genes. Consequently, the actual involvement of the candidate gene in most cases remains to be confirmed by genetic and physical mapping, positional cloning, expression analysis, or genetic transformation experiments. Cost-saving alternatives to large genome-wide and population-wide analyses with minimal loss of information have been proposed: analysing pooled samples of phenotypically extreme members of the population or concentrating on genotypically-selected individuals[29]. Decoding cotton genomes will be yet another prospect for improved understanding of the functional and agronomic significance of polyploidy and the genome size variation within the *Gossypium* genus[2]. It should also be noted that there is no single cotton cultivar that has been grown globally. Findings made in one variety should not be generalized for the rest of the varieties since each cultivar differ in its character, habit, performance etc., When there is no similarity among the transcriptome and proteome of cell types, then we cannot expect an identical functional role for the same gene or protein among the cultivars. Hence, before making any final conclusion a comprehensive comparison of the multidisciplinary approaches is required.

Conclusion

Global gene expression analysis is an important tool for unravelling genetic architecture and the connections between genotypic and phenotypic variation, but the results of such studies require careful interpretation. Despite the wealth of efforts in genetic mapping, transcriptomics and DNA sequencing to decipher the molecular determinants for the quality of the cotton fibre, it is difficult to predict in what aspects these studies will eventually impact on breeding processes. The next major task at hand will be the functional analysis of unannotated genes using reverse genetic approaches, which is much more promising in light of recent advances in cotton transformation and regeneration technology [8]. In addition to the potential for bioengineering fibre properties in the future, significant headway is being made to exploit the fibre ESTs to genetically map the fibre transcriptome as a step toward marker-assisted selection by molecular breeding programmes[28, 31], although this will remain challenging because of the need to assay expression levels in large numbers of genotypic samples. The compilation through meta-analysis of fibre QTL data from various studies (since the majority of the markers used are cross referenced in other populations) and the integration of QTL data with expression data (eQTL) would help to identify chromosomal regions important for fibre quality as well as important candidate genes influencing fibre quality, and ultimately facilitate the breeding of superior genotypes[2]. A good parallel approach may be to search for candidates in commercial cultivars that are having naturally superior fibre qualities. Several studies performed to compare the structural differences in the genomes have shown that the difference is in the expression pattern, rather than in the presence or absence of particular genes. Hence, the comparison of gene expression profiling between contrasting genotypes with respect to fibre quality can be extended to transcription profiling at the QTL level, and the genes identified at such QTLs may potentially be better candidates for superior fibre quality. Progress has been initiated in this direction at this laboratory in a recently sanctioned project on the genetic improvement of cotton for drought tolerance and superior fibre quality through marker aided selection using a mapping population developed from commercial cotton cultivars adapted to the target environment. Harnessing the full potential of the functional genomics will require a multidisciplinary approach and integrated knowledge of the molecular and other biological processes of fibre development. This information has immediate applications in breeding programmes geared toward the genetic improvement of the cotton yield and fibre quality which is the ultimate aim and outreach of these efforts.

References

1. Arpat *et al,* , *Plant Mol Biol,* 54 (2004) 911–929.

2. Food and Agriculture Organization of the United Nations, Agriculture Data, 2000. http:// http://apps.fao.org/page/collections?subset=agriculture.

3. Barwale *et al., AgBioForum* 7(1&2) (2004) 23-26.

4. Sundaram *et al,* In *Handbook of cotton in India,* (Indian Society for Cotton Improvement, Mumbai, India) 1999, 56-69.

5. Fryxell P A, *The Natural History of the Cotton Tribe* (Texas A&M University Press, College Station, TX) 1979.

6. Meredith J W R, In *Proceedings of the World Cotton Research Conference II*, Athens, Greece, 2000, 97–101.
7. Taliercio E W & D Boykin, *BMC Plant Biol* 7 (2007) 22.
8. Wilkins T A & Arpat A B, *Physiologia Plantarum* 124 (2005) 295–300.
9. Udall *et al*, *BMC Genomics* 8 (2007) 81.
10. Udall *et al*, *Genome Res* 16 (2006) 441-450.
11. Wu *et al*, *Plant Cell Physiol* 47 (2006) 107–127.
12. Allison *et al.*, *Nature Rev Genet* 7 (2006) 55-65.
13. Jamet *et al.*, *Trends Plant Sci* 11 (2006) 33-39.
14. Basra A S & Malik C P, *Int Rev Cytology* 89 (1984) 65-113.
15. Tiwari S C & Wilkins T A, *Canadian J Bot* 73 (1995) 746–757.
16. Ruan *et al.*, *Plant Cell* 13 (2001) 47.
17. Seagull R W, *J Cell Sci* 101 (1992) 561–577.
18. Hsieh Y L, In *Structural development of cotton fibres and linkages to fibre quality* (Hawthorne Press Inc., New York) 1999.
19. Carpita N C & Gibeau D M, *Plant J* 3 (1993) 1–30.
20. Marx-Figini M, *et al.*, *Nature* 210 (1966) 747–755.
21. Wilkins T, 2007. In *World Cotton Research Conference* 4, held on September 10-14, 2007 (Lubbock, Texas, USA) 2007.
22. Wu *et al*, *Mol Genet Genomics* 274 (2005) 477-493.
23. Shi *et al*, *Plant Cell* 18 (2006) 651–664.
24. Hovav *et al*, *Planta* 227(2) (2008) 319-329.
25. Yao *et al.*,, *Electrophoresis* 27 (2006) 4559-4569.
26. Vij S & Tyagi A K, *Plant Biotech J* 5 (2007) 361-380.
27. Jansen R C & Nap J P, , *Trends Genet* 7 (2001) 388-391.
28. Lacape *et al*, In *World Cotton Research Conference* 4, held on September 10-14, 2007 (Lubbock, Texas, USA).
29. Xu Z, Zou F & Vision T, *Genet* 170 (2005) 401-408.
30. Chen *et al*, *Plant Physiol*, 145 (2007) 1303-1310.
31. Cedroni *et al.*, *Plant Mol Biol* 51 (2003) 313-325.

13

Low Cost Component Protocol for *In vitro* Micro Tuber Seed Production of Potato: Review and Future Strategy

ANOOP BADONI AND J. S. CHAUHAN

Introduction

The potato is one of the most important food crops both in the developed as well as in the developing countries. Due to its diversified use in developed countries as food, feed raw material for producing starch, the potato was generally regarded as a crop suited for the western world. The Potato is next only to rice, wheat and maize in cultivation in India. Next to cereals the potato is the only crop which could supplement the need of the food of the country. It is potentially a crop, which can be harvested, and the tuber can be consumed any time after sixty days of planting. As a source of energy it surpasses cereals like wheat and rice (Das; 1999). In India the potato is grown in almost all the states. Nearly 80% of the crop is grown in Indo-Gangetic plains comprising Punjab, Haryana, Uttar Pradesh, Bihar and West Bengal. Its world average yield is 16.1 tonnes/ha and per caput availability is 50.5 kg/year (Prasad; 2004). The successful cultivation of the seed potato depends upon the availability of disease free seed, soil, moisture, plant protection measures, low temperature and short day's conditions during the tuberization phase, resulting in the rapid bulking rate. The Potato plant is very sensitive to ecological factors such as temperature, rainfall and the photo period (Singh; 2002). Two tuber bearing species *Andigena* have been exploited worldwide for commercial cultivation. Starting from 1958 to date, many numbers of varieties have been released for different agro climatic situations in the country (Shekhawat et al; 1992). Potato varieties are distinguished on their habit, pigmentation on the stem, structure of leaf, flower and fruit (berry) color and tuber like shape, size and color, depth of eyes and flesh color etc. (Thamburaj and Singh; 2001). Some of the varieties are Kufri Kundan (1958), Kufri Red (1959), Kufri Safed (1958), Kufri Alankar (1968), Kufri Chamatkar (1968), Kufri Chandramukhi (1968), Kufri Badshah (1979), Kufri Jyoti (1968). Kufri Giriraj (SM/85-45), Kufri Pushkar (JE/JC-166), Kufri Chipsona-I, Kufri Chipsona-II etc. (Arya; 2002).

The Indian Council of Agriculture Research (ICAR) has identified two new hybrid varieties of potato Kufri Chipsona-3 and Kufri Himalini (Anonyms 2005). Nearly 8% of the total potato area namely in the country lies in the hills, where it is an important cash crop. Late blight disease (*Phytophthora infestans*) appears in the hills in epiphytotic form every year and it is the most important impediment in potato cultivation. Losses up to 74% in the North-Eastern hills and up to 65% in the North-Western hills have been recorded in susceptible potato varieties (Joseph *et. al.*; 1998) The ICAR has identified Kufri Himalini for commercial cultivation in hilly regions. Late blight disease has intensified over the last few years, and the resistance to the disease has been decreasing in existing varieties such as Kufri Jyoti and Kufri Giriraj. The new variety, with a medium maturity of 110-120 days has been recommended for cultivation in the north- western and eastern hills during summer. Kufri Himalini provides a yield advantage of over 10% over Kufri Jyoti and Kufri Giriraj. In the plains its keeping quality is better than all the cultivars developed so far for hill regions (The Hindu, 2005).

Potato cultivars are tetraploid vegetatively propagated crops and pose several problems in seed production. Generally the tuber of a potato is used as a seed. Due to progressive accommodation of viral diseases in seed stock, the availability of good quality seed is a major constraint in potato production, which is approximately 50% of the total production cost. Besides the high costs of seed potato, the propagation is also characterized by the low multiplication rate of only 4-6 times. Hence to produce the disease free planting material and for decreasing the production cost, new methods of propagation are to be derived and adopted. In India, the systematic work on potato tissue and cell culture was initiated in 1972 at the Central Potato Research Institute, Shimla.

To enlarge the production of clonal material i.e., to produce the uniform, identical seed material of potato, micro propagation is the better alternative over the conventional propagation of the potato. The in vitro propagation method is a most suitable alternative to produce micro tuber seed material for the potato. By using the technique, which involves low cost components, the large scale clonal material can be achieved in a short time duration. The use of micro propagation for commercial seed production has moved the potato from the test tubes to the field (Wang and Hu; 1982). The advances are the beginning of the second "Green Revolution" in agriculture and are expected to make farming more efficient, profitable and environmentally safe (Dhingra *et. al.*; 1992). Micro propagation is a sophisticated technique of regenerating plants using small pieces of plants (so called explants) that are proliferated on an artificial medium under sterile conditions. The importance of micro propagation lies in very fast clonal multiplication of vegetable crops. Micro propagation is used mainly for getting disease- free plants of superior vigour and productivity (Singh; 1997).

Two new hybrid potato varieties:

1- Kufri Chipsona-3

2- Kufri Himalini

The Indian Council of Agriculture Research (ICAR) has identified two new hybrid varieties of potato. The 'Kufri Chipsona-3' can be used by the food processing industry to make chips and flakes. Kufri Chipsona-3 has a higher protein, potassium and calcium content and is suitable for people with high blood pressure. It is identified for the north Indian plain, gives higher yields at 333 q/ha compared with earlier varieties of kufri chipsona-1 (278 q/ha) and kufri chipsona-2 (272 q/ha) (The Tribune, 2005). The ICAR identified 'Kufri Himalini' for commercial cultivation in hilly regions. Late blight disease has intensified over the last few years, and resistance to the disease has been decreasing in maturity of 110-120 days and, has been recommended for cultivation in the North-Western and North-Eastern hill during summer (The Hindu, 2005).

Problems in conventional potato cultivation

In India, farmers use the traditional methods for potato cultivation and they face many problems in these conventional methods and by these problems the farmers get affected economically. The most severe problem faced by farmers, regarding the non-availability of foundation and certified seed, (Thakur & Moorti; 1991) is that the farmers use pieces of potato or a whole potato tuber, as a seed and by using these conventional methods, a large quantity of food material is used for cultivation. Some major problems with conventional potato cultivation are as follows:

- Wastage of a large quantity of food material.
- Disease and Insect problems.
- Absence of uniformity and quantity and quality of the product may be decreased.
- Availability of good quality seed is a major constraint in potato production, due to progressive accommodation of viral disease in seed stock.
- High cost of potato seed.
- Low multiplication rate.

Need of *in vitro* method for production of potato micro tuber seed material

Obtaining quantities of clean planting material has been a major barrier to increase potato production in many developing tropical countries (Uyaen *et. al.;* 1983). Tissue culture can be derived from a variety of plant organs (viz. meristems-tip, shoots, roots, leaves anthers etc.). Initiation of culture involves; selection, preparation and sterilization of explants and inoculation. In India, Professor P. Maheshwari, realized the significance of the in vitro culture technique quite early. He initiated investigation with complex tissue such as ovary, ovule and control of fertilizers. The technique has developed around the concept that a cell is a totipotent that is it has the capacity and ability to develop into a whole organism. (Malvee; 2007). A Systematic effort to promote Biotechnological Research in India began with the establishment of the National Biotechnological Board' under the Department of Science and Technology in the early eighties. Biotechnology has rapidly emerged as an area of activity having a marked reality as well as having a potential impact on virtually all human welfare, ranging from food processing, protecting the environment, to human health. As a result it now plays a very important role in employment, production and productivity, trades, economics and economy, human health, and the quality of human life through out the world. (Singh; 1993).

Some major advantages of *in vitro* methods

To produce the disease free planting material and for decreasing the production cost, new methods of propagation are to be devised and adopted. In India, the systematic work on potato tissue and cell culture was initiated in 1972 at the Central Potato Research Institute, Shimla (Dhingra *et. al.;* 1992).

To create a large production of clonal material i.e. to produce the uniform identical seed material of potato, micro propagation is the alternative to conventional propagation methods. The *in vitro* propagation is a most suitable alternative to produce micro tuber seed material for potatoes. By this technique which involves low cost components, the large scale clonal material can be achieved in a short time duration. The use of micro propagation for commercial seed production has moved the potato from the test tube to the field (Wang and Hu, 1982). The advances are the beginning of a second "Green Revolution" in agriculture and are expected to make farming more efficient, profitable and environmentally safe. In vitro regenerated micro tubers have become an important mode of rapid multiplication of potato germplasm and prebasic stock in tuber seed multiplication schemes as well as germplasm exchange. (Dhingra, *et. al.;* 1992). Meristem culture procedure utilizes the smallest part of the shoot tip as the explants, including the meristem dome and a few subtending leaf primordia. The primary reason for this procedure is to produce rooted micro plants that are free from systematic viruses, virus like organisms, and superficial fungi and bacteria (Hartmann, Kester, Davies and Geneve; 1997).

Low cost media for micro tuber production

For the commercialization and acceptance of *in the vitro* propagation technique, the cost of production has to be competitive with conventionally propagated plant material, therefore the reduction of the cost of media becomes inevitable. In the recent past some research work has been initiated on this aspect and the same is being reviewed here. A low cost medium was used for the propagation of potatoes made by using V_8juice, table sugar, an NPK fertilizer, $CaCo_3$and agar (V_8juice medium). The plants grown on the V_8 medium were slower to develop roots but after 3 weeks of cultivation in Magenta medium vessels at 22° C, there was no difference in plant height and the number of nodes per plant between plants repeatedly grown on an MS medium and V_8 medium (Bains, 1991). Chandra et al., (1991) found that the inorganic constituents of the MS medium and sucrose could be replaced with commonly available fertilizers like CAN, SSP, and MOP in the proportion of 6.5, 1.0, g/l respectively with 3% ordinary sugar and tap water (potable) for *the in vitro* propagation of potato using single node in liquid medium over filter paper bridges. Different concentrations of ordinary sugar i.e. 8, 10, 12, and 14% obtained from open markets did not have any statistical differences over 231 mM sucrose AR for the micro tubers formed in cultivar Kufri Jyoti, Kufri Badshah, and Kufri Sinduri. Chandra (1992) recommended that the lowest concentration of sugar i.e. 8% could replace 23mM sucrose AR in the micro tuber induction medium. Purnima (1997) replaced expensive ingredients like analytical grade chemicals and sucrose with commercial grade chemicals and table sugar, distilled water was substituted with tap water without any adverse impact on the efficiency for economy of *the in vitro* propagation process. Nene and Sheila (1997) have used a low cost medium in which tapioca was used as a substitute for agar for propagating chick peas.

J. Gopal *et.al.;* (1998), has carried out work on the "Microtuberization in Potato (*Solanum tuberosum* L.)" and described that cultures maintained under a short photoperiod (10 h of 6-12 wmol m^{-2} S^{-1}) and low temperatures (day 20^0 -2^0 C and night 18^0-2^0 C) had both a higher yield (255 mg/plantlet) and a greater number (2/paltlet) of micro tubers than those maintained under long days (16h of 38-50 wmol $m^{-2}s^{-1}$) combined with high temperatures (day 28^0-2^0C and night 25^0-2^0C) (yield 207 mg/plantlet; micro tuber number, 0.9/plantlet), over a wide range of genotypes. After the plantlets had been cultured under long days for an initial period of 60 days, continuous darkness advanced microtuberization by 2-3 months in various genotypes. Under short day and low temperature conditions the addition of 6-benzylaminopurine increased micro tuber yield from 255 mg/plantlet to 645 mg/plantlet and the average micro tuber weight from 115 mg to 364 mg. a similar pattern was observed under conditions of long days and high temperature. Micro tuber produced under light had a greater number of eyes (maximum average: 5.96/ micro tuber) than those produced in the dark (maximum average: 3.50/plantlet). The genotype 2 cultural conditions interactions were significant indicating the importance of developing genotype-specific protocols to maximize microtuberization. J. Gopal *et.al.;* (2003), worked with the objectives of using micro tubers for the conservation of potato germplasm, the main effects of genotype, abscisic acid (ABA), and sucrose level, and their interactions on biomass production, microtuberization, micro tuber dormancy, and dry mater content, were studied. Higher concentrations of ABA decreased both micro tuber production and micro tuber dormancy, whereas higher concentrations (60-80 gl^{-1}) of sucrose promoted biomass production, micro tuber production as well as micro tuber dry matter content.

Scope and Conclusion

As one of the principal cash crops, the potato contributes to the national economy in many ways. It gives handsome returns to the growers. On the average, the net returns are about Rs. 4,700.00/hac. In other terms; 1 rupee of investment gives a net return of Rs. 1.40. in a short period of 80-100 days. The Potato is both a labor and a capital-intensive crop (Shekhawat, *et. al.;* 1992). There have been several problems in conventional potato cultivation, which have already been discussed the production of the micro tuber of the potato is the best way to solve these problems. Some major advantages of micro tuber as a seed are as follows:

- Saving a large quantity of food material.
- Disease free planting material.
- Large production of clonal material.
- Low cost of seed material.
- High multiplication rate with good quality seed material.

Potato cvs are a tetraploid vegetatively propagated crop. It poses several problems in seed production. Generally the tuber of a potato is used as a seed. Due to progressive accommodation of viral diseases in seed stock, the availability of good quality seed is a major constraint in potato production, which is approx. 50% of the total production cost. Besides high cost of seed potato, propagation is also characterized by a low multiplication rate of only 4-6 times.

Hence to produce the disease free planting material and for decreasing the production cost new methods of propagation are to be derived and adopted. To the large production clonal material i.e. to produce the uniform, identical seed the material of potato, micro propagation is the better alternative over the conventional propagation of the potato. The *in vitro* propagation method is a most suitable alternative to produce micro tuber seed material for a potato. By using this technique, which involves low cost components the large scale clonal material can be achieved in the short time duration. The use of micro propagation for commercial seed production has moved the potato from the test tubes to the field. The advances are the beginning of the second "Green Revolution" in agriculture and are expected to make farming more efficient, profitable and environmentally safe.

14

Evaluation of Stres Response of *Gmelina Arborea* to Coalmine Overburden Stress

N. Pathak, J. Gogoi, H. P Deka Boruah and A. K. Handique

Introduction

Plants cannot avoid abiotic environmental stress. They can face different stresses by making necessary adjustments, but the nature of response of a plant species is varied from each other as plants species respond differently to a particular stress according to their genetically determined reaction norm. Different factors like age, adaptability, seasonal and even diurnal activity of plants influence plants responses (Darwin 1880, Larcher 2003, Rizhsky *et al.*, 2002). Mine OB materials devoid of true soil characters and are highly acedic or basic in nature for which plants have to adapted lot (Chaoji 2002, Deka Boruah 2006). Tolerance towards coal mine stress by different plants has been reported by many workers (Gogoi *et al.*, 2007, Pathak *et al.*, 2006 Maiti 2007, Li 2006). However, tolerant plants may show changes to their physiological, biochemical, morphological and anatomical nature during the period of stress adaptation. Salisbury 1985 reported that if a plant develops tolerance, it can endure the adverse environment. A high tissue mass density is generally associated with a species characteristic of a stressed environment such as roots of some species which tend to have thick cell walls and a large proportion of stele and sclerenchyma (Craine *et al.*, 2001, Wahl and Rysor 2000). Stress may also have an influence in enzymatic reactions. The fluctuations of enzyme activity in plants occurs in drought and other abiotic stressful situations (Gale 1975; Ayala *et al.*, 1996; Koyro 1997).

The morphological change of coalmine OB tolerant plant species *G. arborea, D. sisso, S. rostrata* and *C. streata* were found to change appreciably (Gogoi *et al.*, 2007, Pathak *et al.*, 2006 and Deka Boruah *et al.*, 2008, Maiti 2007). The importance of stomatal behaviour, which is only the means of plant communication to an external environment, was reviewed in detail with respect to environmental change (Hetherington *et al.*, 2003; Assamma *et al.*, 2001). Many studies also depicted the accumulation of osmoprotectant (Iyer *et al.*, 1998, Walden and Teare 1974). To overcome this oxidative stress a higher plant converted reactive oxygen to H_2O_2 and H_2O thereby saved the plant species from stress (Meloni *et al.*, 2003). Proline and soluble sugars are key osmolytes contributing towards osmotic adjustment (Yoshiba *et al.*, 1997, Mundree *et al.*, 2002).

Similarly ion exchange plays a significant role in maintaining the membrane stability during stress. Water stressed plants showed greater adaptation to water deficits at higher K levels (Premachandra *et al.*, 1991). Here the physiological response of the plant species *Gmelina arborea*, the family verbenaceae tolerant to coalmine OB stress was evaluated at morph physiological and biochemical level.

Materials and Methods

Stumps of six month old Gomari (*Gamelina arborea*) were taken for experimental purposes. To perform the experiment the earthen pots (size 30×30) were filled up with mine OB materials. The soil mix was prepared by mixing Mine OB and Un-mine soil at the ratio of 1:1. Un-mine soil was taken and as a control to it stumps of *G. arborea* were planted and allowed to grow under normal climatic conditions in the polyhouse. Morphological data was recorded according to the method described by Gardner *et al.*, (1985) and Misra (1992). Plant height, girth and leaf area were recorded after one year for comparison. Normal height was measured by taking the height of the plant from the base of the plant to the top. Girth of the plant was measured by $ðr^2$ where ð (constant) = 3.14 and r = Radius of the plant. Leaf areas were measured by multiplying the universal constant.

Leaf area = L × B × 0.66
Where L = Length; B = Breadth of an individual leaf;
0.660 is the universal constant.

Leaf area index was calculated by the formula LAI = LA/P

Where, LA = Leaf area and P = Total land area occupied by one plant.

Enzyme activity responsible for growth and stress i.e., nitrate reductase (NR), catalase and peroxidase were determined according to Ram *et al.*, (1999). NR activity of leaves was estimated by homogenizing 100 mg of leaf samples in a mixture of phosphate buffer (pH-7): 0.02 M KNO_3: 5% propanol (4:1:1). The mixture was then separated by centrifugation at 10,000 rpm, for 10 min at 4°c and supernatant was collected. From it aliquot (0.4 ml) was transferred to a clean test tube and to it 0.3 ml sulphanilamide and 0.3 ml N-1 napthylethylent diamine hydrochloride was added. The enzyme mixture was diluted with distilled H_2O up to 4 ml and amount of NH_3 ion released was measured spectrophotemetrically at 540 nm and quantified from standard curve.

For an estimation of catalase activity to the enzyme extract (100 mg of fresh plant leaves extracted with 1M phosphate buffer by grinding and separated by centrifugation) following reagent were mixed in different proportions for both test and blank.

Reagent	Test	Blank
H_2O_2(0.2)	1.25ml	-
Enzyme extract	0.50ml	0.50ml
Phosphate buffer	3.25ml	4.50ml

After 3 min 2 ml reaction mixtures mixed with 2 ml potassium di-chromate-acetic acid reagent (1:1). After 10 minute of incubation OD was taken at 570 nm.

For peroxidase enzyme activity, to the enzyme extract following reagent was added in different proportion to prepare test and blank sample.

Reagent	Test	Blank
Enzyme extract	2.0ml	2.0ml
Phosphate buffer	2.0ml	3.2ml
Pyrogallol	1.0ml	-
H_20_2	0.2ml	-

The solution was kept at room temperature for 3 min and activity was measured at 430 nm.

The total enzyme activity was quantified from the standard curve prepared for the specific enzymes viz., release of H_2O_2 for Catalase, perpurogallin for peroxidase, NH_3 for Nitrate reductase, respectively.

Total soluble sugar was determined by the anthron method (Thimmaiah, 1999). For this a 100 mg leaf sample was macerated in 5ml of softened distilled water which was hydrolyzed in 2.5 N HCL and finally neutralized by $NaCO_3$.The green coloured intensity was measured spectrophotometerically at 630nm.

Proline was estimated by following the method of (Thimmaiah, 1999). A leaf sample was homogenized in 10 ml of 3% sulphosalicylic acid. After the filtration was over using wahtmen using filter paper no 42, the filtrate 2 ml was mixed with 2ml glacial acetic acid and 2ml acid-ninhydrin. After boiling the samples for 1 h the reaction was terminated by placing the tubes in an ice bath. Then toluene was added and stirred for 20 to 30 sec. After the separation of the toluene the red layer coloured intensity was measured at 520nm.

Leaf permeability status was measured by the release of ions Na^+, $K^+$and Ca^+ from the leaf samples of *G. arborea.* For this, measured amount of leaf sample was immersed in double distilled water for 2 h. The amount of Ca^+, Na^+ and K^+ was measured in the liquid by an Atomic absorption spectrophotometer.

Mine OB **mineOB: un-mined soil** **Un-mined soil**

Plate 1: Control experiment of *Gmelina arborea* grown in different substrate in polyhouse conditions

Results and Discussion

It was found that a plant grown in a mine OB showed statistically significant less plant height, leaf area and girth (Table 1). Approximately 40%, 70% and 50% less plant height, respectively less leaf area and girth was found for the mineOB grown plants. These results were also supported by the report of (Gogoi *et al.*, 2007) where the author has found the stunted behaviour of the plant.

*Table*1.Comparison of growth of *Gmelina arborea* grown in different substrate

Substrate	Height (cm)	Leaf area (cm^2)	Girth (cm^2)
Un-mine	117.5**	132***	30.1**
Mix OB	73.5**	132.1***	42.9*
Mine OB	48	96.9	15.19

NB:*P>0.05 ; **Statistically significant at p>0.01; ***similar letter not significant different

Table 2. Comparison of leaf area one month old *Gmelina arborea* at weekly intervals grown in different substrate

Substrate	Leaf area in cm^2				
	1st week	2nd week	3rd week	4th week	5th week
Un-mined soil	59.30±58.4**	100±71.5**	103±71.3**	103.90±71.4**	107.00±74.4**
Mix OB	32.18±16.09	37.30±18.65	39.06±19.53	38.96±19.48	38.90±19.45
MineOB	17.90±8.95	12.30±6.25	12.50±6.25	18.80±9.45	18.90±9.4

Reduction of leaf size of *G. arborea* was also found for the 2 months old saplings (Table-2).This depicted the affect of mine OB on leaf size.

Table 3. Comparison of leaf area index (LAI) in *Gmelina arborea* at weekly intervals grown in different substrates

Substrate	Leaf area index(LAI) in cm^2				
	1st week	2nd week	3rd week	4th week	5th week
Un-mined soil	0.080±0.08**	0.142±0.10*	0.145±0.10*	0.146±0.09*	0.137±0.09
Mix OB	0.035±0.01	0.041±0.01	0.040±0.01	0.043±0.02	0.043±0.01
Mine OB	0.019±0.01*	0.013±0.01	0.013±0.01	0.021±0.01	0.020±0.00

The leaf enzyme activity (Catalase, Peroxidase, Nitrate reductase) of the plants grown in mine OB soil was also found to have significantly less than that of the plant grown in Un-mined soil (Table 3). The enzymes NR, peroxidase and catalase all played a critical role in the plant growth and development (Pessarakli 1999).Though reports are available on increase of peroxidase and catalase activity when plants were grown in abiotic stress (Balakumar T, 2000) no appreciably high accumulation of catalase and peroxidase depictedin was found to be in the present study. This may be due to the fair tolerant nature of the plant towards mine OB Stress.

Table 4. Comparison of enzyme activity of *Gmelina arborea* grown in different substrates.

Substrate	Enzyme activity in mg/g/h		
	Peroxidase	Catalase	Nitrate reductase
Mine OB	0.015 ± 0.010	0.010±0.007	0.066±0.046
Mix OB	0.014 ± 0.009	0.009±0.006	0.070±0.049
Un-mined soil	0.024 ± 0.016	0.014 ±0.009	0.104±0.780

The accumulation of proline was found to be about two times higher in the plants of mine OB condition than control (Table-5). Osmoprotectectant proline accumulates significantly in a higher concentration when plants are allowed to grow in abiotic stress (Vessllues et al., 1999, Yamchi *et al.*, 2007, Iyer and Caplan 1998).This supports our present observations. Compared with Unmined soil *Gmelina arborea* showed a higher amount of soluble sugar while young but slowly a less sugar content in a one year old plant (Table 7). This may be due to an osmotic adjustment of the young seedling of *G.arborea* for survival.

The accumulation of higher soluble sugar, proline in stress adapted plant species was reported for many plant species (Thomas and James 1999; Dekankova et al., 2004).This suggested that there is a requirement of a plant species to an osmotic and membrane stability adjustment for an adaptation to adverse edaphic situations.

Table 5. Comparison of proline accumulation by *Gmelina arborea* grown in different substares

Substrate	Proline accumulation in μ mol/g tissue
Un-mined	30.92
Mix OB	27.69
Mine OB	57.10

Ion exchange capacity specifically Na and K was found to be higher in the plants of Mine OB (Table6). On an osmotic stress condition the permeability status of leaf and root may change (Valentonic *et al.*, 2006).

Table 6. Membrane permeability of *Gmelina arborea* grown in different substrates

Substrate	Ion exchange (mg/lit)		
	Ca	Na	K
Un-mined soil	19	0.14	0.15
Mix OB	8	0.88**	0.65**
Mine OB	11	0.44**	0.24**

Table 7. Comparison of total soluble sugar of *Gmelina arborea* grown in different substrates

Substrate	Total sugar content in %	
	6 month old	1 year old
Un-mined soil	0.192 ± 0.13	0.81±0.57
Mix OB	0.105±0.074	0.72±0.50
Mine OB	0.245±0.173	0.61±0.43

Conclusion

Gmelina arborea showed adaptability towards mine OB stressful situations. Though a stunted growth i.e., reduced leaf size, LAI, they will have to change their morph physiological as well as biochemical nature for their adjustment. Further study will be helpful for stress physiological research.

Acknowledgement

Authors are thankful to Dr. P. G. Rao, Director, NEIST, Jorhat, Dr. B. K. Gogoi, Head, Biotechnology Division, NEIST, Jorhat, Assam for their support and to carry ing out this research successfully. Authors are also thankful to CMPDI, Ranchi for financial support.

References

1. Aasamaa K., Sober A. and Rahi M., (2001). Leaf anatomical characteristics associated with shoot hydraulic conductance, stomatal conductance and stomatal sensitivity to changes of leaf water status in temperate deciduous trees. *Aust. J. Plant Physiol.* 28, 765-774
2. Ayala F., O'Leroy J. W. and Schumaker K. S., (1996). Increased vacuolar and plasma
3. membrane H^+-ATP ase activities in Salicornia bigelovii Torr in response to NaCl. *J.Exp.Bot.*, 47, 25-32
4. Bradford K. J., and Hsiao T. C. (1982). Physiological responses to moderate water stress. InO.L Lange,P S Nobel, C B Osmond, and H Ziegler (eds.), Physiological Plant Ecology II,

Water relations and carbon assimilation. Pages 263-324 in A. Pirson, M H. Zimmermann (eds.), *Encyclopedia of plant physiology,* New series.12B. Springer-Verlag, Berlin, Heidelberg, New York

5. Balakumar T. 2000. Mechanism of aluminium toxicity and tolerance in plant perspective and prospective Ch. 13 . In Advances in plant physiology 3, pp 321-333 Scientific Publishers, Jodhpur, edited by. Hemantarajen, A..
6. Chaoji S.V., (2002). Environmental challenges and the future of Indian coal. *Journal of Mines, Metals and Fuels,* 257
7. Craine J M., Froehle J., Tilman D. G., Weldin D. A., Chapin F. S. III (2001). The relationship among root and leaf traits of 76 grassland species and relative abundance along fertility and disturbance gradients. *Oikos.,* 93, 274-285
8. Darwin R., (1880). The power of movement in plants. John Murry, London.
9. Deka Boruah H. P., (2006). North Eastern Coal and Environment, an Overview. Proceedings on Workshop-cum-Interactive meet on Characterzation and Gainful Utilization of North East Coal, Proceedings, Published by RRL, Jorhat 28-33
10. Deka Boruah H. P., Rabha B. K., Pathak N., Gogoi J (2008). Non-uniform patchy
11. stomatal closure of a plant is a strong determinant of plant growth under stressful situation :a case study, *Current science,* 94 1310-1314
12. Dekankova K., Luxova M., Gasparikova O., Kolarovic L (2004). Response of maize plants to water stress. *Biologia 59/Suppl.,* 13, 151-155
13. Gali J., (1975). The combined effect of environmental factors and salinity on plant growth.In: Pants in saline environments, edited by Poljakwff-Mayber, H & Gale, J. New York: Springer Verlag, 186-192
14. Gardner F. P., Pearce B. R. and Mitchell R. L. (1985). *Physiology of crop plants.* Iwa State University Press.
15. Gogoi J., Pathak N., Duara J., Deka Boruah H. P., (2007). *In-Situ* selection of tree species in environmental restoration of opencast coalmine wasteland, Proceedings of International Seminar, MPT Mumbai
16. Hetherington U. M., Woodward F. I., Woodward F. I., (2003). The role of stomata in sensing and driving environmental change, *Nature,* 424, 901-908
17. Iyer S. and Caplan A., (1998). Products of proline catabolism can induce osmotically regulated genes in rice. *Plant Physiol,* 116, 203-211
18. Koyro H. W., (1997). Ultrastructural and physiological changes in root cells of Sorghum plants(Sorghum bicolor X S.sudanensis C V sweet Sioux induced by NaCl. *J. Exp.Bot.,* 48, 693-706
19. Larcher W., (2003). *Physiological Plant Ecology,* 4th Edition. Springer.
20. Li M. S. (2006). Ecological restoration of mineland with particular reference to the metalliferous mine wasteland in China: A review of research and practice Science of The Total Environment. 357(1-3), 38-53.

21. Maiti S. K., (2007). Bioremediation of coalmine overburden dumps-with specisl empasis on micronutrients and heavy metals accumulation in tree species. *Environ Monit Assess,* 125,111-122.

22. Misra K. C., (1992) Mannual of plant ecology , Oxford & IBH Publishing Co. Pvt.Ltd.

23. Meloni D. A., Olivia M. A., Martinz C. A., Cambraia J., (2003). Photosynthesis and activity of superoxide dismutase, peroxidase and glutathione reductase in cotton under salt stress. *Environmental and Experimental Botany,* 49, 69-76.

24. Mundree S. G., Baker B., Moula S., Peters S., Marais S., Wilingen C. V., Govender K., Meredza A., Muyanga S., Farrant J. M., Thomson J. A., (2002). Physiological and molecular insights into drought tolerance. *Afr. J Biotechnol* 1 28-38

25. Passarakli M., (1999). Handbook of plant and crop stresses, CRC press

26. Pathak N., Gogoi J., Saikia N., Gohain R., Deka Boruah H. P., (2006). Biodiversity

27. prospecting for phytoremadiation of coalmine tailings dumping site in the environment. In Souvenir cum abstract of *Value addition to bioresources of NE India Post harvest technology and cold chain* on the occasion of National Technology Day.112-113

28. Premachandra G S., Saneoka H., and Ogata S., (1991). Cell membrane stability and leaf water relations as affected by potassium nutrition of water stressed maize. *Journal of Experimental* Botany 42. (6), 739-745

29. Rizhsky L., Liang H and Mittler R., (2002). The combined effect of drought stress and heat shock on gene expression in Tobacco., *Plant Physiology,* 130, 1143-1151

30. Ram P. C., Lal R. K., Singh A. K., Singh U., Chatmvedi G. S., (1999). *Technique in plant metabolism, Mannual,* N.D University of Agriculture& Technology , Faizabad (UP)

31. Salisbury., Frank B., and Marinos N. G., (1985).The ecological role of plant growth substances in R P Pharisand D M Reid (eds), *Encyclopedia of plant physiology* Vol11. Springer-Verlag, Berlin, Heidelberg, New York.

32. Thimmaiah S. R., (1999). *Standered methods of biochemical analysis,* Kalyani publishers, New Delhi

33. Thomas H., James A. R., (1999). Partitioning of sugars in Lolium perenne (perennial ryegrass) during drought and rewatering. *New Phytol,* 142, 295-305

34. Valentovic P., Luxova M., Kolarovic L., Gasparikova O., (2006). Effect of osmotic stress on compatible solutes content, membrane stability and water relations in two maize cultivers. *Plant soil environment.* 52, 186-191

35. Versleues E. P and Sharp R. E., (1999). Proline accumulation in Maize (Zea mays L.) Primary roots at low water potentials.11.Metabolic source of increased proline deposition in the elongation zone. *Plant Physiolog,y* 119, 1349-1360

36. Wahl S., Ryser P., Edward P. J., (2001). Phenotypic plasticity of grass root anatomy in response to light intensity and nutrient supply. *Ann Bot* 88: 1071-107

37. Walton D. C., (1980). Biochemistry and physiology of abscisic acid. *Annual Review of Plant Physiologt* 31 453-485.

15

Mycorrhizal Inoculation of Some Medicinal Plants : Field Trials

YUDHVIR K. BHOON

Introduction

We would like to share with the farming community who are cultivating the medicinal plants on a commercial scale, the results of our investigations on the inoculation of various medicinal plants with VAM (Vesicular Arbuscular Mycorrhiza). These investigations were carried out under the National Medicinal Plant Board sponsored Research Project No.87/2002 "Mycorrhiza" is made up two words-mycor (Fungus) and rhiza-(root) and literally means root fungus and these soil microorganisms ie . the fungus are thought to be as old as our Mother Earth and are found distributed all over the earth. Mycorrhizae are the plant phytosymbioant i.e. there is symbiosis between the root and the fungus. The fungus cannot make its own food and multiplies in the root of the plants (feeder roots only). It taken carbon as carbohydrates from the plants for its survival and in turn gives a lot of things back to the plant for its overall growth and protection from various pathogens. The various functions of the mycorrhiza (VAM) which it does for the plants (except Chenopodium, Brassica etc as these are not infected by this fungus) are as follow.

a) The main function of the Mycorrhiza is to dissolve the fixed phosphate available as insoluble phosphate in the rhizosphere zone ie. the zone in the soil surrounding the roots and make it available to the plants through the hyphel roots which are tube like structures which these fungi leave in and the rhizosphera zone through which the dissolved phosphate reaches the roots. Easy availability of the organic phosphate in the dissolved form, being the major nutrient will result in rapid growth, early flowering, maturing and other stages of the life of the plant.

b) It also helps in the dissolution of the trace elements which are in the form of insoluble compounds due to high alkalinity (pH more than 7.6) and make them available to the plants. The soil due to a continuous use of the area over the years ever since the mid sixties when the Green Revolution started has become more alkaline and the trace elements already available in the soil will be in the form of insoluble hydroxides/other amines. Easy availability of the trace elements like Zn, Mn, Fe, Co, Ni, Mo etc results in better growth, flowering, fruit setting etc. There is an over all growth resulting in better yields.

c) These fungi synthesize certain chemicals like HCN etc and release them in the rhizosphere zone which protect the feeder roots of the plants from the attack of the various pathogens in the rhizosphere zone and the roots spread faster and this helps in the rapid growth of the plants.

d) In the transplanted crops, because of the mechanism explained in steps a,b,c, if inoculation is done of the seedlings, there plants stabilise at a much faster rate than the control and if the inoculation is carried out right in the beds where seedlings are raised from the seeds, the transplanted seedlings will stablise more quickly and grow faster and the mortality rate can also be lowered. We have also observed that inoculated seedlings of Phyllanthus niruri/P.amarus, Achyranthes aspera, Ocimum sanctum stablise more quickly than the control seedling even during the scorching summer months of May and June.

e) The Mycorrhiza also helps in the nitrogen fixation because the phosphate requirement in the nitrogen fixation is fulfilled through mycorrhizal activity and hence Mycorrhiza can be used with Rhizobium.

f) There are a number of Mycorrhiza benefitting Bacteria like Azotobacter, Phosphate Solubilizing Bacteria, Azospirilium. Actinomycetes etc which all work in an harmonious way and there is no competition among these microorganisms and better results are obtained if Mycorrhiza is used along with other microorganisms mentioned above which make available the nitrogen and phosphorous (by nonsymbiotic way)

Results of VAM inoculation of some selected Medicinal Plants in the Pots and Fields

1. Solanum nigrum

The bio mass in the inoculated plants was found to be double the control plants. The size of the berries was slightly bigger than the control and the weight of the berries was found to be 1.9-2.1 times more than the control.

2. Aspargus racemosus

The weight of the tubers which are of medicinal importace was found to be almost double the control. The flowering was observed the same year and one can get the seeds for raising more planting material.

3. Ocimum basilicum,O.gratissimum OC-14, O.canum,OC-12 and O.sanctum

In all the ocimum varieties mentioned above, the biomass was higher than the control. The size of the leaves was higher by 10-15% in the inoculated plants which gives more yield of the fresh leaves and more oil production. The inflorecense were found to be longer in the inoculated plants, more branching and this results into higher yields of the seeds. However, there was no increase in the weight of the seeds in case of O.basilicum.

4. Andrographis paniculata

Like other plants, this species has also given us encouraging results. The biomass was 1.5-1.8 times more in the inoculated plants. The spread over area of the green biomass was also more in the inoculated plants. The results of the HPLC analysis carried out on the control and mycorrhizal plants(aerial parts) indicated the % of Andrographolide to be 3.4 and there was no change in terms of the contents of the active ingredients as a result of the mycorrhizal inoculation.

5 Aloe vera.

The size and thickness of the leaves was more in the inoculated plants. The inoculated plants gave a greater number of the suckers than the control plants. These suckers are the planting material to cover more area under plantation.

6. Chlorophytum borivilianum

The incidence of root rot was minimized in the fields inoculated with mycorhhiza. The weight of the bunch of the fingers dug out from the fields inoculated with mycorrhiza was at a par with those grown on chemical fertilizer doses.

7. Bacopa monnieri

The spread over area due to vegetative growth was much more and faster in the inoculated plants than the control.

The majority of the Medicinal plants belong to the families of the plants which are easily colonized by Mycorrhiza and hence VAM can be successfully used in the cultivation practices. The work to determine the % of active ingredients in the inoculated med. plants is in progress. We wish that you could try also the use of Mycorrhiza in the various medicinal plants on trial in the current season to get more benefits and through the use of Mycorrhiza, you will be increasing the fertility of the soils of your fields and the plants produced will be exclusively organic.

8. Silybum marianum

The experiment was conducted in a view of the extensive use of silybum marianum in the allopathic system for the treatment of Hepatitis B, C which are virus based and is being adopted by the Indian System of Medicine also and also because of the fact that the plant is being cultivated commercially in India to meet the international market demand. The growth of the plants raised from direct sown seeds which were inoculated with Ecorrhiza at the time of the sowing of the seeds, was more satisfactory than non inoculated plants. The total height of the plants, the leaf area, the branching, and the collar girth of the plants were all higher in inoculated plants. The following was observed 10 days earlier in the inoculated plants. The colour of the seeds was dark black in the inoculated plants while in the control, it was light brown. The seeds of the inoculated and control plants were subjected to an HPLC Analysis for the silymarin contents and the seeds from the inoculated plants were found to have a silymarin contents of 1.88 % while the control seeds were having 1.77 %. There is a marginal increase in the silymarin contents as a result of inoculation. The weight of 1000 seeds from inoculated plants was higher by 14.25% compared with the to than the control plants and the amount of the seeds harvested from 10 plants was 25% more than that of the control plants which will be beneficial to the farmers in taking 25% more production than the control plants. During October, 2004 the seeds collected from the inoculated and non inoculated plants were sown and the plants raised from the seeds of the inoculated plants were in a flowering stage during the 3rd week of March 2005 while the plants raised from the seeds of the non inoculated plants were under going vegetative growth only. The results will be presented in the Final Report to be submitted in September, 05.

9. Clitoria ternetea

One of the most interesting observations made while doing the inoculation of Clitoria ternetea was that the leaf area of the leaves of the plants raised from the seeds obtained from the inoculated plants was practically 150 t more than the plants raised from the seeds of the non inoculated plants during the monsoon of 2004. The plants after completing their dormancy in Dec-Jan (2004-05), are again sprouting and the leaf area of the plants, either inoculated with Ecorrhiza or raised from the seeds of the inoculated plants, was 100-200 t more than the control plants. More results pertaining to Phytochemical studies and other parameters will be presented in the final report when the data will be available during June 2005 after the flowering is over and the seeds are harvested.

10. Cassia occidentalis

The seedlings of Cassia Occidentalis raised from the seeds collected from the plants growing in the campus of Sri Venkateswara College, were inoculated both in the pots and the beds in an effort to undertake the field trials. Like the results obtained with other plants, the growth of the inoculated plants was much faster than the non inoculated plants. The average collar girth of the inoculated plants was found to be 92% more than the non inoculated plants during December 2004. However, when the measurements were undertaken during March-April, 2004, the collar girth was found to be only 13.20% higher in the inoculated plants.

The increase in the height of the plants was noted during the vegetative growth period in June and July 04. there was an increase of 8.53% in the height of the inoculated plants in mid June while a 30.62% increase in the height of the inoculated plants was observed during mid July 04. The Collar girth in the inoculated plants was higher by 30% than the non inoculated plants, more seedpods were formed in inoculated plants and the quantity of the seeds collected from 10 plants was 35% more than non inoculated plants. The leaves and the seeds of the non inoculated and inoculated plants were subjected to the total anthraqinane contentes and the inoculated plants were having a higher % of the anthraquinone both in the seeds and the leaves in comparison with the non inoculated plants.

Totalanthraquinones: Seed,wild:0.088+0.006,Control:0,084+0.001,VAM:0.103+0.003: Leaf:Cont.0.067+0.007,VAM: 0.079+0.005,

11. Acalypha indica

While doing the inoculation of Ecalypha indica, some of the interesting observations are as follow:

i. The growth of the inoculated plants was much faster

ii. The leaf viral attack on the inoculated plants was much less

iii. The total height of the inoculated plants was 15-20% more than the non inoculated plants

iv. The flowering was early

v. The branching in the inoculated plants was more than the non inoculated plants.

The total biomass in the inoculated plants was higher by 20% than in the non inoculated Plants. The average height of the plants inoculated with Ecorrhiza (taken for 8 plants) as observed on 12.6.04 and 17.7.04 was found to be 39.75 and 78.63 cm for control plants and 47.41 and 89.5 cm for the inoculated plants (average taken for 12 plants). The increase in the average height in the inoculated plants was found to be 19.27% and 13.86% more at those dates as compared with the non inoculated plants. However, at the maturity stage after 3 months of plantation, the average height of the inoculated plants was 114 cm and the non inoculated plant was 110.0 cm.

The total average biomass of the arial plants was 123.5g(Inoculated) and the non inoculated was 101.5g, so there was an increase of 42.68% in the biomass as a result of Ecorrhizal inoculation. The increase in the root weight was 26.47% more in inoculated plants than in the control plants. It is interesting to note that the panchang (whole plant) is used in the IndianSystemofMedicine.

12. Phyllanthus amarus

The results of inoculation in Phyllanthus amarus are equally encouraging. The inoculated plants were longer in length, collar girth and total biomass. It is significant because, the whole of the plant is used in the Indian system of medicine. The increase in the biomass in the inoculated plants was found to be 34.3% more than in the non inoculated plants. The collar thickness was found to be 70% more in the inoculated plants. However, per unit mass, the % of the active ingredient Phyllanthin (1.7%) was found to be exactly the same. However, per unit area, because there is 34.5% increase in the biomass in inoculated plants, one should expect higher yields.

13. Catheranthes roseus

The results of the Mycorrihzal inoculation in Catheranthes roseus are also equally encouraging. The average collar girth of the inoculated plants was found to be 30% more than the non inoculated plants. The average biomass in the inoculated plants was found to be 92.03% more than the non inoculated plants. The roots and the leaves are with Natural Product Division (Prof. K.K. Bhutani) NIPER, Mohali for the analysis.

14. Lepidium sativum and Linum uritatissimum

The weight of the 1000 seeds from inoculated Lepidium sativum was found to 4.5 grams and Control Plants were only 3.6g and there was an increase of 25% in size of the inoculated seeds while in the case of Linum uritatissimum, the 1000 inoculated seeds were weighing 19.15% more than the non inoculated plants. The oil contents in case of Linum usitatissimum is to be analyzed . An increase of 25% in the weight of the inoculated seeds will result in an overall higher yield per acre.

15. Asparagus racemosus

After the preliminary results of inoculation on the seedlings of Asparagus racemosus, the field trials were undertaken both in the college campus and at Surjeevan Farm, Village Besar Akbar Pur, Distt. Gurgaon, where the soils are sandy. The plants were uprooted after 1.5 years of plantation and comparative morphological studies indicated the following results.

Three and four branches of the inoculated and non inoculated plants were taken for the comparative studies. There was a 97.94% increase in the average weight of the tubers from the inoculated plants which are of medicinal importance.

In the bunch of the tubers of the inoculated plants, the thickness of the 168 tubers which were obtained varied from 0.6 cm - 24 mm while in the control plants, the 148 tubers which were obtained, the thickness was in the range of only 0.8mm to 10 mm. The length of the tubes in the inoculated bunch was in the range 5cm to 44cm while in the control plants, it was in the range of 4cm - 33cm.

The tubers harvested after 1.5 years of plantation were dried after peeling off the thin layer and were subjected to HPTLC (M/s Natural Remedies, Pvt Ltd) while in the inoculated plants, Shatavarin IV was found to be <0.1 %. Shatarvarin was so low in the noninoculated plants that it could not be detected indicating that the Ecorrhizal inoculation helps in the easy synthesis of the higher % of the active ingredient.

16. Andrographis paniculata

The inoculation of 3 months old saplings was done with Ecorrhiza directly in the fields at the Surjeevan Farm during Feb'03 and the plants were allowed to grow and the arial parts of the plants were harvested during Nov'04. The inoculated plants in the 5'x20' beds showed good branching and the flowering was earlier by 10 days compared with the non inoculated plants. The foliage area spread over in the inoculated plants was 20-25% more than the non inoculated plants. The leaves and twigs were analyzed for Andrographolide and there was no change in the % of the active ingredient per unit mass as a result of the inoculation. However, the biomass which is of a medicinal value was found to be 26% more in the inoculated plants, so the inoculation with Ecorrhiza can give a higher yield and hence may be incorporated in the cultural practices.

Inoculation Methods

VAM inoculum is now commercially available in India due to the successful and extensive research carried out by Dr. Alok Adholeya, Fellow, Microbial Biotechnology Division (now

Mycorrhiza Research Centre, The Energy Research Institute, New Delhi) over two decades. The research was sponsored by Deptt. Biotechnology, GOI in developing the protocols for mass scale production of mycorrhiza by a root organ culture technique. This is the first technique developed in India and the technology has been sold to M/s KCP Sugars and Industries Pvt Ltd, Vuyyuru AP and Cadila (Agro Div.) Ahmadabad and Majestic AgronomicsPvt Ltd,A-1/289 Janak PuriNew Delhi-110028 under the "Transfer of Technology"pograme. Mycorrhiza-AM is a consortium of three AMF spores while the other two inoculum contains only one species of mycorrhiza. VAM is now available in 1kg pack costing Rs. 100 under the trade name "Ecorrhiza" , "Josh" and "Mycorrhiza-AM" respectively.

4-5 Kg of VAM inoculum is sufficient for one acre of the land which can be mixed with 200 Kg of powdered cow dung manure/FYM/soil and mixed in the fields uniformly while preparing the fields. The field can be ploughed two to three times so that mixture gets distributed in the soils. When the no. of the plants per acre are more than 40,000, 5-10 Kg of the inoculum will be required as the requirement of the plants will increase with the increased number of plants. Even the seedlings may be dipped in the water suspension of the inoculum containing some jaggery or molasses before transplantation. Alternatively, the seed beds may be inoculated with VAM inoculum raising the nursery. The seedlings so raised will be all inoculated with mycorrhiza. Yet another method is to put 2 g of the inoculum in the micropits to a depth of 3-4" below the soil level while transferring the seedlings in the fields. Even if the seedlings have been stablized after the trasplantation, the inoculation with VAM can still be carried out by making 5" deep holes with a hollow pipe (1-2" diameter) 1-2" away from the stems of the plants and putting 2g of VAM inoculum in these holes and then fill the holes with the soil. The inoculum should reach the level of the softer and thinner feeder roots as the spores of the Mycorrhiza in the inoculum will attack only the feeder roots and not the thicker roots.

Mycoorrhiza which is a potential biofertilizer is the correct answer to undertake organic farming without any fear in mind that organic farming if undertaken will result in lower yields. By the use of nitrogen fixing Biofertilizers (nonsymbionant) with mycorrhiza, PSB and organic manure, you will be increasing the concentration of the micro flora responsible to increase the fertility of the soils, you will be increasing C:N and ultimately, the production will go on increasing. High quality organically grown medicinal plants or any agricultural crops can be harvested through the use of Biofertilizers viz, VAM .Nitrogen fixing bacteria/Phosphate solubilizing Bacteria etc.

References

1. Feldmann, F., 1998: The strain-inherent variability of arbuscular mycorrhizal effectiveness: II. Effectiveness of single spores. Symbiosis, 25, 131-143.
2. Feldmann, F., 1999: Mykorrhizaeinsatz im Pflanzenbau. Deutscher Gartenbau 17, 24-26.
3. Feldmann, F., Idczak, E., 1994: Inoculum production of VA-mycorrhizal fungi. In: Norris, J. R.; Read, D.J.; Varma, A. K. (eds.): Techniques for mycorrhizal research, Academic Press, San Diego, 799-817.
4. Jakobsen, I., 1995: Transport of Phosphorus and Carbon in VA Mycorrhizas. In: Varma A. und Hock, B.(eds): Mycorrhiza. Springer, Berlin, 297-324.

16

Compost Tea and its effect on Growth of Tomato Seedlings

R. SUNITHA AND DR. P. SUBRAMANIAN

Introduction

Compost tea, in modern terminology, is a compost extract brewed with microbial food sources *viz.*, molasses, rock dust, humic and fulvic acids. The compost tea brewing technique is an aerobic process, involving the extract and growth of beneficial microorganisms. Compost teas are distinguished from compost extracts both in method of production and the way in which they are used. Teas are actively brewed with microbial food and catalyst sources are added to the solution, and a sump pump bubbles and aerates the solution, supplying plenty of much needed oxygen. The compost tea provides the source of microbes, the microbial food and the catalytic amendments which promote the growth and multiplication of microbes in the tea (US EPA, 1993).

There is a huge potential available in the beneficial micro organisms as well as compost tea, which play a major role of good seedling establishment, crop growth and production. At present a liquid based organic source of nutrients is required to use in drip/ fertigation system.

Methodology

Preparation of Compost tea

Vermicompost and water was taken in 1:4 ratio. To this mixture, different carbon sources like jaggary (10 %) and molasses (10 %) were added separately. The vermicompost-water mixture, vermicompost-water + jaggary and vermicompost-water +, molasses mixtures were taken in separate container and then incubated for 3 days by providing sufficient aeration with aerators. The frothing was removed periodically. The contents were filtered separately and the filtrate was the compost tea. The compost tea was stored in a sterilized plastic can container for further use.

Growth estimation of the microbial inoculants

Azospirillum lipoferum

The growth estimation of *the Azospirillum lipoferum* was done by serial dilution and most probable number technique as described by a Dobereiner (1980). Compost tea was serially diluted by transferring one ml of culture to 9 ml of sterile water blank which corresponds to 10^{-1} dilution. Likewise serial dilutions upto 10^{-9} were made. The tubes were incubated for 6 days for the growth of *Azospirillum lipoferum*. The positive test tubes showing the growth of the culture as subsurface pellicles was scored and the population of inoculants was estimated using the Most Probable Number (MPN) table.

Phosphobacteria, *Pseudomonas fluorescens* and *Azotobacter chroococcum*

The growth estimation of Phosphobacteria, *Pseudomonas fluorescens* and *Azotobacter chroococcum* in terms of colony forming units per ml was estimated through serial dilution and plate count method described by Allen (1953).

Culture broth was serially diluted as per the procedure mentioned in the above paragraph. From appropriate dilutions, one ml was used for plating. Enumeration of Phosphobacteria was done by plating in Pikovskya medium (Appendix-1), *Pseudomonas fluorescens* in King's B medium (Appendix-1) and *Azotobacter chroococcum* in Waksman 77 medium (Appendix-1).

Physico-chemical characterisation of compost tea

pH balane (p Germen potenz=power. H= symbol for Hydrogen)

About 50ml of compost tea solution was taken in a 100 ml beaker and the electrode of the pH meter was immersed in the solution, and the readings were recorded (Jackson, 1973).

Total nitrogen

Total nitrogen was estimated using the micro Kjeldahl's method as given by Humphries (1956).

Ten ml of the compost tea was digested using 15 ml of the diacid (concentrated sulphuric acid and perchloric acid in the ratio of 5:2). The extract of the digested samples were then made upto 100 ml using distilled water. 10 ml of the aliquot was taken in distillation flask and 25 ml of 40 percent Sodium hydroxide was added. 25 ml of Boric acid with double indicator were added in a beaker and placed at the delivery end. The distillation was carried out for 3 minutes. The liberated ammonia collected in the beaker was titrated against 0.02 N sulphuric acid till the blue colour changes to the red colour. From the titre values (x), the total nitrogen was calculated.

Total nitrogen in the sample (percent) = $(X \times 0.00028 \times V \times 100) / (10 \times W)$

V- Volume of 0.02N sulphuric acid used for sample titration

W-Weight of the sample taken (g)

Total phosphorus

Total Phosphorus content was estimated from the triacid extracts, using photoelectric colorimeter at 420nm (Jackson, 1973)

Ten ml of the sample was digested with 15 ml of the triacid (9:2:1 ratio of nitric acid, sulphuric acid and perchloric) until the contents become colourless. The digested samples were then filtered through Whatman No.41 filter paper and filtrate was made upto 100 ml using distilled water. It was then preserved for further use.

5 ml of the triple acid extract was pipetted into a 25 ml volumetric flask and 5 ml of Barton's reagent (Appendix II) was added and made upto 25 ml using distilled water. The set up was kept for 30 minutes for the yellow colour development. The intensity of the yellow colour was measured with the photoelectric colorimeter at 420 nm. The concentration of phosphorus in the sample (X) was calculated from the Standard curve plotted with standard concentration in the X-axis and OD value read from photoelectric colorimeter the in the Y-axis

Total phosphorus content of the sample (percent) = $(X/10^6)x(25/5) \times (100/W)x\ 100$

W-Weight of the sample taken (g)

Total potassium

Total potassium was estimated from the triple acid extracts of the sample using the flame photometer (Jackson, 1973). The extract was directly fed into the flame photometer and the readings were recorded.

Total potassium content of the sample (percent) = $[(X/W)/10^6] \times 100$

W-Weight of the sample taken (g)

Total micronutrients

The micronutrients zinc, copper, manganese and iron were estimated from the triple acid extracts prepared as described in 3.3. The extracts were fed into the Atomic Absorption Spectrophotometer (AAS) and the readings obtained for the total micronutrients were recorded (Lindsay and Norvell, 1978).

Plant growth promoters

Indole Acetic Acid (IAA) production

Extraction of IAA (Chandramohan and Mahadevan, 1968)

The sample was adjusted to a pH balance of 2.8 with 1N Hydrochloric Acid and an equal volume of ice cold (4°C) diethyl ether was added. The contents were shaken, covered with black paper and allowed to stand for 4 h at 4°C with intermittent shaking. Using a separating funnel, the aqueous phase was separated from the organic phase and the extraction was repeated 2-3 times. Discarding the aqueous phase, the organic phases were pooled and evaporated to dryness in the dark. The residue was dissolved in two ml of absolute methanol.

Spectrophotometric estimation of IAA (Gorden and Paleg, 1957)

The quantity of IAA produced was estimated using salper's reagent (1 ml of 0.5 N ferric chloride mixed in 50 ml of 35% perchloric acid). A quantity of 0.5ml methanol fraction was added to 1.5 ml of distilled water and 4.0 ml of salper's reagent and incubated in darkness for 1 hour at 28°C.

The intensity of the pink colour developed was read in a spectrophotometer at 535nm against a solvent reagent to a standard graph of IAA, prepared from a series of IAA solutions of known concentrations *viz.*, 5 µg, 40µg, 60µg, 80µg and 100µg.

Gibberellic acid production

Extraction of gibberellins (Borrow *et al.*, 1955)

One ml of the compost tea was acidified to a pH balance of 2.0 with equal volumes of ethyl acetate. The ethyl acetate phase was evaporated at 32°C and the residue was redissolved in 2ml of distilled water containing 0.05 percent of Tween 80.

Spectrophotometric estimation of gibberellins (GA3)

Gibberellic acid was estimated spectrophotometrically as given by Mahadevan and Sridhar (1982).

Two ml of zinc acetate solution (Appendix II) was added to the dissolved residue. After two ml of Potassium Ferro cyanide solution (Appendix II) was added, the mixture was centrifuged at 10000 rpm for 10 minutes. Five ml of supernatant was added to 5 ml of 30 percent hydrochloric acid and the mixture was incubated at 20°C for 75 minutes. The blank was prepared with 5 percent hydrochloric acid. The absorbance was measured at 254 nm in spectrophotometer.

Effects of compost tea on tomato Seedlings

A plastic pot size 48 x 48 cm was filled with 5 kg soil. The tomato seedlings from the control treatment of the protray were transplanted to the soil in the pots. The compost tea prepared from different substrates was sprayed over the seedlings on the 15th day after transplanting as per the treatments given below.

T_1- Control

T_2- Foliar spray

T_3- Root application

T_4- Root + Foliar application

For the foliar spray 15 ml was applied. For the Root and foliar application, 30 ml was taken. Regular watering was done. On the 30th day, plant biometric parameters such as the number of laterals and plant yield parameters *viz.*, number of flowers and fruits were recorded.

Results

Studies on chemical and biological properties of compost tea

Macronutrients

Total nitrogen

The total nitrogen content recorded at the highest level was in the compost tea with jaggery with 0.182 per cent. This was followed by compost tea with molasses with 0.168 per cent and compost tea alone with 0.072 per cent (**Table.1**).

Total phosphorus

The total phosphorus content was recorded in compost tea. The maximum phosphorus content was in the compost tea with jaggery with 0.186 per cent, followed by compost tea with molasses with 0.166 per cent and compost tea with 0.002 per cent (**Table.1**).

Total potassium

The maximum potassium content was recorded in the compost tea with molasses (0.2145 per cent), followed by the compost tea with jaggery (0.1435 per cent) and compost tea alone (1.2 per cent) (**Table.1**).

Table.1 Studies on the estimation of macronutrient content in compost tea

Samples	Total N (%)	Total P (%)	Total K (%)
Compost tea with molasses (10%)	0.168	0.168	0.215
Compost Tea with Jaggery (10%)	0.182	0.186	0.144
Compost Tea	0.072	0.002	0.120

Micronutrients

Total copper and total zinc

The maximum copper content was recorded in compost tea with jaggery (1.45 mg l^{-1}), followed by the compost tea with molasses (1.44 mg l^{-1}) and compost tea alone 0.85 mg l^{-1} (**Table.2**).

The zinc content also showed an increasing trend. The maximum zinc content was recorded in compost tea with molasses (16.0 mg l^{-1}), followed by compost tea with jaggery (7.5 mg l^{-1}) and the compost tea alone (1.2 mg l^{-1}) (**Table.2**).

Total iron and total manganese

The total iron content was recorded in compost tea with jaggery (5.27 mg l^{-1}), followed by the compost tea with molasses (3.25 mg l^{-1}) and compost tea (3.17 mg l^{-1}) (**Table.2**).

The total manganese content was recorded in the compost tea with jaggery (13 mg l^{-1}, followed by the compost tea with molasses (10 mg l^{-1}) and compost tea alone (1.1 mg l^{-1}) (**Table.2**).

Table.2 Studies on the estimation of micronutrient content in compost tea

Samples	Cu (mg l^{-1})	Fe (mg l^{-1})	Mn (mg l^{-1})	Zn (mg l^{-1})
Compost tea with molasses (10%)	1.44	3.25	10.00	16.00
Compost Tea with Jaggery (10%)	1.45	5.27	13.00	7.50
Compost Tea	0.85	3.17	1.10	1.20

Beneficial microbial population

Azospirillum

The maximum *Azospirillum* population was recorded in the compost tea with molasses with an 8.1

x 10^6 cfu ml^{-1}, followed by compost tea with jaggery (3.9 x 10^5 cfu ml^{-1}) and compost tea (1.6 x 10^4 cfu ml^{-1}) (Table 3).

Azotobacter

The maximum *Azotobacter* population was recorded in the compost tea with molasses with a (1.7 x 10^6 cfu ml^{-1}) followed by compost tea with jaggery (4.6 x 10^5 cfu ml^{-1}) and compost tea - control (2.7 x 10^4 cfu ml^{-1}).

Phosphobacteria and *Pseudomonas*

The Phosphobacteria population was recorded in compost tea with molasses with a (2.5 x 10^6 cfu ml^{-1}) followed by compost tea with jaggery (5.3 x 10^5 cfu ml^{-1}) and compost tea (1.5 x 10^4 cfu ml^{-1}).

The *Pseudomonas* population was recorded in compost tea with molasses with a (5.7 x 10^6 cfu ml^{-1}) followed by compost tea with jaggery (2.9 x 10^5 cfu ml^{-1}) and compost tea (2 x 10^4 cfu ml^{-1}).

Table. 3 Studies on the enumeration of beneficial microorganisms in compost Tea (5th day after extraction)

Samples	*Azospirillum lipoferum* (cfu ml^{-1})	*Azotobacter chroococcum* (cfu ml^{-1})	Phosphobacteria (cfu ml^{-1})	*Pseudomonas fluorescens* (cfu ml^{-1})
Compost tea with molasses (10%)	8.1 x 10^6	1.7 x 10^6	2.5 x 10^6	5.7 x 10^6
Compost Tea with Jaggery (10%)	3.9 x 10^5	4.6 x 10^5	5.3 x 10^5	2.9 x 10^5
Compost Tea	1.6 x 10^4	2.7 x 10^4	1.5 x 10^4	2.0 x 10^4

Changes in Indole Acetic Acid (IAA) and Gibberellic Acid (GA)

The changes in the Indole Acetic Acid content during the development of compost tea were presented in Table 20. The IAA content varied when the compost tea prepared with molasses (78 µg ml^{-1}), compost tea with jaggery (72 µg ml^{-1}) and compost tea (60 µg ml^{-1}) **(Table.4)**.

Similarly, the GA content also had in compost tea prepared with molasses (42 µg ml^{-1}), compost tea with jaggery (38 µg ml^{-1}) and compost tea (42 µg ml^{-1}).

Table.4 Studies on estimation of growth promoters in compost Tea

Samples	IndoleAceticAcid (µg ml^{-1})	Gibberellicacid (µg ml^{-1})
Compost Tea with Molasses (10%)	78	42
Compost Tea with Jaggery (10%)	72	38
Compost Tea	60	42

EFFECTS OF COMPOST TEA ON TOMATO SEEDLING UNDER POT CULTURE EXPERIMENT

ROOT APPLICATION

FOLIAR APPLICATION

CONTROL FOLIAR APPLICATION

PREPARATION OF COMPOST TEA

COMPOST TEA WITH JAGGERY

COMPOST TEA WITH MOLASSES

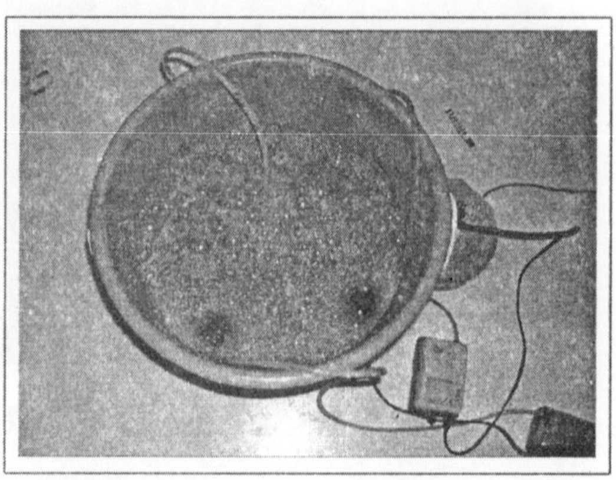

PREPARATION OF COMPOST TEA

EFFECTS OF COMPOST TEA ON TOMATO SEEDLING UNDER POT CULTURE EXPERIMENT

CONTROL ROOT APPLICATION

CONTROL FOLIAR + ROOT APPLICATION

Studies on the effect of compost tea on plant growth and flower setting on Tomato Seedlings Compost tea prepared with molasses

Number of branches

The maximum number of branches 8.04 was recorded in the treatment T_3 (root application) on the 15th day after transplanting. This is followed by T_4 (foliar + root application), T_1 (control) and T_2 (foliar application) with the values of 4.06, 3.98 and 3.48 numbers respectively.

Number of flowers

The number of flowers increased significantly in the treatments on the 15th day after transplanting. In the treatments, a T_3 (root application) recorded a maximum number of flowers with 8.64, which was on par with the treatment with T_4 (foliar and root application), T_1 (control) with the values of 4.68 and 1.58 numbers respectively.

Number of fruits

The maximum number of fruits was recorded in the treatment T_3 (root application) with the value of 4.12, which was on par with the treatment with T_4 (foliar and root application) with the value of 1.08 numbers.

Compost tea with jaggery

Number of branches

The number of branches increased significantly in all the treatments on the 15th day after transplanting. In the treatments, the T_3 (root application) had the highest number of branches with 13.22 numbers, which was on par with the treatment on the T_4 (foliar and root application) and T_1 (control) with the numbers of 9.32 and 6.06.

Number of flowers

The number of flowers increased significantly in all the treatments. In the treatments, the T_3 (root application) had the highest numbers with 11.64, which was on par with the treatment T_4 (foliar and root application) with 6.52 numbers.

Compost tea (Control)

Number of branches

The total number of branches was increased significantly in the various treatments on the 15th day after transplanting. Among the treatments, the T_3 (root application) recorded highest numbers with 6.32, followed the treatments of T_1 (control), T-4- (foliar and root application), and T_2 (foliar application) with 3.36, 3.2 numbers and 3.16 numbers respectively.

Number of flowers

The number of flowers increased significantly in various treatments on the 15th day after transplanting. In the treatments, the T_3 (root application) recorded highest numbers with 2.84, which was on par with the treatment of the T_4 (foliar and root application) with 2.04 numbers.

Table.5 Studies on the effect of compost tea on the growth and yield parameters on tomato on 15 DAT

	Compost Tea with Molasses (10%)			Compost Tea with Jaggery* (10%)		Compost Tea (control)*	
Treatment	No. of branches per plant	No. of flowers per plant	No. of fruits per plant	No. of branches per plant	No. of flowers per plant	No. of branches per plant	No. of flowers per plant
T_1 - Control	3.980	1.580	0.380	6.060	3.260	3.358	0.728
T_2 - Foliar application	3.480	1.120	0.180	4.340	2.860	3.160	1.062
T_3 - Root application	8.040	8.640	4.120	13.220	11.640	6.320	2.840
T_4 - Foliar + Root application	4.060	4.680	1.080	9.320	6.520	3.200	2.040
SEd	0.3376	0.2154	0.2045	0.2665	0.2347	0.3008	0.2135
CD (0.05)	0.7158	0.4566	0.4334	0.5649	0.4976	0.6377	0.4527

*Fruits were not initiated at 15 DAT

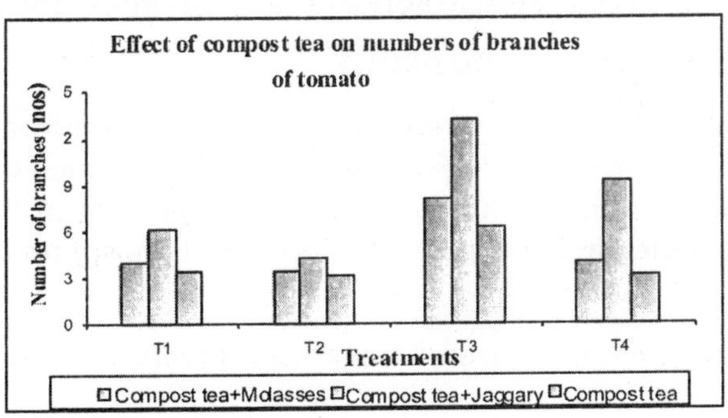

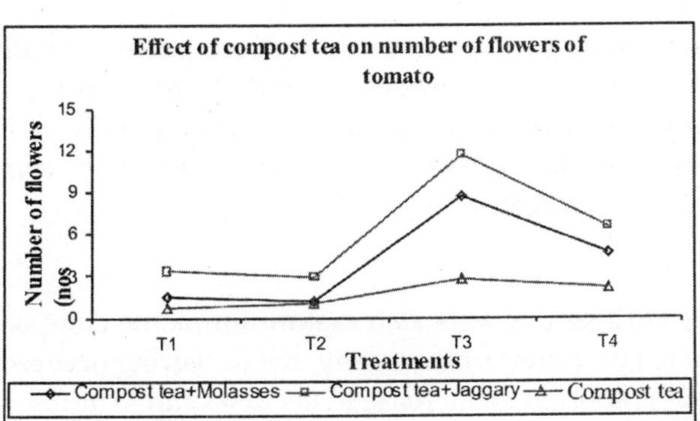

T_1- Control
T_2- Foliar spray
T_3- Root application
T_4- Root + Foliar application

DISCUSSION

In the compost tea prepared from molasses and jaggery the population of *Azospirillum lipoferum, Azotobacter chroococcum,* Phosphobacteria and *Pseudomonas fluorescens,* was higher than compost tea prepared without these carbon sources. The sugar content in molasses and jaggery as well as the other nutrients supported the maximum growth of the organism. The continuous aeration with additional carbon sources provided the quick cell division of the organism. The Organic Farming Research Foundation Project Report (1998) reported that the organic tea produced through "active" or aerated (an organic tea system that receives a boost of oxygen with the use of mechanical mixing, packed column or forced air), extracts a higher quantity and quality of nutrients and microbes which can be drawn from the organic feed stock. This report also indicated that aerobic teas made from aged or suppressive compost contain more beneficial microbes, especially if they have been well aerated and allowed to brew for several days.

The plant growth promoting substances of IAA and the GA content in compost tea is in a moderate quantity only, but in compost tea with molasses recorded a higher IAA and GA content. The source for IAA and GA are from rhizobacteria only. The substrate which contains more *Pseudomonas fluorescens* population with recorded a maximum plant growth promoting substances. The addition of molasses in compost tea increased the level of *Pseudomonas* population, which in turn increases the IAA and GA content. Minakshi (2005) reported that bacteria belonging to the genera *Arthrobacter, Azospirillum, Bacillus, Enterobacter, Serratia* and *Pseudomonas* have been found to have enormous potential as plant growth promoting rhizobacteria (PGPR) and are now used in agriculture as bio-inoculants.

Macro and Micronutrient content

The compost tea prepared from jaggary recorded the maximum nitrogen and phosphorous content. The addition of carbon source with active aeration accelerated the growth of nitrogen fixing organism and phosphorous solubilizing organism present in the compost tea. This increase in population resulted in more nitrogen content and phosphorous content in compost tea.

The maximum content of Copper, Iron, Manganese and Zinc were recorded in this compost tea.

Plant growth parameters

The compost tea application to the tomato seedling plants, with the root application of compost tea recorded the highest plant growth parameters, when compared with the foliar application. The application of the compost tea in the root zone favoured the microbial activity in the rhizosphere as well as the nutrient content and plant growth promoters present in the compost tea are available in the root zone for immediate absorption.

Yield parameters

In the compost tea application also the fruit setting was also maximum in the compost tea prepared from molasses. In the compost tea prepared from jaggery, the flowering occured and the fruit setting was delayed. Welke (2001) reported an increase in the marketable number and weight of strawberries and also the increased weight of broccoli heads due to the application of compost tea from cattle manure composts.

Summary and Conclusion

In the present investigation, the compost tea was developed from vermicompost enriched with molasses and continuously aerated for 72 hours. This product was evaluated for its quality by utilizing it in a pot culture experiment with tomato seedling. The results of the study are summarized here below.

The present study revealed that compost tea prepared with molasses and jaggery was rich in its nutritive value. The total macronutrients (N, P) found in a maximum level in compost tea were prepared with jaggery (0.182 per cent, 0.166 per cent) and the potassium content was higher in compost tea prepared with molasses (0.214 per cent).The micronutrients (Cu, Mn and Fe) content found in the maximum level in compost tea were prepared with jaggery (1.45 mg l^{-1} and 13 mg l^{-1}, 5.27 mg l^{-1} respectively) and a higher zinc content of 16.0 mg l^{-1} found in the compost tea was prepared with molasses. The beneficial microorganisms viz., *Azospirillum lipoferum*, *Azotobacter chroococcum*, Phosphobacteria and *Pseudomonas fluorescens* (8.1 x 10^6 cfu ml^{-1}, 1.7 x 10^6 cfu ml^{-1}, 2.5 x 10^6 cfu ml^{-1} and 5.7 x 10^6 cfu ml^{-1} respectively) were found to be maximum in compost tea prepared with molasses. The growth promoting substances like IAA and GA (78 µg ml^{-1} and 42 µg ml^{-1}) were found to be comparatively higher in compost tea prepared with molasses. The number of branches, number of flowers and number of fruits were found in a maximum level in the treatment in which the compost tea prepared with molasses was applied as a root application (4.06, 4.68, 4.12 respectively). The number of branches and number of flowers (13.22, 11.64) were higher in the compost tea prepared with jaggery applied tomato crops to as a root application.

It also contained growth promoters (IAA and GA) and beneficial microbial population further enhancing its nutritive value for application to crops. The application of compost tea in tomato crop proved its effectiveness in increasing the plant productivity. Hence, compost tea can be recommended to be used as source of plant nutrients and growth promoters which can supplement inorganic fertilizers effectively and enhance plant yield in an eco friendly manner.

References

1. Borrow, A., P.W. Drian, V.E. Chester, P.J.Curtis and N. Radley. 1955. Gibberellic acid a metabolic product of the fungus *Gibberella fujikuroi*. Some observations on its production and isolation. **J.Sci.Food. Agric., 6**: 340-348.

2. Chandramohan, D. and A. Mahadevan. 1968. Indole Acetic Acid metabolism in soil. **Curr.Sci., 37** : 112-113.

3. Gorden, S.A. and L.G.Paleg. 1957. Quantitative measurement of Indole Acetic Acid. **Physiol.**

4. Humphries, E.C. 1956. Mineral components and ash analysis **Modern methods of plant analysis.** Springer- Verlag, Berlin. pp. 468-502 **Planta, 10**: 112-113.

5. Jackson, M.L. 1973. **Soil chemical analysis.** Prentice Hall of India, Newdelhi. P. 498

6. Lindsay, W.L. and W.A.Norvell. 1978. Development of DTPA Soil test for Zinc, iron, manganese and coper, **Soil sci. Soc. Amer.J.,42**: 421-428.

7. Mahadevan,A. and R.Sridhar. 1982. Hydrolytic enzymes. Methods in Physiological Plant Pathology, second edition, Sivakami Publishers, Madras, pp: 24-44.

8. Minakshi,A.,K. Sexena and N.K. Matta. 2005. selection of culturable PGPR from diverse pool of bacteria inhabiting pigeonpea rhizosphere. **Indian J. of Microbiol., 45(1)**: 21-26

17

Biodiversity in Arid Lands of Rajasthan

AMRITA BAJAJ AND RAJESH KUMAR ABHAY

Introduction

Arid lands have an immense scientific, economical and social value. They are the habitat and source of livelihood for about one-quarter of the earth's population. It is estimated that these ecosystems cover one-third of the earth's total land surface and about half of this area is in an economically productive use as agricultural land(CCD Secretariat, 1997). Arid ecosystems cover extensive land areas stretching across more than one-third of the earth's land surface. They cover a variety of terrestrial biomes which are extremely heterogeneous with wide variations in topography, climatic, geological and biological conditions. Arid ecosystems also contain a variety of native animal, plant and microbial species that have developed special strategies to cope with the low and sporadic rainfall, and extreme variability in temperatures that prevail in these ecosystems. Such adaptive traits have global importance, especially in the context of predicted climate change.

Arid pastoralists and farmers have developed efficient pastoral and mixed cropping systems adapted to the difficult conditions of arid lands These systems have sustained the livelihoods of generations of arid land people. Furthermore, arid land pastoralists and farmers have successfully created and maintained high levels of agro biodiversity of crops and livestock breeds. Yet, global awareness about the great value of arid lands remains frustratingly low. Compared with tropical rainforests, for example, the wealth of arid land biodiversity and indigenous knowledge is less well documented, it and has received much less support and advocacy in the conservation media.

Although remnants of healthy arid land biodiversity and indigenous knowledge still exist at various locations, arid lands face increasing threats of further degradation. Already, as sources quoted by the SBSTTA (1999) report, it is estimated that 60 percent of arid lands is already degraded resulting in an estimated annual economic loss of USD 42 billion world-wide. Thus, the continued degradation of arid lands is a major threat to the ecological functions of arid lands and to the species and genes living in these ecosystems and thus to human welfare. Arid lands need urgent attention. Efforts in the past to address arid land issues have achieved much less than expected. Thus new paradigms are needed to go beyond the status quo with imagination and courage.

The word biodiversity has been coined by W.G. Rosen in 1985 while planning the National Forum on Biological Diversity organized by the National Research Council (NRC) which was to be held in 1986, and first appeared in a publication in 1988 when the entomologist E. O. Wilson used it as the title of the proceedings of that forum The word biodiversity was deemed more effective in terms of communication than biological diversity.

Since 1986, the terms and the concepts have achieved widespread use among biologists, environmentalists, political leaders, and concerned citizens worldwide. It is generally used to equate to a concern for the natural environment and natural conservation. This use has coincided with the expansion of concern over extinction observed in the last decades of the 20th century. Today, as ever, human beings are dependent for their sustenance, health, well-being and enjoyment of life on fundamental biological systems and processes. Humanity derives all of its food and many medicines and industrial products from the wild and domesticated components of biological diversity. Biotic resources also serve recreation and tourism, and underpin the ecosystems which provide us with many services.

While the benefits of such resources are considerable, the value of biological diversity is not restricted to these. The enormous diversity of life in itself is of crucial value, probably giving greater resilience to ecosystems and organisms. Biodiversity also has important social and cultural values. Diversity within the natural environment is important. It provides variety that people enjoy, both in species and landscape. Species variety plays a dual role of ensuring and signalling the vitality of the natural environment. Protecting biodiversity protects the health of the natural environment and this enables it to provide services upon which people depend.

Conceptual Framework

Various scholars have tried to define biodiversity according to their own understanding and perceptions. "A definition of biodiversity that is altogether simple, comprehensive, and fully operational ... is unlikely to be found", (Noss, 1990). There are several different definitions used by resource managers and ecologists. Together, they allow us to develop an understanding of the broad concept of biodiversity.

"Biodiversity" is defined as the variability among living organisms from all sources, including, 'inter alia', terrestrial, marine and other aquatic ecosystems and the ecological complexes of which they are part. This includes diversity within species, between species and of ecosystems (United Nations Convention on Biological Diversity, 1992).

"Biological diversity is the variety and variability among living organisms and the ecological complexes in which they occur" (US Congress Office of Technology Assessment, 1987)

"Biodiversity is the totality of genes, species, and ecosystems in a region". Biodiversity can be divided into three hierarchical categories — genes, species, and ecosystems that describe quite different aspects of living systems and that scientists measure in different ways.

Genetic diversity refers to the variation of genes within a species. This covers distinct populations of the same species (such as the thousands of traditional rice varieties in India) or genetic variation within a population (high among Indian rhinos, and very low among cheetahs).

Theme	Lovejoy, 1980 Definition:	Norse and Manus, 1980	Norse et al., 1986	Wilson, 1988	Flint, 1991	Reakka Kudla et al., 1991
Inventory	global species total	Definition: global species total	Definition: global species total	Definition: global species total; habitat-specific totals	Definition: global species total	Definition: genetic species total and ecosystem level; linking ecosystem processes to variety and richness
Loss rates	Estimated losses due to forest destruction	Causes (settlement transport, fragmentation, agriculture, forestry, over-exploitation, introduced species)		Global and habitat-specific estimates; geological patterns case studies	Habitat examples; causes (development, market failure,interventions, habitat loss, over-exploitation)	Estimated global extinction rates, species and site specific case studies
Value	Potential uses	Potential uses (food, energy, chemicals, raw materials, medicine)	Products (actual and potential); ecosystem services; forestry; psychological well-being	Medicine industrial, food, potential	Use and non-use values potential, habitat examples	Specific examples of value to biotechnology and agriculture; value as a scientific tool and to raise awareness
Economic	Valuation of ecosystem function	Gene banks, botanical gardens, species specific schemes, ecosystem management, legislation	Timber production	Market Failure to price biodiversity, new economic appro-aches, pricing bio-diversity	Economic approaches to value biodiversity	
Conversation			Concepts; conservation and manage-ment of forests and rare species; laws	Priorities, case studies, technologies, reintroductions	Aid policy procedures for priority sites, habitat examples, criteria	Strategic and case-specific examples, marine and terrestrial; genetic, species and ecosystem level; emphasis on infra structure
Attitudes		Philosophical, psychological, affinity to life, right to exist	Ethics and stewardship	Green movements, society links to ecosystem, morals, religion, philosophy	Ethical value of biodiversity	Brief review of history in science

Source: Jeffries, M.J. (1997) Biodiversity and Conservation, pp. 8-9.

Species diversity refers to the variety of species within a region. Such diversity can be measured in many ways and scientists have not settled on a single best method. The number of species in a region — its species "richness" — is one often- used measure, but a more precise measurement, "taxonomic diversity", also considers the relationship of different species to each other. For example, an island with two species of birds and one species of lizard has a greater taxonomic diversity than an island with three species of birds but no lizards.

Ecosystem diversity is harder to measure than species or genetic diversity because the "boundaries" of communities — associations of species — and ecosystems are elusive. Nevertheless, as long as a consistent set of criteria is used to define communities and ecosystems, their numbers and distribution can be measured..." (World Resources Institute, 1992).

Composition, structure, and function determine, and in fact constitute, the biodiversity of an area. Composition has to do with the identity and variety of elements in a collection, and includes species lists and measures of species diversity and genetic diversity. Structure is the physical organization or pattern of a system, from habitat complexity as measured within communities to the

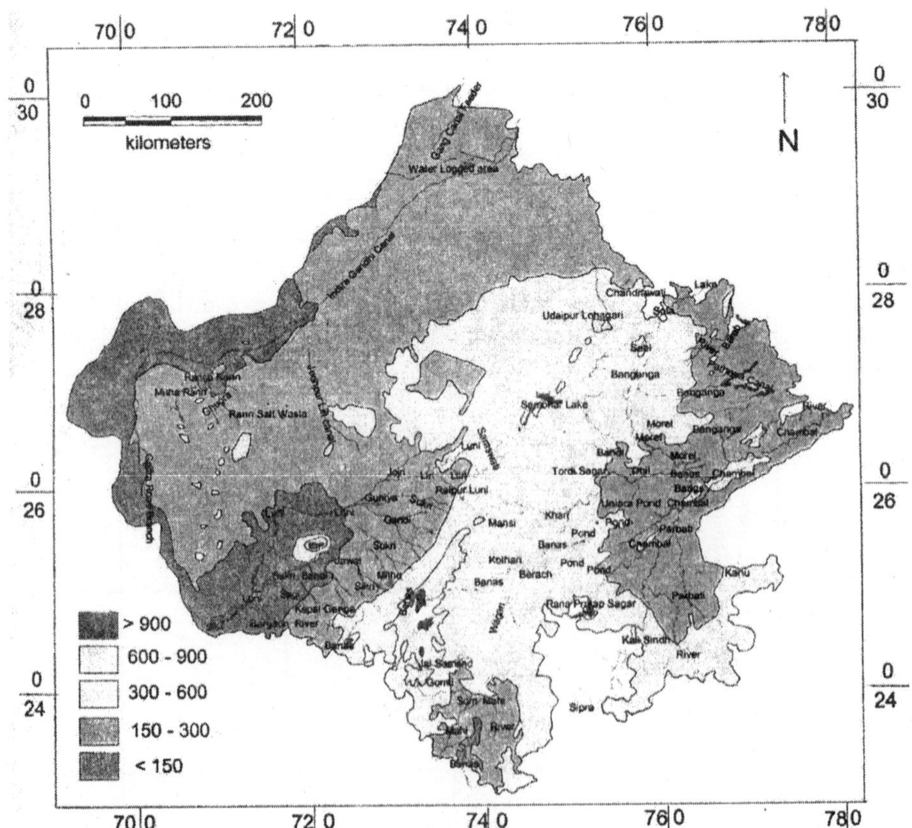

Figure 1: Study Area

pattern of patches and other elements at a landscape scale. Function involves ecological and evolutionary processes, including gene flow, disturbances, and nutrient cycling" (Noss, 1990).

The wide scientific, social and philosophical horizons incorporated by the concept of biodiversity are evident in much of early literature (Table 1.1)

Aims and Objectives

The purposes of this paper are:

i. to draw greater attention to the potentials of arid lands, and
ii. to examine the variety of flora and fauna in the study area,
iii. to examine the possible threats to biodiversity and their causes
iv. to suggest options for conservation and sustainable use of arid land biodiversity.

Another powerful selection agent is human population. Local people have developed complex pastoral and cropping systems for which they have selected and maintained the biological diversity they value most in domestic livestock and crops.

Study Area: Thar Desert

Physiography and geology

There are three principal landforms in the desert region i.e. the predominantly sand covered Thar Plains with hills including the central dune free country (Fig. 1).

Hills

It is a desolate country where sand is piled up into huge wind blown dunes (technically this is known as an erg). The sand dunes are of three types viz., longitudinal parabolic, transverse and barchans. The first type, running NNE-SSW, i.e. parallel to the prevailing winds, occurs to the south and west of the Thar. The transverse dunes, aligned across the wind direction, go to the east and south of the Thar and barchans, with the conclave sides facing the wind in the interior, predominant in the Central Thar. On the whole the Thar Desert slopes imperceptibly towards the Indus Plain and the surface unevenness is due mainly to sand dunes. The dunes in the south are higher, rising sometimes to 152 m whereas in the north they are lower and rise to 16 m above the ground level.

The Aravalli Range forms the main landmark to the south-east of Thar Desert. The more humid conditions that prevail near the Aravallis prevent the extension of the Thar Desert towards the east and the Ganges Valley. In the heart of the sand covered area, the bare, dune free country of Barmer, Jaisalmer and Bikaner present an anomaly.

Desert soils

The soils of the Arid Zone are generally sandy to sandy-loam in texture. The consistency and depth vary according to the topographical features. The low-lying loams are heavier and may have a hard pan of clay, calcium carbonate ($CaCO_3$) or gypsum. The pH balance varies between 7 and 9.5. The soils improve in fertility from west and northwest to east and northeast. Desert soils are regosoils of wind blown sand and sandy fertile deposits, derived from the disintegration of rock in the subjacent areas and blown in from the coastal region and the Indus

Valley. The desert soils occupy the districts of Jodhpur, Bikaner, Churu, Ganganagar, Barmer, Jaisalmer, and Jalore. The Thar consists mainly of the wind-blown sand. The area is covered not only by sheets of sand but also of rocky projections of low elevations which constitute the older rocks of the country. Water is scarce and occurs at great depths, from 30 to 120 m below the ground level. Some of these soils contain a high percentage of soluble salts in the lower horizons, turning water in the wells poisonous. Being poor in organic matter the soil became infertile because it contained varying amounts of calcium carbonate.

Agriculture

The main occupation of people in the desert is agriculture and animal husbandry. Agriculture is not a dependable proposition in this area— after the rainy season, at least 33% of crops will definitely fail. Animal husbandry, trees and grasses, intercropped with vegetables or fruit trees, is the most viable model for arid, drought-prone regions. Arid regions of Western Rajasthan, Gujarat, and Southern parts of Haryana receive low precipitation (under 400 mm normally) with high evaporation due to high solar radiation and wind speed. The region faces frequent droughts. Overgrazing due to the high animal population, wind and water erosion, mining and other industries are serious land degradation of processes.

Livestock

In the last 15-20 years, the Rajasthan desert has seen many changes, including a manifold increase of both the human and animal population. Animal husbandry has become popular due to the difficult farming conditions. At present, there are ten times more animals per person in Rajasthan than the national average and overgrazing is also a factor affecting climatic and drought conditions.

Agro-forestry

Forestry has an important part to play in the amelioration of the conditions in semi-arid and arid lands. If properly planned forestry can make an important contribution to the general welfare of the people living in desert areas, the living standard of the people in the desert would not be so low. They cannot afford other fuels like gas, kerosene etc. Fire-wood is their main fuel and of the total consumption of wood about 75 percent is firewood. The forest cover in the desert is low. The forest is insufficient to fulfill the need of firewood. This fact diverts the much needed cattle dung from the field to the hearth and this in turn results the decrease in agricultural production.

The scientists of Central Arid Zone Research Institute (CAZRI), have successfully developed and improved dozens of traditional and non-traditional crops/fruits, such as Ber trees (like plums) that produce much larger fruits than before (lemon-size) and can thrive with a minimal rainfall. These trees have become a profitable option for farmers. One example from a case study of horticulture showed that in situation of budding in 35 plants of Ber and Guar (Gola, Seb and Mundia variety developed in CAZRI), using only one hectare of land, the yielded 10,000 kg. of Ber and 250 kg. of Guar, which translates into a double or even a triple profit. The most important tree species, in Agro-forestry, providing livelihood support in Thar Desert is the Prosopis cineraria.

Prosopis cineraria provides wood of a construction class. It is used for house-building, chiefly for rafters, posts, doors and windows, and for well construction for water pipes, upright posts of Persian wheels, agricultural implements and shafts spokes for the yokes of buffalo carts. It can be used also for small turning work and tool-handles. Container manufacturing is another important wood based industry, which depends heavily on desert grown trees. Prosopis cineraria is much valued as a fodder tree. The trees are heavily lopped particularly during the winter months when no other green fodder is available in the dry tracts. There is a popular saying that death will not visit a man, even at the time of a famine, if he has a Prosopis cineraria, a goat and a camel, since the three together are somewhat said to sustain a man even under the most trying conditions. The forage yield per tree varies a great deal. On an average, the yield of green forage from a full grown tree is expected to be about 60 kg with complete lopping having only the central leading shoot, 30 kg when the lower two third crown is lopped and 20 kg when the lower one third crown is lopped. The leaves are of high nutritive value. The feeding with leaves during the winter when no other green fodder is generally available in rain-fed areas, is thus profitable. The pods are a sweetish pulp and are also used as a fodder for livestock.

The Tecomella undulata is the tree species, locally known as the Rohida, found in the Thar Desert regions of northwest and western India, and is another important medium sized tree of great use in Agro-forestry, that produces quality timber and is the main source of timber amongst the indigenous tree species of desert regions. The trade name of the tree species is Desert teak or Marwar teak. It is mainly used as a source of timber. Its wood is strong, tough and durable. It takes a fine finish. Heartwood contains quinoid. The wood is excellent for firewood and charcoal. Cattle and goats eat leaves of the tree. Camels, goats and sheep consume flowers and pods.

The Tecomella undulata plays an important role in environmental conservation. It acts as a soil-binding tree by spreading a network of lateral roots on the top surface of the soil. It acts as a windbreak and helps in stabilizing shifting sand dunes. It is considered as a home of birds and provides shelter for other desert wildlife. The shade of the tree crown gives shelter to the cattle, goats and sheep during the summer days. The Tecomella undulata has got medicinal properties as well. The bark obtained from the stem is used as a remedy for syphilis. It is also used in curing urinary disorders, the enlargement of the spleen, gonorrhoea, leucoderma and liver diseases. The seeds are used against abscesses.

Eco-Tourism

Desert safaris on camels have become increasingly popular around Jaisalmer. Both foreigners as well as the rapidly growing Indian middle (and upper) class frequent the desert seeking adventure on camels for anything from a day to several days. This industry ranges from cheaper backpacker treks to plush Arabian night style campsites replete with banquets and cultural performances. During the treks the tourists are able to view the fragile and beautiful ecosystem of the Thar desert. This form of tourism provides income to many operators and camel owners in Jaisalmer as well as employment for many camel trekkers in the desert villages nearby.

Natural vegetation

The natural vegetation is classed as the Northern Desert Thorn Forest. This desert Thorn Forest occurs in small clumps scattered in a more or less open form. The density and size of patches increases from the West to the East following an increase in rainfall the natural vegetation of the Thar Desert is composed of the following tree, shrub and herb species.

Tree species: Acacia jacquemontii, Acacia leucophloea, Acacia senegal, Anogeissus rotundifolia, Prosopis cineraria, Salvadora oleoides, Tecomella undulata, Tamarix articulata

Small trees and shrubs: Calligonum polygonoides, Acacia jacquemontii, Balanites roxburghii, Ziziphus zizyphus, Ziziphus nummularia, Calotropis procera, Suaeda fruticosa, Crotalaria burhia, Aerva tomentosa, Clerodendrum multiflorum, Leptadenia pyrotechnica, Lycium barbarum, Grewia populifolia, Commiphora mukul, Euphorbia neriifolia, Cordia rothii, Maytenus emarginata, Capparis decidua.

Herbs: Eleusine compressa, Dactyloctenium scindicum, Cenchrus biflorus, Cenchrus setigerus, Lasiurus hirsutus, Cynodon dactylon, Panicum turgidum, Panicum antidotale, Dichanthium annulatum, Sporobolus marginatus, Saccharum spontaneum, Cenchrus ciliaris, Desmostachya bipinnata, Cyperus arenarius, Erogrostis species, Ergamopagan species, Phragmitis species, Typha species.

The Driving Forces of Biodiversification in arid lands

Dryland ecosystems cover a variety of terrestrial biomes (i.e. arid steppes, grasslands at various altitudes and latitudes; tropical and sub-tropical savannahs; dry forest ecosystems and coastal areas) which are extremely heterogeneous. A major driving force of the biological diversification in a dryland environment is the relative aridity. Within each arid zone, significant variations between sites are introduced by topography and geology as mentioned earlier, and also by variations in the most limiting factors - i.e. water and soil nutrients. The resulting patchwork of habitats determines the distribution of living organisms.

In addition to habitat differentiation, other driving forces of biodiversification in drylands include the seasonal pattern of rainfall, fires and herbivorous pressure. Types and intensities of these environmental stresses combine to determine the main selection pressures in drylands: low and highly variable rainfall in time and space; recurrent but unpredictable droughts that may persist for several consecutive years; high temperatures; inherently low soil fertility; high incidence of salinity; prevailing herbivorous pressure; and fires. These stresses have been selected for a large diversity of adaptive traits.

The seasonal pattern of rainfall, for example, has been selected by a number of plant and animal species and micro-organisms which are able to develop rapidly and complete their life cycle in a very short period of time. Some species have mechanisms to escape drought whereas other species have appropriate organs to resist drought. Plant species, for example, have large below ground systems to store water and nutrient or corky bark to insulate living cells from desiccation and fire burning (Medina et al., 1992; SBSTTA, 1999).

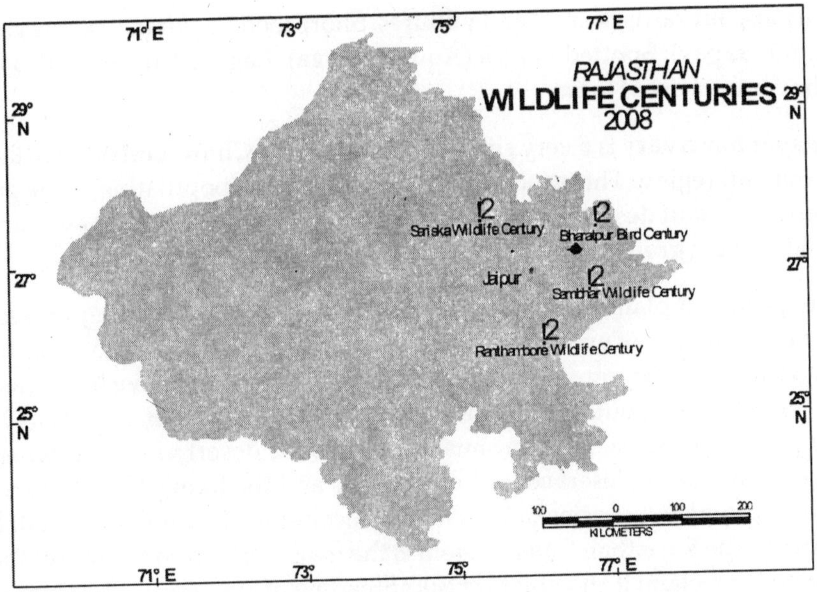

Figure 2: Wildlife Centuries in Rajasthan

Biodiversity

Stretches of sand in the desert are interspersed by hillocks and sandy and gravel plains. Due to the diversified habitat, the vegetation and animal life in this arid region is very rich. About 23 species of lizard and 25 species of snakes are found here and several of them are endemic to the region. There are four major wildlife sanctuaries in Rajasthan mainly located in the eastern part of the state (Fig. 2).

Some wildlife species, which are fast vanishing in other parts of India, are found in the desert in large numbers such as the Great Indian Bustard (Ardeotis nigriceps), the Blackbuck (Antilope cervicapra), the Indian Gazelle (Gazella bennettii) and the Indian Wild Ass (Equus hemionus khur) in the Rann of Kutch. They have evolved excellent survival strategies as their size is smaller than other similar animals living under different conditions and they are mainly nocturnal. There are certain other factors responsible for the survival of these animals in the desert. Due to the lack of water in this region the transformation of the grasslands into cropland has been very slow. The protection provided to them by a local community, the Bishnois, is also a factor.

The Desert National Park at Jaisalmer, is spread over an area of 3162 km^2, and is an excellent example of the ecosystem of the Thar Desert, and its diverse fauna. Great Indian Bustard, Blackbuck, chinkara, desert fox, Bengal fox, wolf, desert cat etc. can be seen here easily. Seashells and massive fossilized tree trunks in this park record the geological history of the desert. The region is a haven for migratory and resident birds of the desert. One can see many eagles, harriers, falcons, buzzards, kestrel and vultures. Short-toed Eagles (Circaetus gallicus), Tawny Eagles (Aquila rapax), Spotted Eagles (Aquila clanga), Laggar Falcons (Falco jugger) and kestrels are the commonest of these.

The Tal Chhapar Sanctuary is a very small sanctuary in the Churu District, 210 km from Jaipur, in the Shekhawati region. This sanctuary is home to a large population of the graceful Blackbuck. The desert Fox and desert cat can also be spotted along with typical avifauna such as the partridge and and the grouse.

The local tree species for planting in the desert region are not many and they are also very slow growing. The introduction of exotic tree species in the desert for plantation, has become necessary. Many species of Eucalyptus, Acacia, Cassia and other genera have been tried from Israel, Australia, USA, Russia, Southern Rhodesia, Chile, Peru, Sudan etc. In Thar Desert, Acacia tortilis has proved to be the most promising species for desert afforestation and has helped in checking the spread of desert soils to other areas and the fixing of sand dunes. The jojoba tree is another promising species of tree with an economical value found suitable for planting in these areas. The Rajasthan Canal system is the major irrigation scheme of the Thar Desert and is conceived to reclaim it and also to check spreading of the desert to fertile areas.The river Luni is the only natural water source that drains inside a lake in the desert.

The habitat is greatly influenced by the extreme climate. The sparse vegetation consists of xerophilious grasslands of Eragrostis spp. Aristida adscensionis, Cenchrus biflorus, Cympogon spp., Cyperus spp., Eleusine spp., Panicum spp., Lasiurus scindicus, Aeluropus lagopoides, and Sporobolus spp. Scrub vegetation consists of low trees such as Acacia nilotica, Prosopis cineraria, P. juliflora, Tamrix aphylla, Zizyphus mauritiana, Capparis decidua, and shrubs such as Calligonum polygonoides, Calotropis spp., Aerva spp., Crotalaria spp., and Haloxylon salicornicum. Haloxylon recurvum is also present.

Despite the climate, several species have evolved to survive the extreme conditions here. Among the mammal fauna, the blackbuck (Antilope cervicapra), chinkara (Gazella bennettii), caracal (Felis caracal), and desert fox (Vulpes bengalensis) inhabit the open plains, grasslands, and saline depressions known as chappar or rann in the core area of the desert. The overall mammal fauna consists of forty-one species. None are endemic to the ecoregion, but the blackbuck is a threatened species whose populations take refuge in this harsh environment.

Among the 141 birds known in this ecoregion, the great Indian bustard (Chirotis nigricaps) is a globally threatened species whose populations in this ecoregion have rebounded in recent years. A migration flyway used by cranes (Grus grus, Anthropoides virgo) and flamingos (Phoenicopterus spp.) on their way to the Rann of Kutch further south crosses this ecoregion. Chaudhry et al. reported eleven reptile species from ten genera from the Cholistan desert in the western Thar.

There are eleven protected areas that cover almost 44,000 kilometers2, representing about 18 percent of the ecoregion's area (Table 2). These include several reserves that are quite large, with one (Cholistan) exceeding 20,000 km2, and two (Nara Desert, Rann of Kutch) being over 5,000 km2.

Table 2: WCMC (1997) Protected Areas that Overlap with the Ecoregion.

S.No.	Protected Area	Area (km2)	IUCN Category
1	Nara Desert	7,290	IV
2	Desert	2,710	II
3	Tando Mitha Khan	220	UA
4	Rann of Kutch	10,540	IV
5	Nara	600	UA
6	Cholistan	21,830	UA
7	Rahri Bungalow	50	UA
8	Abbasia	100	UA
9	Ramgarh Bundi	300	IV
10	Lal Suhanra	70	V
11	Tal Chapar	80	IV
	Total	43,790	

Climatic Conditions and Biodiversity in the Thar Desert

The Great Indian Desert, the Thar Desert, extends over 285 000km2, of which 196,000 km2 is in Rajasthan, comprising 61% of the total geographical area of the state. The terrain condition varies from plains to sand-dunes and rocky hill outcrops. The area consists of dry, undulating plain of hardened sand and gravelly soil and rolling plains of loose sand. The sand dunes vary from 2 to 10 km in length and from 20 to 80 m in height. The mean annual rainfall in the region varies from 100 mm in the northwestern sector of the Jaisalmer district to 458 mm in the eastern part of the desert. Wide variations in temperature are characteristic of the Thar Desert. The mean maximum temperature during the summer is 40 °C but at times 50 °C is reached. The winter season temperature varies from 14 to 16 °C but at times the lowest temperature goes below freezing point during the month of January. High wind velocities during the summer are another characteristic feature of the region. During the months of June and July winds usually blow at 20±30 kmh but wind speeds as high as 130 kmh, have been experienced. The soils are poor in organic matter with very low productivity. The Thar Desert is one of the most densely populated desert regions in the world (87 persons per km2). The density of population varies from 264 persons per km2 in Jhunjhunun to 9 persons per km2 in Jaisalmer.

The total population in the 11 desert districts was 17.43 million in 1991, as against 13.48 million in 1981. The growth rate during this period was 9%. The total livestock population of the region as per the 1988 census was 17.82 million as against 23.35 million in 1983, whereas in 1972 it was only 16.3 million. The vegetation, which is predominantly xerophytic and quite sparse, occurs in great variety, there being as many as 682 species amongst which different grasses alone account for 107 species. The commonest plant is sewan grass (Lasiurus sindicus). The other prominent shrubs are sinia (Crotalaria burhia), khip (Leptadenia pyrotechnica), phog (Calligonum polygonoides), banvli (Acacia jacquemontii), bui (Aerva tome-ntosa), murut (Panicum turgidum), lana (Halo-xylon salicornicum), ker (Capparis decidua), lamp (Aristida funiculata), bhurat (Cenchrus biorus), and beker (Indigofera cordifolia). Khejri (Prosopis cineraria) is the lifeline of the desert but the plant that provides timber to the desert is rohira (Tecomella undulata). Some more common plant species are ber (Zizyphus nummul-aria), kumat (Acacia senegal), jal (Salvadoraoleoides), thor (Euphorbia caducifolia), gugal (Commiphora wightii), tantia (Eleusine compressa) and ganthia (Dactyloctneum sindicum). Of the 682 plant species of the desert, 209 are common to Africa and the Thar Desert. Sixty of the 209 are distributed only in the Thar Desert while 149 extend to other parts of the Indian subcontinent. Migration from the eastern side of the Thar desert accounts for 123 species and the numbers from Malayan, Sino-Indian and Australian Indian areas are 12, 11 and 15, respectively (Bhandari,1993). The overall distributional pattern of the desert biota reflects the migratory routes of the desert plants in the past. It seems that west to east migration could have started from the Sahara Desert. Many plant species such as Barleriaa canthoides, Dipterygium glaucum, Dignethia hirtella and L. pyrotechnica are restricted up to the Thar Desert, while a few others, such as Arnebiahispidissima, T. undulata and Orobanche cernuavar, nepalensis have extended to most parts of India. A similar pattern of migration and distribution is met within the forms speculated to have migrated from the Iranian region. Species such as Aristida hystricula, Aristida pogonoptila, Blepharissindica, C. polygonoides, Dipcadi erythraeum, H. salicornicum and Tamarix aphylla are examples of this type. These plants have their eastern limit in the Thar Desert. The past history of the wildlife inhabiting the Thar Desert is fascinating. The Asiatic lion, now found only in the Gir Forest of India, was found in fair numbers until the 1870s on the plains of Rajasthan and Punjab. Leopards were common in the Siwana Hills in the Barmer District. The cheetah, now rare in India, was at one time found in the Khathiawar region. The wild ass, now restricted in the little Rann of Kutch was well distributed in parts of the Rajasthan Desert. Over 60 species of mammals have been recorded from the Thar Desert. The chinkara (Gazella gazella), desert cat (Felis libyca), desert fox (Vulpes vulpes), jackal (Canis aureus), wolf (Canis lupus), and desert hare (Lepus nigricollis) are quite common in the area. Approximately 300 species of birds have been reported from the Thar Desert. The most important of all is perhaps the great Indian bustard (Choriotes nigriceps) which breeds in the area and migrates locally in different seasons. Other bird species are the grey partridge (Francolinuspondicerianus), peafowl (Pavo cristatus), commons and grouse (Pterodesexutus), common quail (Coturnix sp.), Indian courser (Cursorious coromandalicus), doves, pigeons, bulbuls, larks, babblers etc. The rich insect fauna of the desert consists of termites (Psammotermes rajasthanicus), the white grub (Holotrichia spp.) locusts (Schistocerca gregaria) and grass hoppers (Hrilidiaanis), etc. The amphibian fauna of the desert is restricted to one species of toad (Bufo and ersonii) and five species of frogs. Among reptiles, the monitor liz-ard (Varanus griseus) and spiny tailed lizard (Ur-omastix hardwickii) are prominent. The desert is quite rich in several snake species of which the sawscaled viper (Echis carinatus) is noteworthy. The snakes which occur near villages and towns in the desert are Eryx conicus, Eryx johni and Naja naja.

Biological diversity is the foundation upon which human civilizations are built. Thousands of species of plants and animals supported the development of early societies, providing the basis for the evolution from a hunting and gathering subsistence to agricultural and industrial organizations. Not only are genetic resources important for crop species, but hundreds of cases have been documented concerning the medical or industrial value of the plant and animal species. There is a great scope for the systematic exploitation of biological diversity, provided that this resource is maintained until human technological capacity can make use of it. In agriculture alone, it is estimated that prospective breakthrough in genetic engineering will generate products, worth some $50±100 billion per year, before the end of the present century, through contributions to new vaccines, hormones, bacterial feeds, anabolic steroids, pesticides, nitro-xed osmoregulation and nitro-xedcereal enzymes, etc. These will enable humankind to achieve a quantum leap in its capacity to feed hungry people. In the Indian desert, there are numerous populations of genera such as the Tephrosia, Citrullus, Cucumis, Tribulus, Convolvulus and Lasiurus. Their genetic diversity could be used to improve these plants for their better utilization. For example, there are a large number of strains of the Citrullus colocynthis, locally known as tumba, an important sand dune stabilizer found throughout the desert, spreading on loose sand. However, in the past 15 years, ever since its seeds began to be collected for oil, it has become a threatened plant in the desert. Moreover, a natural hybrid between C. colocynthis and Citrullus lunatus, the watermelon, has been found in the desert (Bhandari and Shirangi, 1986). There is the possibility of transferring disease resistant and hardiness genes to the water melon. The genetic diversity of the desert adapted plants is likely to be of immense value in increasing the yields and for coping with changing environmental conditions. The opportunity and scope provided by genetic engineering should further greatly enhance agricultural productivity (Bhandari, 1988). There are as many as 200 species of plants of minor medicinal value in the desert and there are,it is believed, as many others about which their medicinal utility has yet to be discovered. Co-mmiphora wightii, Withania somnifera, Urgineaindica and Solanum surattense are found in the desert and these species are very well known for their medicinal properties. The bulbs of U. indicate that they are rich in glycoside contents (Gupta, 1988).

Benefits

The three categories mentioned can all be related to economical impacts. These impacts can be directly in the sense that they have direct economical value, like medicine, wood, tourism, or, indirectly in the sense that the economical value follows from a better economy and often sustainable infrastructure, like research, education and protection measures. The benefits of biodiversity conservation can be categorized into three categories as given in table 3:

Table 3: Benefits of Biodiversity

Benefits		
Biological Resources	Ecosystem Services	Social Benefits
1. Food for humans and for cultivated animals,	1. Protection of water resources,	1. Research, education and monitoring,
2. Medicinal and pharmaceutical resources	2. Soils formation and protection,	2. Recreation & tourism,
3. Breeding stocks, population reservoirs,	3. Nutrient storage and cycling,	3. Cultural values
4. Resources not yet identified (future resources),	4. Pollution breakdown and absorption,	
5. Wood products,	5. Contribution to climate stability,	
6. Ornamental plants and animals.	6. Maintenance of ecosystems ,	
7. Potential agents for crop improvement or biological control.	7. Recovery from unpredictable events	

There are a multitude of anthropocentric benefits of biodiversity in the areas of agriculture, science and medicine, industrial materials, ecological services, in leisure, and in cultural,

Figure 3: Checking of shifting sand dunes through plantations of Acacia tortilis near Laxmangarh town (Sikar),

aesthetic and intellectual value. Biodiversity is also central to an ecocentric philosophy. It is important for the contemporary population to understand the reasons for believing in the conservation of biodiversity. One way to identify the reasons of why we believe in it is to look at what we get from biological diversity and the things that we lose as a result of any species extinction, which has already taken place over the last 600 years. Mass extinction is the direct result of human activity and not of any natural phenomenon which is the perception of many modern day thinkers. There are many benefits that are obtained from natural ecosystem processes. Some ecosystem services that benefit society are air quality, climate (both global CO2 sequestration and regional and local), water purification, disease control, biological pest control, pollination and prevention of erosion. Alongwith those come non- material benefits that are obtained from ecosystems which are spiritual and aesthetic values, knowledge systems and the value of education that we obtain today. However, the public remains unaware of the crisis in sustaining biodiversity. Biodiversity takes a look into the importance to life itself and provides modern people with a clear understanding of the current threat to life on Earth.

An important benefit of biodiversity in arid areas is the checking of the sand dunes (Fig.3).The soil of the Thar Desert remains dry for a long period of the year and is therefore most prone to wind erosion. Due to the high velocity of winds in these areas the dust is blown off with it and deposited on fertile lands causing a problem of shifting sand dunes. The movement of the shifting sand buries fences and blocks roads and railway tracks. A permanent solution to this problem of shifting sand dunes can be provided by the fixation of the shifting sand dunes with suitable plant species and the planting of windbreaks and shelterbelts. They also provide protection from hot or cold and desiccating winds and the invasion of sand.

Problems Related to Biodiversity

The Thar probably is the world's most densely populated desert. The grazing of livestock, mostly sheep and goats, is intensive, affecting soil fertility and destroying native vegetation. Many palatable perennial species are being replaced with inedible annual species, thus, changing the vegetation composition and the ecosystem dynamics. The availability of water since the completion of the Indira Gandhi Canal Scheme has provided for water irrigation to the once non-arable desert, thus attracting farmers to the area. Salt pans for commercial salt production will have serious impacts on the Sambhar Lake in Rajasthan. Together with recent climatic changes, these pressures combine to degrade and destroy the fragile desert ecosystems.

The extreme conditions found in the drylands mean that the natural biodiversity is often fragile, easily affected by local human-induced habitat conversion, overgrazing or over-harvesting, the introduction of exotic species or changes in water availability, to name only a few sources of disruption. One result is the acceleration of soil erosion and nutrient loss leading to land degradation in major dryland areas.

Land degradation is further affected by demographic trends, a lack of adequate social and economic wellbeing, an absence of innovative technologies that enable people to live more sustainably, cultural factors and belief systems, and unresponsive institutions and policies. For example, the dry Seridó of northeast Brazil is becoming degraded through deforestation and forest fires set by local poor people to clear land, which results in the uncontrolled withdrawal of water and the discharge of pollutants into reservoirs.

Drylands are also under threat from irreversible human-induced global climatic changes. This threatens the dryland biodiversity through changes in the precipitation patterns and in

Figure 4: Sheep overgrazing the Thar Desert area of Rajasthan, lead to the Depletion of Plant Species.

local or regional temperatures, as well as through increases in the incidence of diseases like malaria or dengue fever. Unfortunately, scientific methods that can accurately predict these kinds of changes are limited.

Over the past century the population in the Thar desert has increased dramatically and it is now the most populous desert in the world with a human density of 84 per square km. The inhabitants main occupations are agriculture and animal husbandry which has in turn caused a decrease in nutritious land for farming due to overgrazing. It is a matter concerning forage demand and supply, and high demands meeting low supplies have affected the environment negatively, making life in the desert more hostile for the inhabitants. This is a problem that is only on the upturn and unless some course of action is taken, the human effect on the Thar deserts biodiversity will just become more destructive.

Species are disappearing at an unprecedented rate. This will amount to an irreversible loss of unique natural resources. Biological impoverishment is not usually caused by the deliberate overexploitation of a species but by in virtually all cases, it results from the degradation or destruction of a species habitat by some human activity such as agriculture. Indeed, although biological diversity contributes to such vital human activities as the production of food, fibre and energy, those same activities along with population growth are the most frequent causes of the depletion of biological diversity. In the Thar Desert, biological diversity is subjected to such interference. Trees, shrubs and even their roots are mercilessly removed by human beings for fuel, fodder, fencing and construction purposes. The requirement for woody biomass for fuel in the region has increased from 1.85 million tonnes in 1951 to approximately 4.0 million t (Bhandari, 1991). Prosopis cineraria, Acacia nilotica, A. senegal and C. polygonoides are exploited for fuel purposes. Commiphora wightii, W. somnifera,S. surattense and U. indica are exploited for medicinal purposes.

Livestock levels in the Thar Desert are much beyond the land's carrying capacity and there is a continual depletion of the biological diversity due to grazing (Fig.4). Besides the livestock, the natural fauna of the Indian desert is sufficiently rich to exert substantial pressure on the vegetation. Among large animals, the ungulates are the major consumers of grass and crops, e.g. the blue bull (Boselaphus tragocamelus), blackbuck (Antelope cervicapra rajputanae), gazelle (G. gazella benetti) and wild boar (Sus scrofa). Hares (L. nigricollis), langurs (Presbytis entellus), peacocks (Pavo cristatus) and squirrels (Fun-ambulus pennanti) are other herbivores with fairly large population sizes (Saxena and Prakash,1992).

Habitat destruction, overexploitation, and species introduction are the major causes of biodiversity loss in the Thar desert of India.. Other factors include fires, which adversely affect regeneration in some cases, and such natural calamities as droughts, diseases etc. Habitat destruction, decimation of species, and the fragmentation of large contiguous populations into isolated, small, and scattered ones has rendered them increasingly vulnerable to inbreeding depression, high infant mortality, and susceptibility to an environmental catastrophe and, in the long run, possibly to extinction.

Besides these, the failure to stem this tide of destruction results from an amalgamation of lacunae in economical, political, institutional and governance systems. Among others, these include, management with limited local community participation and involvement and inadequate implementation of eco-developmental programmes; poor implementation of the Wildlife Protection Act of 1972 as amended in 1991.

Poor conviction rates of wildlife cases are due to the inadequate legal competence in the forestry department together with the lack a daisical approach of the courts with the cases result that pending for years. The Protected Area Network comprises the National Parks and Sanctuaries which cover a mere 4.2% of the land area and is inadequate in protecting such ecologically important and fragile ecosystems such as wetlands, mangroves, and grasslands that lie outside such protected areas. The protected areas themselves are susceptible to denotification and further reduction in extent due to other pressures emanating from the industrial commercial-political combination. The root cause of any biodiversity loss can nicely be explained from the following figure:

Endemism and Endangered Species in the Thar Desert

The degree of endemism of the plant species in the Thar desert is higher than in the Sahara Desert. Of the total plant species found in the Thar desert only 6.4% are endemic as opposed to only 3% endemic species in the Sahara Desert. The most important aspect is that endemic species require conservation; however, other species are important from the point of view of biomass and plant cover. However, many endemic species require immediate attention to save them from the dangers of extinction. The densities of phog (C. polygon-oides), rohida (T. undulata), khejri (P. cineraria), babool (A. nilotica), kumat (A. senegal), jal (S. oleoides), sevan (L. sindicus), ker (C. decidua), nagori ashgandh (W. somnifera), guggal (C.wightii) and S. surattense are reducing very fast. These will very soon become endangered species of the Thar Desert. The chinkara (G. gazella), bustard (C. nigriceps) and black buck (A. cervicapra) have already been declared endangered animal species of the Thar Desert.

Major Biodiversity Threats

1. Habitat destruction,
2. Extension of agriculture,
3. Filling up of wetlands,
4. Conversion of rich bio-diversity site for human settlement and industrial development.

Major Problems with Biodiversity Conservation

Biodiversity conservation in India is also impeded by a lack of knowledge of the magnitude, patterns, causes, and rates of deforestation and biodiversity loss at the ecosystem and landscape level. Poaching and trading in wildlife species are among the most important concerns in the management of protected areas today but information on poaching, trade, and trade routes is sketchy and current wildlife protection and law enforcement measures are inadequate and inefficient. The major problems related to biodiversity in the study area are listed below:

1. Low priority for the conservation of living natural resources,
2. Exploitation of living natural resources for monetary gains,
3. Values and knowledge about the species and ecosystems inadequately known,
4. Unplanned urbanization and uncontrolled industrialization.

Conservation Efforts

The conservation of biological diversity requires a major shift in thinking at all levels. It will involve a commitment to increasing the levels of human and financial support. Comprehensive and reliable information is particularly important to enable biodiversity to be fully integrated into all levels of decision making. The most effective way to con-serve biological diversity is through the in situ conservation of ecosystems; this respects the ecosystemic interactions within and between wildlife populations as well as between different species. The convention on biological diversity, in preparation since 1988, was channelized just before the United Nations Conference on Environment and Development held in June 1992 in Rio de Janeiro, Brazil. By mid-December 1993, a total of 167 countries had participated and 36 had formally signed at the convention. India is one of the mega diversity' centres of the world. It has contributed very many economical plants, such as rice, sugar cane, millets, lentils, brinjal a number of cucumbers and a number of medicinal and ornamental plants to world use. The main repositories of

India's genetic wealth are the 410 sanctuaries and 69 national parks, constituting 3% of the land of the country. These are dwindling day-by-day. Many of the 115,000 known Indian plant and animal species have become endangered with the shrinking forest cover. Population pressure has led to encroachments and illegal exploitation of the forest resources. Immense pressure is mounted on protected areas by the increasing demand for fodder, fuel wood and other plant produce. Nearly 72% of the sanctuaries and 56% of the national parks have people living within their boundaries. Approximately 250 million tonnes of fuel wood is chopped in India every year. If the human population continues to increase at the current rate, there will be no forest cover left by the middle of the next century. Major economical, social and political reforms are needed to tackle the population poverty cycle. The spread of industrial activity into protected areas has caused havoc. For example, all the political parties of Rajasthan are proposing to develop marble, dolomite, barytes, soapstone, feldspar, quartzite, clay and granite mining in the Sariska tiger project region of Rajasthan. The recent announcement by the Indian Union Minister of State for Forest and Environmental Matters, Shri Kamal Nath, for umbrella legislation for biodiversity, the formation of a 'cats' crisis cell and the setting up of a strike force in all Project Tiger areas is particularly welcome. However, it is more important to implement and tighten the legal system. There are no environmental courts and the introduction of the Indian National Environmental Tribunal Bill has been delayed for more than 2 years. The most urgent and crucial thing is to make a directory of all plant and animal species; only then can the rate of disappearance of species be evaluated. In the Thar Desert around 10% of the existing species are endangered due to agriculture, fodder, fuel wood, industrial and medicinal needs and grazing by animals (livestock and wildlife T.I. KHAN, 1997). The following recommendation can be made for the conservation and management of biodiversity in the Thar desert:

1. An environmental impact assessment should be a mandatory requirement to check ecological refuge, deforestation and loss of biological diversity whenever a change of landuse is considered.

2. People should be made aware of the importance of their local fauna and flora.

3. Environmental education should be imparted to 'ecosystem people' and to those who are directly dependent upon the biomass for their basic needs.

4. Environmental ethics and religious means should be used to develop greater in situ conservation of biological diversity. Village grazing lands (Orans), permanent pastures (Jhors), long fallows and lands in the names of gods and goddesses should be given importance for protection, development and expansion. Orans are of religious and ceremonial significance and should provide protection to the native species. They now function as centres for the dispersal of genetic material.

5. The practice of replacing indigenous species with exotic species should be discouraged.

6. Merely declaring that an area be known as a sanctuary oran for a national park is insufficient. Local people should participate in the sanctuary and its conservation.

7. There is a lack of awareness amongst the village people of the serious importance of conservation. Such in situ conservation is critical.

8. Existing sanctuaries, Tiger Project areas and national parks should not be removed from protection.

Further steps

There are a few more things which can be done to sustain biodiversity not only in arid areas but in other areas as well. These are discussed below:

1. Documentation of biodiversity is an urgent requirement as the latest statistics and data on the floral and faunal biodiversity of India have not been compiled and documented.
2. The information and data should be made available to the scientific and socio-economic agencies to support the evaluation and revision of the policies.
3. Lack of knowledge of the magnitude, patterns, causes and rates of deforestation and biodiversity laws at the ecosystem and landscape level.

Policy recommendations

1. Most of the legal provisions pertain mainly to the use and exploitation of biological resources, rather than their conservation. Even the Wildlife Protection Act 1972, focuses on protection rather than conservation. Protection under the Wildlife Protection Act is largely directed towards large animal species (charismatic terrestrial species) rather than the large spectrum of fauna and flora also found in the marine realm.
2. The existing laws relating to biodiversity shall be examined in order to bring them in line with the provisions of convention to reflect the current understanding of biodiversity conservation.
3. Need for comprehensive legislation on biodiversity conservation and use especially fisheries policies, which are generally ignored.
4. Formulation of policies for the protection of wetlands, grasslands, sacred groves, marine flora, fauna and other areas is significant from the point of view of biodiversity.
5. Improving policies on the environment.
6. Passage of biodiversity bill.
7. Increase of the allocation of financial resources for the conservation of biodiversity.
8. Integrating conservation with development.
9. Incentives and disincentives for the improper use of biodiversity.
10. There should be a continuous monitoring of biodiversity use for a review of the results of implementation of policies and programmes.
11. Perform more strategic environmental and social assessments.
12. Involvement of all stakeholders, e.g., local to national governments, private sector, NGOs, farmers, consumers

Shortcomings in Policy Efforts

The lack of policies for the protection of grasslands, sacred groves and other areas significant from the point of view of biodiversity:

1. Inadequate implementation of eco-development programmes,
2. Need for enhanced role of NGOs and other institutions,
3. Need for political commitment and goodwill,
4. Need for providing Institutional Structure,
5. Human resource development – limited local community participation.

Conclusion

In the present world where everybody is living a fast lifestyle and is too busy even to take out time for themselves, our biodiversity is in a state of neglect.Therefore, before this neglect becomes a more dangerous situation hampering the survival of human beings which is already happenning man, has to control his destructive activities and take time out to look into the matter seriously. It does not matter he is acting as a government official or an NGO activist, a local individual or an international association member. He has to bring all these roles together to save biodiversity which is already becoming an alarming problem.

References

1. Ahmedullah, M. and Nayar, M.P. (1987) Endemic Plants of the Indian Region, Calcutta: Botanical Survey of India.
2. Alexander, V., Markov and Andrey V. K. (2007) "Phanerozoic marine biodiversity follows a hyperbolic trend", Palaeoworld, 16(4): 311-318.
3. Alroy, J.C.R. et al. (2001) "Effect of sampling standardization on estimates of Phanerozonic marine diversification", Proceedings of the National Academy of Science, USA , 98: 6261-6266.
4. Altieri, M.A. (1990) "How common is our common future?", Conservation Biology, 4(1): 102-103.
5. Arroyo, M.T.K, Raven P.H. and Sarukhan, S. (1991) "Biodiversity", Invited Communication at the International Conference on an Agenda of Science for Environment and Development into the 21st Century, Vienna, Austria, p. 17.
6. Barrow, E., Gichohi, H. and Infield, M. (2000) "Summary and key lessons from a comparative review and analysis of community conservation in eastern Africa. IUCN Eastern Africa", Working Paper No. 2, p. 25.
7. Bhandari, M.M. (1988) "Life support systems for heat tolerance", in National Bureau of Plant Genetic Resources (eds) Life support plant species, New Delhi: National Bureau of Plant Genetic Resources publications, 60-65.
8. Bhandari, M.M. and Shirangi, O.P. (1986) "Conservation of threatened plants of western Rajasthan", in S.M. Mehno and M.M. Bhandari (eds) Environmental Degradation in Western Rajasthan, Jodhpur: Scientific Publishers, 129-132.

18

Floral Distribution Pattern in Surroundings of Sidculife-Pantnagar

T. BANERJEE, R.K. SRIVASTAVA AND U. MELKANIA

Introduction

Increased levels of industrial activity to meet the Sdemands of humanity have caused irreversible damage to our ecosystem. Environmental pollution always exists due to some developmental activities. The industrial pollution due to its nature has the potential to cause irreversible reactions in the environment and hence it is posing a major threat to sustainable development. Therefore, due to the building of this SIDCUL Integrated Industrial Estate-Pantnagar, the environmental degradation can not be ignored in the surroundings which include the G.B. Pant University of Agriculture & Technology, Crop Research Centre, Horticulture Research Centre and Vegetable Research Centre etc as well as other research centres of the university which are within a 20 km radius from the IIE-Pantnagar.

The present investigation was carried out to determine the status of floral pattern in the surrounding areas of the SIDCUL. The determination of floral patterns in the adjacent areas and any impact on floral patterns by SIDCUL are crucial aspects to determine the sustainability of existing industrial estates. The baseline data on terrestrial ecology was derived by adopting the quadrant method of survey to determine frequency, density, abundance, cover etc. within the region.

2. Study Area

Uttarakhand is a new state created from the state of Uttar Pradesh in the year 2000 and now it is developing very fast with a rapid boom in industrialization. State Industrial Development Corporation of Uttarakhand Limited (SIDCUL) is responsible for the establishment of industries in the industrial estate of Pantnagar, Haridwar and Sitargang. SIDCUL Integrated Industrial Estate-Pantnagar (with latitude between $28^0$59′51″ N and $29^0$01′12″ N and longitude between $79^0$24′9″ E and $79^0$26′15″ E) in the District of Udham Singh Nagar of Uttarakhand has attracted

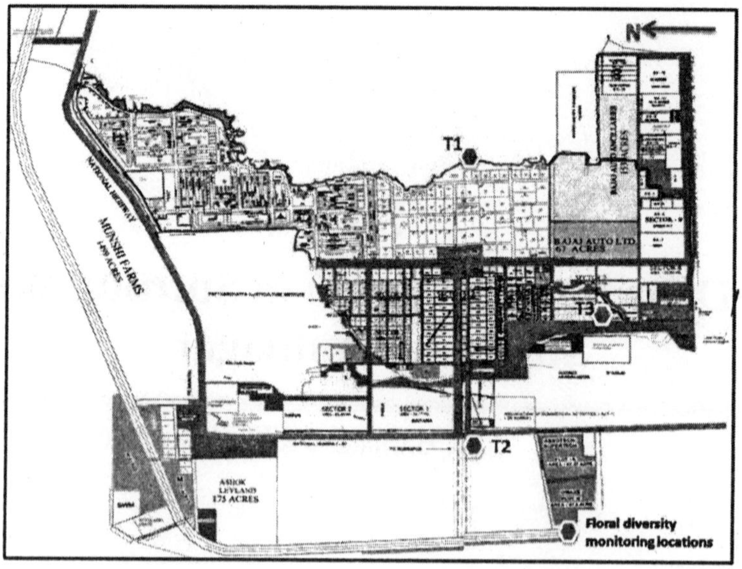

a large number of industrialists for setting up of their industries due to the available infrastructure facilities and suitable Uttarakhand industrial polices. A total of 3339 acre of land is earmarked for the setting up of various industries on prime agriculture land, which earlier belonged as agricultural farm land to the G.B. Pant University of Agriculture and Technology, Pantnagar (U.S. Nagar). The monitoring of floral patterns was carried out in the vicinity of to the SIDCUL IIE-Pantnagar to quantify existing floral diversity in the particular area. Floral pattern monitoring locations inside the SIDCUL are shown in the Fig. 1.

3. Methodology

The quadrat (1 sq meter in size) method was used to determine floral pattern in the surroundings of the SIDCUL. The quadrat is a square sample plot or unit used for the detailed analysis of vegetation. In vegetation analysis, a quadrat of any size, number and arrangements may be used depending upon the study area. The encountered vegetation of the study area was analyzed to calculate density, frequency & abundance **(Curtis and McIntosh, 1950)** with standard methods as outlined in the Ecology Work Book (**Mishra, 1968**). Relevant data was used to the calculate Importance Value Index of a particular species **(Phillips, 1959)**. Shannon-Wiener Index and Simpson's Index were also measured to compare diversity indices between different monitoring locations.

3.1. *Frequency*

Frequency indicates the number of sampling units in which a given species occurs, and thus expresses the distribution or dispersion of various species in a community. Percentage frequency is therefore calculated as

$$\text{Frequency}=\frac{\text{Number of quadrats in which species occured}}{\text{Total number of quadrats studied}}\ 100$$

$$\text{Relative frequency} = \frac{\text{Frequency of a species}}{\text{Total frequency of all species}} \times 100$$

After determining the percentage frequency of all species, various species are distributed among Raunkiaer's five frequency classes depending upon their frequency values as follows:

Frequency %	Frequency class
0-20	A
21-40	B
41-60	C
61-80	D
81-100	E

3.2 *Density (individuals/ unit area)*

The term density represents the numerical strength of a species in the community. The density and frequency taken together are of prime importance in determining community structure.

$$\text{Density} = \frac{\text{Number of individuals of a species in all quadrats}}{\text{Total number of quadrats studied}}$$

$$\text{Relative density} = \frac{\text{Density of the species}}{\text{Total densities of all species in all quadrats}} \times 100$$

3.3 *Abundance (plant/ unit area)*

Abundance is described as the number of individuals per quadrat of occurrence.

$$\text{Abundance} = \frac{\text{Number of individuals of a species in all quadrats}}{\text{Total number of individuals of all species}}$$

$$\text{Relative Abundance} = \frac{\text{Abundance of a species}}{\text{Total abundance of all species}} \times 100$$

The abundance to frequency ratio (A/F) of different species was computed to define the distribution pattern of the species **(Whitford, 1949)**. This ratio indicates regular (<0.025), random (0.025 to 0.050) and contiguous (>0.050) distribution **(Curtis and Cottam, 1956).**

Importance Value Index (IVI) = Relative density + Relative frequency + Relative Abundance

IVI of a species in the community gives an idea of its relative importance in the community. The two main factors taken into account when measuring diversity are richness and evenness.

3.4 Species Richness

Richness is a measure of the number of different kinds of organisms present in a particular area. For example, species richness is the total number of different species present in a community. Some communities may be simple enough to allow complete species counts to determine species richness but in a complex ecosystem, species richness is determined by using formulae of **Margalef (1958).**

$$\text{Species Richness} = \frac{S - 1}{\ln N}$$

where, S = total no of species in the community

N = total density of all the species

3.5 Evenness

Evenness is a measure of the relative abundance of the different species making up the richness of an area. A community dominated by one or two species is considered to be less even than one in which several different species have a similar abundance.

3.6 Shannon and Wiener Species Diversity Index (H′)

The Shannon index (also known as the Shannon-Wiener Index), is one of several diversity indices used to measure diversity in categorical data. It is simply the information entropy of the distribution, treating species as symbols and their relative population sizes as the probability **(Shannon and Wiener, 1963).** In ecology, a diversity index is a statistic which is intended to measure the biodiversity of an ecosystem. Moreover, diversity indices can be used to assess the diversity of any population in which each member belongs to a unique species. The advantage of this index is that it takes into account the number of species and the evenness of the species. The index is increased either by having additional unique species, or by having greater species evenness.

The proportion of species 'i' relative to the total number of species (Pi) is calculated, and then multiplied by the natural logarithm of this proportion (In Pi). The resulting product is summed across species, and multiplied by -1.

$$H' = -\text{SUM}\{P_i * (\ln P_i)\}$$

3.7 Simpson's Index (D)

Simpson's Index is a measure of diversity. In ecology, it is often used to quantify the biodiversity of a habitat. It takes into account the number of species present, as well as the abundance of each species.

$$D = \frac{\sum n(n-1)}{N(N-1)}$$

Simpson's Index (D) measures the probability that two individuals randomly selected from a sample will belong to the same species (or some category other than species). Here, 'n' signifies the total no. of organisms of a particular species and 'N' signifies the total number of individuals of all the species. With this index, 0 represents infinite diversity and 1, no diversity. That is, the bigger the value of D, the lower the diversity **(Simpson, 1949).**

3.8 Simpson's Index of Diversity (1 – D)

Simpson's index represents the probability that two individuals, randomly selected from a sample will belong to different species. In general, the value of Simpson's index ranges from 0 to 1. The greater the value of Simpson's index, the greater will be the sample diversity.

4. Results and Discussions

The land converted to an industrial estate was basically agricultural land used for intensive agriculturical purposes. Therefore it was a man made ecosystem with dominance of a few particular species. The conversion of land from prime agriculturical land to an industrial area should not affect largely the biodiversity of the particular area, but due to the emissions of pollutants from the industrial area, the existing biodiversity of the particular area may get distressed in the near future.

The land use pattern of the state Uttarakhand has shown that among the total geographical area of 53,483 sq km, the area under forest is 34,622 sq km i.e. 64.81% of the total area, and the net area shown is only 14%, the total irrigated area is 1.54%, of total waste land and fallow land is 7.2% and an area under shrubs is 4%. It signifies that merely 12-15% of the total state area is available for farming. In the case of U. S. the Nagar, total area under forest cover is 36.8% and per capita forest in this district is just 0.0008 sq km. the SIDCUL IIE-Pantnagar situated in the Rudrapur block contains only 0.02% of the forest area **(Forest Department, Uttarakhand, 2006-07)**.

Therefore it was felt necessary to study the floral diversity in the vicinity of the SIDCUL IIE-Pantnagar and represent the information as a baseline status in the initial stage of the development of this industrial estate.

Table 1. Distribution of existing flora in the surroundings of SIDCUL IIE Pantnagar Location-T1 (Near SIDCUL-National highway 87)

Name of the species	% Frequency	Density Sp. No./unit area	Abundance Plants/ unit area	Raunkiaer's Frequency class
Cynodon dactylon	100	97.2	97.2	E
Parthenium sp.	100	5.6	5.6	E
Cannabis sativus	100	5.6	5.6	E
Euphorbia hirta	100	4.4	4.4	E

Name of the species	Relative Frequency	Relative Densitiy	Relative Abudance	Distribution Pattern
Cynodon dactylon	25	86.2	86.2	Contiguous
Parthenium sp.	25	5.0	5.0	Contiguous
Cannabis sativus	25	5.0	5.0	Contiguous
Euphorbia hirta	25	3.8	3.8	Random

Name of the species	IVI	Simpson Index (D)	Special Diversity Index (H)	Special Richness
Cynodon dactylon	197.4			
Parthenium sp.	35			
Cannabis sativus	35	0.75	0.555	0.634
Euphorbia hirta	32.6			

In site-1 (T1), *Cynodon dactylon* showed a maximum IVI with 197.4 and therefore represents the most dominant species in that particular area. All four species represent class 'E' as per Raunkiaer's classifications. Simpson's index in the area is 0.75, therefore Simpson's index of biodiversity in the particular area will be (1 - 0.75) i.e. 0.25, which represents a less floral diversity present in the site-1. The species evenness in the particular site is much smaller signifying that only a few species are dominating in the particular monitoring location. The monitoring site was mostly dominated by the *Cynodon dactylon* with relative density of 86.2. The species richness of the particular area was 0.634.

Table 2. Distribution of existing flora in the surroundingsof SIDCUL IIE-Pantnagar Location-T2 (Inside SIDCUL)

Name of the species	% Frequency	Density Sp. No./unit area	Abundance Plants/ unit area	Raunkiaer's Frequency class
Cynodon dactylon	100	110.8	110.8	E
Lantana camara	40	0.8	2	B
Parthenium sp.	80	8.8	11	D
Sorghum halepense	80	22	27.5	D
Oxalis sp.	60	1.4	2.3	C

Name of the species	Relative Frequency	Relative Densitiy	Relative Abudance	Distribution Pattern
Cynodon dactylon	27.8	77.0	72.1	Contiguous
Lantana camara	11.1	0.56	1.3	Regular
Parthenium sp.	22.2	6.1	7.2	Contiguous
Sorghum halepense	22.2	15.3	9.9	Random
Oxalis sp.	16.7	0.97	1.5	Regular

Name of the species	IVI	Simpson Index (D)	Special Diversity Index (H)	Special Richness
Cynodon dactylon	176.9			
Parthenium sp.	12.9			
Cannabis sativus	35.5	0.62	0.833	0.705
Euphorbia hirta	19.2			

In site-2 (T2), *Cynodon dactylon* shows again a maximum IVI with 176.9 and therefore also represents the most dominant species in that particular area. The species also represent class 'E' as per Raunkiaer's classifications signifying that most frequent species were found on the site. *Parthenium sp.* and *Sorghum halepense* represents class 'D' of Raunkiaer's classifications, representing less frequent species than the *cynodon dactylon* species. Simpson's index in the area is 0.62, therefore Simpson's index of floral diversity in the particular sampling site will be (1 - 0.62) i.e. 0.38, which represents a greater amount of floral diversity present in the site-2. Species evenness in the particular site is much less significant that only a few species are dominating in the particular monitoring location. The site is mostly dominated by the *Cynodon dactylon* with a relative density of 77.0 followed by *Sorghum halepense* with a relative density of 15.3. The species richness of the particular area is 0.805, which is more than the remaining two sites (T1 & T3), signifying more species present in the T2.

Table 3. Distribution of existing flora in the surroundingsof SIDCUL IIE-Pantnagar Location-T2 (Inside SIDCUL)

Name of the species	% Frequency	Density Sp. No./unit area	Abundance Plants/ unit area	Raunkiaer's Frequency class
Cynodon dactylon	100	100.4	100.4	E
Sorghum helipense	80	18.2	22.8	D
Cannabis sativus	60	3.6	6	C
Parthenium sp.	80	4	5	D

Name of the species	Relative Frequency	Relative Densitiy	Relative Abudance	Distribution Pattern
Cynodon dactylon	31.3	79.6	74.8	Contiguous
Parthenium sp.	25	14.4	17.0	Contiguous
Cannabis sativus	18.8	0.57	4.5	Random
Euphorbia hirta	25	3.17	3.7	Random

Name of the species	IVI	Simpson Index (D)	Special Diversity Index (H)	Special Richness
Cynodon dactylon	185.7			
Parthenium sp.	56.4			
Cannabis sativus	23.9	0.65	0.672	0.62
Euphorbia hirta	31.9			

Cynodon dactylon again shows a maximum IVI with 185.7 in case of T3, and therefore represents the most dominant species. Therefore in all the three sites, this particular species

(*Cynodon dactylon*) shows the highest dominance over other species and this species also represents class 'E' as per Raunkiaer's classifications signifying that most frequent species are to be found on the study site. Whereas *Parthenium sp.* and *Sorghum halepense* represent class 'D' of Raunkiaer's classifications, representing less frequence than the *cynodon dactylon* species. Simpson's index in the area is 0.65, so therefore Simpson's index of floral diversity in the study area will be (1 - 0.65) i.e. 0.35. Like the other two sampling locations, the species evenness in this particular site is also much less and again mostly dominated by the *Cynodon dactylon* with relative density of 79.6 followed by the *Parthenium sp.* with relative density of 14.4. Species richness of this particular site is calculated as 0.62.

Simpson's index of diversity for all three sampling sites signifies that floral diversity among weed species in the particular region is much less. but if all the three sites are compared then site-2 (T2) represents the maximum floral diversity over the other two locations. The shannon diversity index is also an important parameter to represent diversity. In site-T1, T2 & T3; The shannon diversity indexes are 0.555, 0.733 and 0.672, respectively, which signifies that the monitoring location T2 is having more diversity than the other two locations. This may be due to more disturbances within the SIDCUL industrial area at the T1 and T3 locations, either by anthropogenic activities or more industrial activities than of the site T2, which is primarily a vacant land outside the industrial estate boundary and lies between road and agriculturical block of field.

Critical analysis signifies that the above identified species present in this industrial estate are weeds or invasive species. However the major threats to exploit biodiversity of a particular area may be considered as bio-invasion. The land converted to the SIDCUL IIE-Pantnagar, was once happened to be prime agricultural land with great productivity, therefore road side fertile land should be effectively managed to reduce the invasion of alien species in this industrial area.

The best way to manage the land is to promote greenery within and surrounding the SIDCUL because it will provide a better barrier between the source of air and noise pollution and the surrounding areas by capturing fugitive emissions and absorbing gaseous pollutants as well as attenuate the noise generated from various industries.

5. Conclusions

The monitoring of the floral distribution in the vicinity of the SIDCUL IIE Pantnagar revealed that, fallow land inside the industrial estate was fully covered by some weed species. All three monitoring locations showed much less floral diversity and it consisted mainly of invasive species. The encountered vegetation consisted of litle species richness with a maximum dominance the of *cynodon dactylon*. After an in-depth analysis of the SIDCUL IIE area, it was recommended that there is scope for the development of a green belt and other plantations which will help only to reduce dispersion of the fugitive emissions and gaseous pollutants but also to be usefulin reducing the growth of any invasive species in the vicinity of the industrial estate.

References

1. Curtis, J.T. and Cottam, G., 1956. Plant ecology work book: Laboratory field reference manual. Burgess Publishing Co., Minnesota. pp: 193.
2. Curtis, J.T., and Mclntosh, R.P., 1950. The interrelation of certain analytic and synthetic phyto-sociological characters. Ecology, 31: 476-496.
3. Forest Department, Uttarakhand, 2006-07. Uttarakhand Forest Statistics. Forest Department, Dehradun, Uttarakhand.
4. Margelf, R., 1958. Information theory in ecology. Gen. Syst, 3: 36-71.
5. Mishra, R., 1968. Ecology work book. Oxford & IBH Publishing Co., New Delhi. pp: 244.
6. Phillips, E.A., 1959. Methods of vegetation study. A Hott Dyreden Book Henry Hold & Co., New York.
7. Shannon, C.E. and Wiener, W., 1963. The mathematical theory of communication. University of Illinois Press, Urabana, U.S. pp: 127.
8. Simpson, E.H., 1949. Measurement of diversity. Nature, 163: 688.
9. Whitford, P.B., 1949. Distribution of woodland plants in relation to succession and growth. Ecology, 30: 199.

19

Isolation, production and partial purification of xylanase from Rhizopus species

R. D. Kamble and A.R. Jadhav

Introduction

Hemicellulose is the second most abundant renewable polysaccharide in nature after cellulose found in a plant cell wall. It is a heteropolymer with backbone of â-1, 4-D-xylanopyranosyl residues and branches of neutral or uronic monosaccharides and oligosaccharide [1]. Xylanases (endo-1, 4-â-D-xylan xylanohydrolase; EC 3.2.1.8) degrade the xylan backbone into small oligomers. Due to the heterogeneity of xylan, the complete enzymatic degradation to its monomer is a complex process requiring the action of a battery of hydrolytic enzymes. They include Endo-(1, 4)-â-xylanase, â -xylosidase, α-L-arabinofuranosidase, α-glucuronidase, acetylxylan esterase and phenolic acid esterases. Endo-(1, 4) - β -xylanase, which attacks the polysaccharide backbone is one of the most crucial enzyme involved in xylan hydrolysis [2, 3].

Various microorganisms such as bacteria, fungi and actinomycetes are reported as xylanase producers. Filamentous fungi have demonstrated a great capability for secreting a wide range of xylanases, with the genera *Aspergillus, Penicillium* and *Trichoderma* being the most extensively studied and reviewed among the xylanase producing fungi. Very little literature is available regarding the fungi belonging to the genus *Rhizopus* for its use in the production of enzymes of biotechnological importance [4, 5, and 6].

Xylanases have considerable potential in several biotechnological applications, which involve the conversion of xylan, which is present in agriculturical waste and the food wastage industry, into xylose. Similarly xylanases could be used for the clarification of juices, for the extraction of coffee, plant oils, starch and for the production of fuel as well as for chemical feed stocks. One of the most important new biotechnological applications of xylanases is their use in pulp bleaching [7, 8].

In this paper, we have isolated a strain of *Rhizopus* showing an xylanase-producing ability and it has been identified by the morphological, cultural and biochemical characteristics. Then the optimum temperature and pH the balance of the partially purified xylanase were determined to investigate their potential in the food industry for the clarification of fruit juices.

2. Materials and Methods

2.1 Sample collection

For the screening of xylanase producing microorganisms soil samples were collected from various composts in and around the Kolhapur district of the Maharashtra state. The soil samples were crushed, sieved, dried and placed in thin plastic bags. These bags were stored at 4^{0}C.

2.2 Screening of xylanase producing microorganisms

2.2.1 Enrichment

Soil samples were enriched using Oat spelt xylan medium (g/L) having following composition: Oat spelt xylan 7.0; Yeast extract 1.0; NaCl 5.0; K_2HPO_4 1.0; $MgSO_4$ 0.2; $CaCl_2$ 0.1; pH 6.5. The inoculated media was incubated in an orbital shaker at 37^{0}C and 150 rpm for 4 days [9].

2.2.2 Isolation

The enriched samples were appropriately diluted and used to inoculate the Oat spelt xylan agar plates for isolation and rapid identification of xylanase producing microorganisms. The plates were incubated (Remi incubator) at 37^{0}C for 72hrs. All the potential micro-organisms were isolated by the clear zone around their growth.

2.2.3 Zymogram analysis

Zymogram analysis was done by preparing a chromogenic substrate i.e. Xylan- Red by using Ten method [10]. The substrate Oat spelt xylan combined with Cibacron Brilliant Red- 3BA is used as a carbon source in the media in place of Oat spelt xylan to observe the colonial growth with a clearance zone around it. The plates were incubated at 37^{0}C for 72 hrs.

2.2.4 Identification of potential isolates

The identification of the xylanase producing isolates will be done using morphological, cultural and biochemical characteristics using Bergey's Manual of Determinative Bacteriology [11].

2.3 Production of extracellular xylanase in Submerged Fermentation

Xylanase production is carried out in submerged fermentation using agro-residues. Potential isolates were grown at 37^{0}C with rotary shaking (150 rpm) (Remi orbital shaker) in 250 mL capacity Erlenmeyer flasks containing 50mL medium having following composition (g/L): Wheat bran 7.0; Yeast extract 1.0; NaCl 5.0; K_2HPO_4 1.0; $MgSO_4$ 0.2; $CaCl_2$ 0.1; pH 6.5.

2.4 Preparation of crude enzyme extract

After fermentation the broth was transferred to centrifuge tubes and clear supernatant was collected after centrifugation (Cooling centrifuge; Remi industries) at 10,000×g for 15 min at

4°C. The supernatant was then used for a further experiment as the crude xylanase extract. All experiments were performed in duplicate.

2.5 Estimation of xylanase activity

Xylanase activity was determined by incubating 1 ml of assay mixture containing 0.5 mL of 1 % xylan and 0.5 mL of suitably diluted enzyme in 50 mM Acetate buffer (pH 4.8) for 30 min at 50°C. Enzyme and reagent blanks were also simultaneously incubated with the test samples. The reducing sugar formed was estimated by Dinitrosalicylic Acid (DNSA) method [12]. One international unit (IU) of enzyme activity for xylanase was defined as the amount of enzyme releasing 1 μmol reducing sugar from xylan per minute using xylose as standard.

The enzyme activity was calculated as follows:

U/ml =BFR × Dilution × Dilution factor x 0.15

BFR = Blank free reading (O.D – Enzyme Blank)

0.15 = μ moles of xylose corresponding to 40 μg/mL xylose

2.6 Determination of total protein concentration

Protein concentration was determined according to the method of Lowry using Bovine Serum Albumin (BSA) as the standard [13].

2.7 Ammonium Sulphate Precipitation

Protein precipitation by salting out technique using Ammonium Sulphate ($NH_4(SO_4)_2$) was carried out with constant gentle stirring. The amount of solid Ammonium Sulphate to be added was calculated according to the formula of Jagannathan et al. [14].

$$X = 50 (S_2-S_1)$$

Where,

X = Amount of Ammonium Sulphate to be added for 100 mL

S_1 = Initial concentration of Ammonium Sulphate

S_2 = Required saturation of Ammonium Sulphate

Solid Ammonium Sulphate (21g in 36mL) was added to the crude enzyme solution to give 90% saturation at 4°C. The mixture was stirred for 60 min and resulting precipitates were collected at 10,000 ×g for 20min at 4°C.

2.8 Dialysis

The precipitate was collected and dissolved in minimum volume of distilled water and dialyzed using 50mM acetate buffer of pH 4.8 for 24hrs at 4°C with periodic change of buffer.

2.9 Characterization of Xylanase

Xylanase will be characterized by determining effect of Temp., pH on enzyme activity.

2.9.1 Determination of effect of temperature on the activity of xylanase:

The activity of the purified xylanase enzyme is determined at different temperatures ranging from 30°C to 70°C.

2.9.2 Determination of the effect of the pH balance on the activity of xylanase:

The activity of the purified xylanase enzyme is determined at different pH values ranging from 4.0 to 11.0.

2.10 Application of partially purified xylanase in clarification of fruit juices:

Freshly collected orange juice was added with 1% partially purified xylanase and tested for reduction in optical density after every 3hrs.

3. Results

3.1 The xylanase producing isolates named as K_1 to K_{10} were isolated from composting soil samples by a screening procedure on the basis of clearance zone and xylanolytic properties on xylan agar plate. Colony characteristics of these 10 isolates were listed in Table No.1.

Table 1: Colony characteristics of isolates K_1 to K_{10}

Isolate no.	Size	Shape	Colour	Margin	Elevation	Consis-tency	Opacity
K_1	4mm	Circular	Chalky white	Regular	Raised	Dry	Opaque
K_2	2mm	Circular	Chalky white	Regular	Raised	Dry	Opaque
K_3	2mm	Circular	Chalky white	Regular	Raised	Dry	Opaque
K_4	3mm	Circular	Chalky white	Regular	Raised	Dry	Opaque
K_5	5mm	Irregular	Grey	Regular	Raised	Dry	Opaque
K_6	3mm	Circular	Grey	Regular	Raised	Dry	Opaque
K_7	3mm	Circular	Grey	Regular	Raised	Dry	Opaque
K_8	2mm	Circular	Chalky white	Regular	Raised	Dry	Opaque
K_9	2mm	Circular	Chalky white	Regular	Raised	Dry	Opaque
K_{10}	2mm	Circular	Chalky white	Regular	Raised	Dry	Opaque

3.2 Xylanase activity in each strain was confirmed by measuring the diameter of zone of clearance showing on the plate containing Oat spelt xylan combined with Cibacron Brilliant Red 3-BA dye (Sigma, Germany). Of these 10 isolates, four were K_2, K_5, K_6 and K_7 selected for further production of xylanase in submerged condition on the basis of diameter of zone of clearance as given in Table No. 2.

Table 2: Diameter of Zone of clearance on xylan-red agar media.

Isolate no.	Diameter of Zone of clearance in mm
K_2	20
K_5	35
K_6	25
K_7	25

3.3 Production of extracellular xylanase was carried by K_2, K_5, K_6 and K_7 in submerged condition on the substrate Wheat bran at 37ºC. The whole broth was kept undisturbed for 12hrs and subjected to centrifugation to collect supernatant as a crude enzyme extract.

3.4 Xylanase activity in each strain was confirmed by measuring the amount of reducing sugars liberated from xylan with Dinitrosalicylic acid reagent using crude extract was found to be maximum in K_5, which was selected for further studies.

Table 3: Xylanase activity of four isolates K_2, K_5, K_6 and K_7 in IU/ml

Isolate No.	Dilution	Aliquot	O.D.	E.B.	BFR	Units	IU/ml	Mean
K_2	1:5	0.05	0.0862	0.0164	0.0698	0.46	23.07	23.76
	1:5	0.1	0.1644	0.0176	0.1468	0.98	24.46	
K_5	1:5	0.05	0.1601	0.0202	0.1399	0.93	46.63	46.00
	1:5	0.1	0.2892	0.0169	0.2723	1.81	45.38	
K_6	1:5	0.05	0.0774	0.0173	0.0601	0.40	20.03	19.89
	1:5	0.1	0.1375	0.0189	0.1186	0.79	19.76	
K_7	1:5	0.05	0.0621	0.0160	0.0461	0.31	15.36	16.72
	1:5	0.1	0.1264	0.0179	0.1085	0.72	18.08	
S.B.	0.0231							

3.5 The strain grows at pH 6.5 and produced a high level of xylanase activity in a solid liquid media containing wheat bran. On the basis of morphological, cultural and biochemical characteristics the isolate was identified as *Rhizopus species.* As it shows stolons, aerial hyphae, grow horizontally and radially over the surface of the substratum and spread the mycelium and rhizoids, rooting hyphae, which arise from each node and anchor the fungus to the substratum and absorb nutrients with sporangiophores.

3.6 Xylanase produced by *Rhizopus sp.* was growth-associated, reaching a maximum after 72 hrs. Enzyme production remained more or less the same up to 96 hrs, while biomass started to gradually decline after 96 hrs.

3.7 Crude enzyme extract was partially purified using Ammonium Sulphate up to 85% and dialyzed. Both crude enzyme extract and dialyzed extract were subjected for determination of the total protein concentration and xylanase activity in U/mg.

Table 4: Total protein concentration and xylanase activity in U/mg shown by *Rhizopus species*

Rhizopus sp.	**Volume (ml)**	**Total protein**	**Total activity (U) (mg)**	**Specific activity (U/mg)**
Crude extract	36 ml	21.6	1620	75
Dialyzed	1.3 ml	1.5567	219.7	141

3.8 The effect of temperature on the activity of partially purified xylanase against xylan obtained from *Rhizopus s species* was examined in the range of 30-70^{0}C. The enzyme showed maximum activity around 50-55^{0}C as shown in Graph No.1.

3.9 The optimal pH of the enzyme activity was 5.5 when the activity was measured over pH range of 4.0-8.0 as shown in Graph No.2.

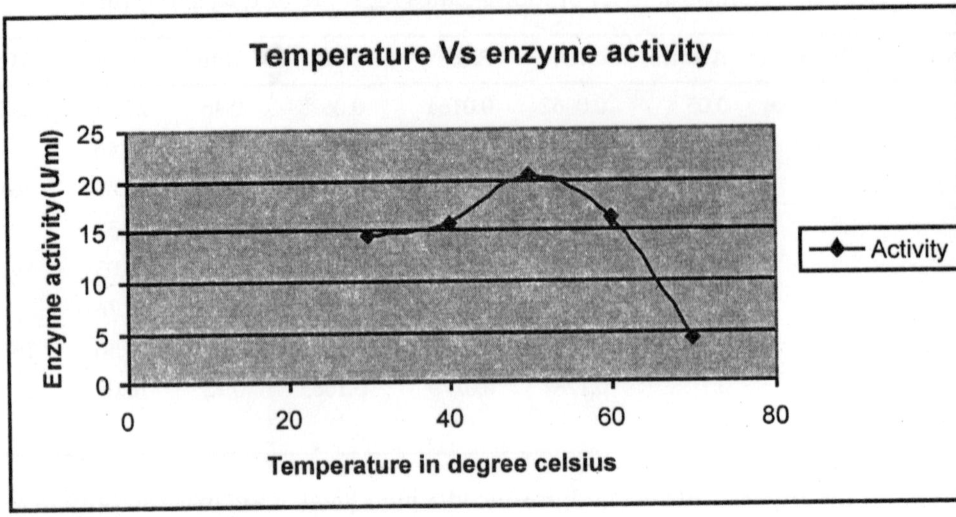

Graph No.1: Effect of temperature on xylanase activity.

3.10 Clarification of orange juice by using partially purified xylanase indicated by gradual decrease in optical density after every 3hrs.

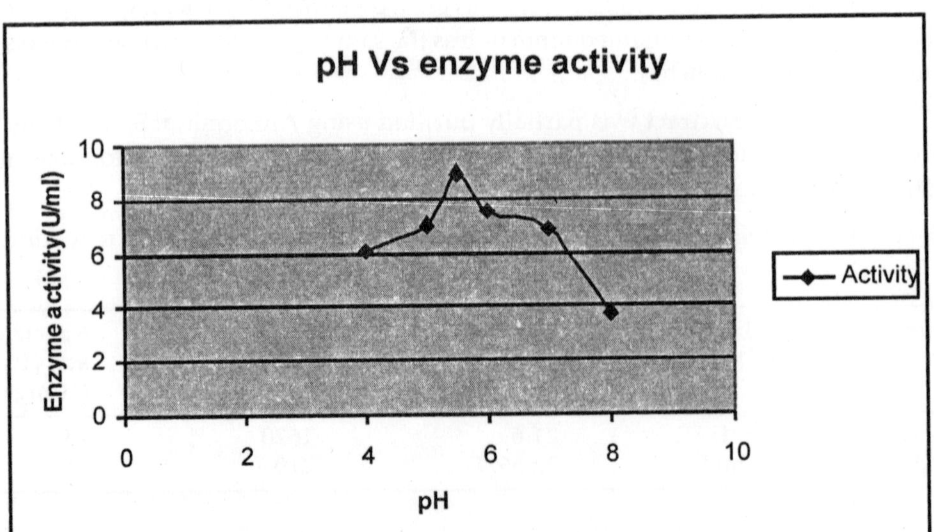

Graph No.2: Effect of pH on xylanase activity.

Table No.5: Decrease in optical density after addition of partially purified xylanase.

Sr. No.	O.D at 540 nm
Before addition of xylanase	2.533
3hrs after addition of xylanase	1.768
6hrs after addition of xylanase	1.102

4. DISCUSSION

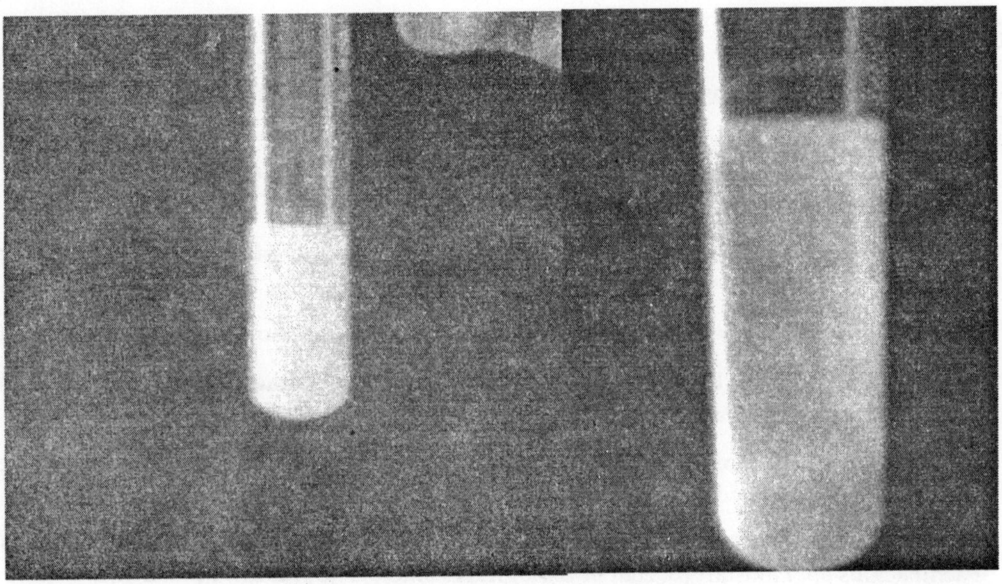

Fig. 2: Orange juice clarification after 6hrs after addition of partially purified xylanase.

Acidic xylanases are considered to have good potential for application in the juice clarification in food industry. This is because the use of such enzymes is expected to greatly reduce the need for pH and temperature readjustments before enzyme addition. Various acid tolerant and thermophilic xylanase producing yeasts and fungi are also reported such as *Streptomyces species S38* by Frederic De Lemos Esteves in 2004, *Penicillium janthinellum* by Luciana A. Oliveira in 2000, *Trichoderma reesei* and *Thermomyces lanuginosus* by Hairong Xiong in 2004, *Aspergillus niger USM AI 1* by Pang Pei Kheng in 2005 shown that it actively degrade xylan [15, 16, 17,18].

Yin Li et al. isolated an acidic *Penicillium sp ZH-30* produced xylanases having an optimum pH of 5.0-6.0 and temperature 50^{0}C [19]. It has been reported that xylanase purified from *Aspergillus ochraceus NG-13* had a specific activity of 0.501 U/mg with a purification fold of 3.0

[20]. Srinivasan and Rele have reported that fungal xylanases had optimum pH of 6.0 to 8.0, temperature of 50 to 60°C [21]. As compared to these observations, the xylanases purified from the culture filtrates of the test fungi in the present study were observed to have higher specific activity. The present work has established the potential of the isolated thermophilic *Rhizopus sp* for xylanase production in submerged fermentation. High-level xylanase production by this strain in submerged fermentation suggests that *Rhizopus species* is an interesting source of new thermostable xylanase and will have important economic advantages. It has been reported that crude and partially purified xylanase from *Rhizopus species* had a specific activity of 75 and 141 U/mg with a purification fold of 2.0. The stability of xylanase at temperature 50-55⁰C and pH 5.5 could be of considerable commercial interest for its potential applications. Further studies on purification and characterization of the xylanase from *Rhizopus species* are in progress.

References

1. Timell T. E., Recent progress in the chemistry of wood hemicelluloses. Wood Sci Technol, (1967), Pp 45-70.
2. Rakshit S. K., Developments in industrially important thermostable enzymes:a review. Bioresource Technology, 89 (2003), Pp 17-34.
3. Lachke A., Biofuel from D xylose- the second most abundant sugar. Resonance, (2002), Pp 50-58.
4. George Sudeep P., Molecular and biochemical aspects of extremophillic actinomycete (2001).
5. Roy N. and A. T. M S. Uddin, Screening, purification and characterization of xylanase from Paenibacillus species. Pakistan journal of biological sciences, 7(3) (2004), Pp 372-379.
6. Rifaat H. M., Production of xylanases by Streptomyces species and their bleaching effect on rice straw pulp, Applied Ecology and environmental research, 4(1): (2005), Pp 151-160.
7. Roy N. and Uddin S. A. T. M., Screening, purification and characterization of Xylanase from Paenibacillus sp. Pakistan Journal of Biological Sciences, 7 (3) (2004), Pp 372-379.
8. Kulkarni, N. and Rao M., Application of xylanase from alkaliphilic thermophilic Bacillus sp. NCIM 59 in bio-bleaching of bagasse pulp. J. Biotechnol, 51 (1996), Pp 167-173.
9. Anuradha, P. K., Vijayalakshmi, N. D. Prasanna and Sridev K, Current science, vol. 92, (9) (2007), Pp1283-1286.
10. Ten L. N., Im W. T., Kim M. K., Kanga M. S., and Lee S. T., Development of a plate technique for screening of polysaccharide-degrading microorganisms by using a mixture of insoluble chromogenic substrates. Journal of Microbiological Methods, 56 (2003), Pp 375 - 382.

20

Molecular diversity of RelA enzyme in marine bacteria- A Bioinformatics analysis

AMBILY NATH I.V*, TRESA REMYA A.T, AKSHATA ALORNEKAR AND NEENA SUSAN VARGHESE

Introduction

The natural environments of marine bacteria are characterized by stresses, such as fluctuations in nutritional availability, The pH balance salin ity, osmolarity, oxygen tension, temperature and inhibitory agents such as drugs, all of which can affect normal bacterial growth. These offer selective pressure to a bacterium, eliciting various adaptive responses for its survival. The adaptive response to nutritional stress encompassing rapid and complex cellular adjustments is called the "stringent response", which has remained the subject of active interest over many years due to its role in growth and the control of gene expression in organisms (Cashel *et al.*, 1996). In such a state, the bacteria adopt a potentially viable but non-culturable (VBNC) state, in which they no longer grow on conventional media. Entry into the VBNC state seems to be more protective as VBNC cells have also been demonstrated to have increased resistance to other stresses such as oxidative damage, mechanical disruption and heat shock.

Cells entering the VBNC state often exhibit dwarfing and during this period a number of major metabolic changes occur including reductions in nutrient transport, respiration rates and macromolecular synthesis (Porter *et al.*, 1995; Oliver, 2000a). In this state, cells are reduced in size, become ovoid in contrast to starved cells. They cannot grow at all on standard laboratory media (Grimes *et al.*, 1986). ATP levels, which decline rapidly in dead and moribund cells, have been found to remain high in VBNC cells (Beumer *et al.*, 1992; Federighi *et al.*, 1998). Further, recent studies have demonstrated continuous gene expression by cells in the VBNC state (Lleo *et al.*, 2000, 2001; Yaron and Matthews, 2002). Extensive modifications in cytoplasmic membrane fatty acid composition that appear to be essential for entry into this state have been described (Day and Oliver, 2004). These are likely necessary for the continuous membrane potential which has been reported (Porter *et al.*, 1995; Tholozan *et al.*, 1999). VBNC cells were also found to have an autolytic capability far higher than that of exponentially growing cells.

The Adoption of this VBNC state is analogous to sporulation in Gram positive bacteria, as it affords protection to bacteria under nutrient stress. The extensive studies conducted on the marine bacterium, *Vibrio vulnificus* shows that the VBNC cells can regain culturability (resuscitation).

It is now abundantly evident that numerous bacteria, both gram-positive and negative, pathogens and nonpathogens are capable of entering into the VBNC state. The VBNC state was first noted with marine *Vibrios*, which were very difficult to culture from marine waters during winter months. VBNC cells remain potentially pathogenic. It has been observed that under starvation, bacteria synthesize Guanosine-3'-diphosphate-5'-diphosphate (ppGpp). The protein responsible for the production of ppGpp is RelA ("relaxed"). RelA is the crucial protein related to the unculturability in bacteria.

The number of species described to enter the VBNC state constantly increases, with approximately 60 now reported to demonstrate this physiological response. These include a number of pathogens *Campylobacter* spp., *E. coli* (including EHEC strains), *Francisella tularensis, Helicobacter pylori, Legionella pneumophila, Listeria monocytogenes, Mycobacterium tuberculosis, Pseudomonas aeruginosa*, several *Salmonella* and *Shigella* spp., and *Vibrio spp.*. *Vibrio vulnificus* has been the most studied as regards VBNC state.

VBNC state and pathogenecity

It has been shown that viable, but non-culturable cells remain potentially pathogenic (Colwell *et al.*, 1985). The VBNC state has important implications for the detection of pathogenic bacteria and monitoring genetically engineered microorganisms. A number of studies conducted on organisms shown that, VBNC cells retain the capacity to establish infections in individuals who are in contact or consume VBNC cells.

It has been observed that the coral pathogen, *Vibrio shiloi* appear to cause coral death while remaining fully nonculturable (ref. Banin *et al.*, 2000; Rosenberg and Ben-Haim, 2002). The key question is, "can cells in the VBNC state resuscitate and initiate infection?" While there is conflicting literature on this point, the answer is clearly 'yes', at least for several pathogens. Studies conducted on *Vibrio cholerae, Vibrio vulnificus* showed that cells in the VBNC state can cause infections. Oliver (2005) has recently conducted an extensive review on the potential public health hazards of VBNC cells present in foods. A few relevant examples are Jones *et al.* (1991), who reported two of four strains of *Campylobacter jejuni* made the VBNC in water cause death in suckling mice. The ability of the *Enterococcus faecalis* cells in the VBNC state to adhere to the cultured heart and urinary tract epithelial cells was shown by Pruzzo *et al.* (2002). Makino *et al.* (2000) studying an *E. coli* O157:H7 outbreak in Japan in 1998, found that culturable cells appeared to have been responsible for this outbreak, but the number was considered to be too low for infection. However, they provided evidence that a larger number of cells likely existed in the VBNC state were the source of the outbreak.

In this context, VBNC has important public health implications, where traditional culture-based methods of microbial assessment may be inadequate to detect VBNC in organisms.

ppGpp and RelA enzyme

One of the factors that provide the bacterium with an ability to survive under hostile conditions is the synthesis of a signal molecule Guanosine-3'-diphosphate-5'-diphosphate (ppGpp) (Cashel, 1975; Chatterji and Ojha, 2001). In 1969, Cashel and Gallant first discovered that bacteria accumulate these molecules during nutrient starvation. The stringent control response which involves a rapid accumulation of ppGpp is triggered when the marine *Vibrio sp.* Strain S14 is subjected to carbon and energy starvation (Jorgen Ostling, Louise Holmquist, Staffan Kjelleberg, 1996). Cells make ppGpp when amino acid levels are low and ppGpp gives a signal to a cell to become dormant until amino acid levels return to normal. The increased level of ppGpp restricts the biosynthetic pathways at the transcriptional level while the amino acids remain unchanged.

ppGpp is produced by the enzyme RelA ("relaxed"). This ribosome-associated Guanosine-3'-diphosphate-5'-diphosphate (ppGpp) synthetase/GTP pyrophosphokinase, belonging to the category spoT/relA family, is the crucial protein related to the unculturability of bacteria. The synthesis of ppGpp involves the transfer of pyrophosphate from ATP to GDP/GTP by RelA enzyme. The mutation of *relA* gene in *V. cholerae* revealed that (p)ppGpp synthesized by RelA plays a significant role both in vitro and in vivo in the regulation of the expression of two principal virulence factors such as cholera toxin (CT) and toxin-coregulated pilus (TCP) (Shruti Haralalka, Suvobroto Nandi and Rupak K. Bhadra, 2003).

RelA enzyme-Pathogenicity-Unculturability

It has been experimentally proved that there is a correlation between the VBNC with unculturability and pathogenicity. Nowadays, the greatest problem microbiologists are facing is the unculturability of certain potential bacterial strains. The problem is related to public health if

they are pathogenic too. A hypothetical approach of either inhibiting or blocking RelA would provide important insights to enhance the culturability of bacteria showing VBNC response.

Molecular Diversity of RelA in marine bacteria

Since the information related to the molecular characteristics of RelA enzyme from marine bacteria is scarce, the available structural details of the enzyme reported from different genera are described here.

The RelA protein from *Escherichia coli* and *Vibrio cholerae* contains two functionally independent domains such as an N-terminal activity domain and a C-terminal regulatory domain. In *E.coli* the relA gene encodes a protein of 744 amino acids of molecular mass of 84kDa. RelA can be dissected both functionally and physically into two domains. The N-terminal domain (NTD) (amino acids [aa] 1 to 455) contains the catalytic site of RelA. The C-terminal domain (CTD) (aa 455 to 744), which is devoid of synthetic activity is involved in regulating RelA activity. Mutational analysis has shown that residues G251 and H354 in the highly conserved NTD are essential for enzyme activity. The extremophile *Thermus thermophilus* HB8 possess a RelA/spot homologue (RSH_{Tt}) encoding a 727-amino acid protein. Based on a DBGET (Integrated Database Retrieval System) search, RSH_{Tt} was found to encode a HD domain (amino acids 40 to 151), which is characteristic of metal-dependent phosphohydrolases, a conserved RelA/SpoT region (248 to 371), a TGS (for threonyl-tRNA synthetase, GTPase, and SpoT) domain (409 to 472), which is predicted to possess a predominant â-sheet structure, and an aspartokinase domain (656 to 726), which binds specifically to a particular amino acid, leading to the regulation of the linked enzyme. The presence of TGS domain suggests a ligand (most likely nucleotide)-binding regulatory role. In *Vibrio cholerae*, a RelA homolog ($relA_{VCH}$) was cloned and sequenced. The $RelA_{VCH}$ gene encodes a 738-amino-acid protein shows functions similar to those of other gram-negative bacteria including *E.coli*. The cloning of the RelA gene in a marine *Vibrio* sp. strain S14 contains a 743-codon open reading frame that encodes a polypeptide that is identical in length and highly homologous to the *E. coli* RelA protein. The amino acid sequences are 64% identical, and they share some completely conserved regions.

Importance of computer-based sequence analysis

Since the X-ray crystallographic structures of the RelA enzymes are not available, an amino acid sequence can be used to predict the functional aspects of these proteins. The polar side chain of amino acids involved in the transfer of phosphoryl groups from ATP to form ppGpp makes them the key molecules of further studies. Even though it is difficult to directly elucidate the structural characteristics of the active site from amino acid sequence alone, the present study aims to give some hints into the characterization of active sites of the enzyme by annotating the physicochemical properties of amino acids in the conserved region. Since the VBNC state is related to the unculturability and often in the pathogenicity of bacteria, the molecular characterization of the RelA enzyme is important to understand its mechanism of action. Hence, a computational approach has been implemented to analyze the RelA sequence of 4 marine bacterial strains which shows a prominent VBNC response.

Approach

The RelA enzyme sequences of 4 marine bacterial strains showing VBNC response have been accessed from NCBI. Among them 3 are pathogenic such as *Escherichia coli* O157:H7 str. Sakai (diarrhoea), *Vibrio vulnificus* (wound infections and primary septicemia), *Campylobacter jejuni* (enteritis) and *Streptomyces coelicolor* A3 (2). The protein sequences have been accessed from NCBI.

The entry details are given below:

1. *Escherichia coli* O157:H7 str. Sakai (Accession No: BAB37067)
2. *Vibrio vulnificus* YJ016 (Accession No: BAC95585)
3. *Streptomyces coelicolor* A3 (2) (Accession No: CAB70915)
4. *Campylobacter jejuni subsp. doylei* 269.97 (Accession No: ABS44393)

Multiple sequence alignment (MSA) was performed with ClustalW2 software.

Results and Discussion

The MSA shows 2 conserved regions, Regions (1) and (2). In both regions, 100 amino acid (a.a) length of each sequence is given. The region, alignment and species details are given below:

Region	Alignment range	Region	Alignment range	Species
(1)	252-351	(2)	389-486	*E.coli*
	250-350		386-483	*V.vulnificus*
	338-437		488-583	*S.coelicolor*
	256-354		392-485	*C.jejuni*

Region-(1)

```
|BAB37067.1|  252 RPKHIYSIWRKMQKKNLAFDELFDVRAVRIVAERLQDCYAALGIVHTHYR 301
|BAC95585.1|  251 RPKHIYSIWRKMQKKSLEFDELFDVRAVRIIADKLQDCYAALGVVHTKYK 300
|CAB70915.1|  338 RPKHYYSVYQKMIVRGRDFAEIYDLVGIRVLVDTVRDCYAALGTVHARWN 387
|ABS44393.1|  256 RIKHSYSIYLKMQRKGIGIEEVLDLLGVRILVEKVSDCYLALGILHTHFN 305

|BAB37067.1|  302 HLPDEFDDYVANPKPNGYQSIHTVVLGPGGKTVEIQIRTKQMHEDAELGV 351
|BAC95585.1|  301 HLPSEFDDYVANPKPNGYQSIHTVVLGPEGKTIEIQIRTKQMHEDSELGV 350
|CAB70915.1|  388 PVPGRFKDYIAMPKFNMYQSLHTTVIGPGGKPVELQIRTFDMHRRAEYGI 437
|ABS44393.1|  306 PLVSRFKDYIALPKQNGYQTIHTTLFDAKS-IIEAQIRTFDMHKIAEFGI 354
```

Region-(2)

```
|BAB37067.1|  389 DSGEMLDEVRSQVFDDRVYVFTPKGDVVDLPAGSTPLDFAYHIHSDVGHR 438
|BAC95585.1|  386 DSGEMLDELRSQVFDDRVYAFTPRGDVVDLPMGATPLDFAYHIHSEVGHR 435
|CAB70915.1|  488 DPGEFLESLRFDLSRNEVFVFTPKGDVIALPAGATPVDFAYAVHTEVGHR 537
|ABS44393.1|  392 NAIELYEYAKDSLYVEDVAVYSPKGEIFTLPRGATVLDFAYEVHTKVGLH 441

|BAB37067.1|  439 CIGAKIGGRIVPFTYQLQMGDQIEIITQKQP-NPSRDWLNPNLGYVTTS 486
|BAC95585.1|  436 CIGAKVAGRIVPFTHKLNMGDQVEIITQKEP-NPSRDWLNPSLGFVTSG 483
|CAB70915.1|  538 TIGARVNGRLVPLESTLDNGDLVEVFTSKAAGAGPSRDWL--GFVKSP 583
|ABS44393.1|  442 AKSAYVNRIKVPLLTELKNGDIVRVVTSNDK-FYRCSWIDS--VKTG 485
```

('*' indicates positions which have a fully conserved residue, ':' indicates the conservation of strong groups of amino acids, '.' Indicates the conservation of weaker group of amino acids)

In the first region, conservation of 74- residues have been identified. Among them 37 are identical (*), with high degree of conservation. Substitution of amino acids belongs to the strongest group of amino acids can be seen in 25 different positions and are denoted by ':'. This indicates residues with very conservative substitutions where one residue is replaced by another residue having the same physicochemical property. For example, one neutral non-polar group is replaced by another neutral non-polar residue, but not identical. The remaining 12 '.'residues indicates conservation of weaker groups with higher degree of substitution. In the 59 a.a conserved second region, 24 are identical, 23 underwent conservative substitutions, 12 show weaker conservation.

Alignments of amino acid sequences are critically important to elucidate the positions and physicochemical properties of the amino acids characterizing the highly conserved regions. Since these properties of amino acids that governs biologically important reactions such as protein-ligand and protein-protein interactions, they are expected to be applicable to predict the molecular mechanism of those reactions involving proteins.

From the ClustalW2 results, an interpretation of the physico-chemical characteristics of amino acids in the highly conserved regions are given:

Most of the residues are polar: They are Arginine (R), Lysine (K), Aspartate (D), Glutamate (E), Asparagine (N), Histidine (H) and Cysteine (C). Moreover, the H residues form the important active site contributors. Cysteine residues play a crucial role in shaping the secondary structure by S-S bonds.

Conclusion

From a Pharmacoinformatics point of view, the study of RelA enzyme is significant. Since the activation of this enzyme produces ppGpp, which is responsible for the crucial biochemical change noticed during the VBNC state, the molecular characterization of conserved catalytic residues would be helpful in predicting the active site of the enzyme. The solubility of a compound depends on the presence of polar residues on the active site of the compound, so that it could interact with other compounds by making or breaking of chemical bonds. If we could locate the active site of RelA enzyme based on the position and properties of the polar residues we would be able to design a potential inhibitor which can lead to new attempts to block the activity of the enzyme thereby avoiding the entry in to VBNC state. This would open a new insight in improving the culturability of bacteria undergoing VBNC state.

References

1. Beumer, R.R., J. de Vries, and F.M. Rombouts (1992). *Campylobacter jejuni* non-culturable cocooid cells. *International Journal of Food Microbiology*. 15, 153-163

2. Chowdhury M.A.R., Ravel, J., Hill, R.T., Huq, A and R.R. Colwell. Physiology and molecular genetics of viable but non-culturable microorganisms.

3. Federighi, M., J.L. Tholozan, J.M. Cappelier, J.P.Tissier, and J.L.Jouve (1998). Evidence of non-coccoid viable but non-culturable Campylobacter jejuni cells in microcosm water by direct viable count,CTC-DAPI double staining, and scanning electron microscopy. *Food Microbiology.* 15,539-550.

4. Flardh K., Axberg T., Albertson, N.H and S. Kjelleberg (1994). Stringent control during carbon starvation of marine *Vibrio* sp. strain S14: molecular cloning, nucleotide sequence, and deletion of the relA gene. *Journal of Bacteriology.* 176 (19), 5949-5957.

5. Gerhard Klebe (2000). Recent developments in structure-based drug design. *Journal of Molecular Medicine.* 78, 269-281 .

6. Gropp Michal, Strausz Yael, Gross Miriam and Gad Glaser (2001). Regulation of Escherichia coli RelA requires oligomerization of the C-terminal domain. *Journal of Bacteriology.* 183 (2), 570-579.

7. Hernandez J., Favos A., Alonso JL and Owen RJ (1996). Ribotypes and AP-PCR fingerprints of thermophilic campylobacters from marine recreational waters. *Journal of Applied Bacteriology.* 80 (2), 157-64.

8. Inaoka Takashi and Kozo Ochi (2002). RelA protein is involved in induction of genetic competence in certain *Bacillus subtilis* strains by moderating the level of intracellular GTP. *Journal of Bacteriology.* 184 (14), 3923-3930.

9. James D. Oliver (2005). The viable but nonculturable state in bacteria. The *Journal of Microbiology.* 43 (S), 93-100.

10. Jain Vikas, Kumar Manish and Dipankar Chatterji (2006). ppGpp: Stringent response and survival. *The Journal of Microbiology.* 44 (10), 1-10.

11. Jorgen Ostling (1997). Global analysis of physiological responses in marine bacteria. *Electrophoresis.* 18, 1441-1450.

12. Jones, K. Melissa, Warner Elizabeth and James D. Oliver (2008). Survival of and in situ gene expression by *Vibrio vulnificus* at varying salinities in estuarine environments. *Applied and Environmental Microbiology.* 74 (1), 182-187.

13. Joux Fabien, Lebaron Philippe and Marc Troussellier (1997). Changes in cellular states of the marine bacterium *Deleya aquamarina* under starvation conditions. *Applied and Environmental Microbiology.* 63 (7), 2686-2694.

14. Miyagi, K., Omura, K., Ogawa, A., Hanafusa, M., Nakano, Y., Morimatsu, S and K. Sano (2001). Survival of Shiga toxin-producing *Escherichia coli* O157 in marine water and frequent detection of the Shiga toxin gene in marine water samples from an estuary port. *Epidemiology and Infection.* 126, 129-133.

15. Nascimento M. Marcelle, Lemos A. Jose, Abranches Jacqueline, Lin K. Vanessa and Robert A. Burne (2008). Role of RelA of *Streptococcus mutans* in global control of gene expression. *Journal of Bacteriology.* 190 (1), 28-36.

16. Nonculturable microorganisms in the environment. (Colwell R. Rita and D. Jay Grimes, Eds.).

21

Relationships between yield components of advanced potato cultivars in irrigation periods by the use of Path analysis and Factor analysis

DAVOUD HASSANPANAH

Introduction

The Potato (Solanum tuberosum L.) is one of the most important crops in Iran and is cultivated on 195,000 ha giving a production of 4,200,000 tonnes. (FAO, 2005). A high yield with good quality is the most important objective in potato breeding. Tuber yield is a complex character associated with many interrelated components. Generally, a path coefficient analysis is needed to clarify relationships between characteristics, because the correlation coefficients describe relationships in a simple manner. The path coefficient analysis shows the extent of direct and indirect effects of the causal components on the response components. In most of the studies involving path analysis, the researchers have considered the predictor characters as the first order variables in order to analyse their effects over a dependent or response variable such as yield (Maity and Chattarzee 1977; Gunel *et al.* 1991; Gopal *et al.* 1994; Yildirim *et al.* 1997; Bhagowati and Saikia 2003; Tuncturk and ÇiftÇi 2005). This approach might result in the multicollinearity for variables, particularly when the correlations among some of the characters are high (Samonte *et al.*, 1998). There may also be difficulties in the interpretation of the actual contribution of each variable, as the effects are mixed or confounded because of collinearity. Samonte *et al.* (1998) adopted a sequential path analysis for determining the relationships between yield and related characters in rice (*Oryza sativa L.*) by organizing and analyzing various predictor variables in the first, second and third order paths. Agrama (1996) and Mohammadi *et al.* (2003) used this model for determining the interrelationships among grain yield and related characters in maize. We used this method to determine the interrelationships among the tuber yield and related characters in potato clones for developing selection criteria for the tuber yield using a sequential path model.

Materials and Methods

This experiment was done by the use of three irrigation treatments (irrigation after 25, 35 and 50 percent discharge of available water) for seven varieties (Satina, Sante, Agria, Marfona, Kennebec, Ceaser and Savalan) in a three replicated split plot design. Every plot was to consist of five lines, and its dimensions were six meters in length and three meters in width. The planting spaces were 75×25 cm in experimental plots. Distances between the plots were 1.5 meters. Water treatments were from irrigation facilities after 25, 35 and 50 percent discharge of the available water. The characters that were measured in the growth period and after harvesting were mainly the stem number, plant height, tuber number, tuber weight per plant, total tuber yield, marketable tuber yield, dry matter percent, marketable tuber number and marketable tuber weight per plant. The analysis of variance was done for noted traits and means compared on the basis of an LSD (Least significant difference) test. Linear correlation coefficients between various pairs of characters were computed. Path and Factor analysis was done for the main stem number, plant height, tuber number, tuber weight in plant, total tuber yield, marketable tuber yield, dry matter percent, marketable tuber number and marketable tuber weight per plant as well as the number and average weight of a mini-tuber per plantlet by Software SPSS.

Results

The results of the analysis of variance showed a significant differences between the effect of irrigation treatments for the number and weight of tubers less than 35mm, between 35-55mm and bigger than 55mm in size, tuber yield, marketable tuber yield, main stem number, tuber number and the tuber weight in plant.

The highest total and marketable tuber yield, the number and weight of tubers per plant, the tuber weight between 35-55mm and bigger than 55mm were produced in irrigation after 25% discharge of available water (Table 1).

Savalan and Ceaser had the highest tuber yield with 40.19 and 39.43 ton/ha, respectively. The highest total and marketable tuber yield, tuber number and weight per plant and tuber number and tuber weight between 35-55mm in each of the three irrigations belonged to the Savalan, Ceaser and Kennebec. Ceaser had the lowest and Sante the highest tuber weight <35mm (Table2).

There was a significant and positive correlation between the total tuber yield with the marketable tuber yield, tuber weight per plant, tuber number and tuber weight bigger than 55mm and the tuber number between 35-55mm (Table 3).

The results of the factor analysis showed that the three main and independent factors accounted for a total variance respect with 91.00, 88.80 and 95.20 percent in conditions of irrigation after 25, 35 and 50% discharge of available water (DAW). The second factor in irrigation after a 25% DAW and the first factor in irrigation after a 35 and 50% DAW named to address of the tuber yield and yield components. This factor was arising from importance of

Table 1. Mean of different traits in irrigation treatments

Irrigation treatments	Tuber Yield (ton/ ha)	Marketable Yield (ton/ ha)	Plant Height (cm)	Number of Main Stem	Tuber Weight per plant (g)	Tuber Number Per Plant	Tuber Number per Plant			Tuber Weight per Plant (g)		
							< 35 mm	35-55 mm	>55 mm	< 35 mm	35-55 mm	>55 mm
Irrigation after 25 percent discharge of available water	39.24 a	36.69 a	43.11 a	4.74 a	729.9 a	8.59 a	1.97 a	5.53 a	1.08 a	47.44 a	430.3 a	252.2 a
Irrigation after 35 percent discharge of available water	35.12 ab	35.12 ab	43.63 a	4.85 a	653.2 ab	7.63 ab	1.59 b	4.95 a	1.09 a	37.97 b	385.8 a	229.4 a
Irrigation after 50 percent discharge of available water	30.02 b	27.02 b	41.37 a	4.64 a	558.4 b	7.47 b	1.69 b	4.99 a	0.80a	38.15 b	353.3 a	167.0 b

Table 2. Mean of different traits in potato cultivars

Irrigation treatments	Tuber Yield (ton/ ha)	Marketable Yield (ton/ ha)	Plant Height (cm)	Number of Main Stem	Tuber Weight per plant (g)	Tuber Number Per Plant	Tuber Number per Plant			Tuber Weight per Plant (g)		
							< 35 mm	35-55 mm	>55 mm	< 35 mm	35-55 mm	>55 mm
Satina	35.97 ab	33.69 bc	45.69 bc	5.56 a	669.2 ab	7.9 b	1.69 bc	5.0 b	1.17 a	42.55 bc	382.6 bc	244.0 bc
Savalan	40.19 a	37.62 ab	37.69 de	4.92 ab	747.6 a	9.7 a	2.03 b	6.7 a	0.91 b	47.92 b	500.5 a	199.2 cd
Agria	32.72 bc	30.89 c	40.97 cd	5.22 ab	608.6 bc	7.5 bc	1.13 de	5.4 b	0.89 b	34.10 cd	405.5 bc	169.0 d
Ceaser	39.43 a	38.00 a	49.97 ab	4.36 bc	733.4 a	7.5 bc	0.99 e	5.2 b	1.25 a	26.48 d	432.7 b	274.1 ab
Marfona	27.12 d	25.08 d	34.67 e	3.75 c	504.4 d	6.5 c	1.56 bcd	4.2 c	0.75 bc	37.94 bcd	295.8 d	170.7 d
Kennebec	37.80 a	35.97 ab	34.06 e	4.03 c	703.2 a	6.4 c	1.22 cde	3.9 c	1.33 a	34.05 cd	351.5 cd	317.7 a
Sante	30.31 cd	26.80 d	55.86 a	5.36 a	563.7 cd	9.8 a	3.65 a	5.6 b	0.61 c	65.25 a	359.9 cd	138.6 d

Table 3. Linear correlation coefficients between different traits in potato cultivars at Irrigation treatments

Correlation Coefficient	Tuber Yield	Marketable Yield	Tuber Number	Tuber Weight	Number of Main Stem	Plant Weight	Tuber Number			Tuber Weight		
							<35 mm	35-55 mm	>55 mm	<35 mm	35-55 mm	>55 mm
Tuber Yield	-											
Marketable Yield	0.98**	-										
Tuber Number	0.33	0.24	-									
Tuber Weight	0.99**	0.99**	0.33	-								
Number of Main Stem	0.08	0.04	0.49*	0.08	-							
Plant Height	0.02	-0.02	0.57*	0.02	0.50*	-						
<35 mm	-0.15	0.81**	0.77**	-0.15	0.41	0.50*	-					
35-55 mm	0.40	0.80**	0.86**	0.40	0.49*	0.31	-0.42	-				
>55 mm	0.72**	-0.09	-0.27	0.72**	-0.07	-0.13	-0.25	0.01	-			
<35 mm	0.03	0.76**	0.82**	0.03	0.40	0.39	0.94**	0.82**	0.93**	-		
35-55 mm	0.82**	0.34	0.62**	0.82**	0.31	0.14	-0.32	0.28	0.50*	-0.51*	-	
>55 mm	0.77**	-0.25	-0.23	0.77**	-0.25	-.017	0.29	0.19	-0.43*	-0.25	0.40	-

Table 4. Path analysis among various characters contributing to tuber yield in potato cultivars at Irrigation treatments

Characters		Indirect Effects					Total
	Direct effect	Tuber Weight between 35-55 mm	Tuber Weight bigger than 55	Tuber Number per plant	Tuber weight per plant	Marke-ting Yield	Correlation coefficients
Tuber Weight between 35-55mm	0.86	-	-0.31	-1.17	0.51	1.60	0.72
Tuber Weight bigger than 55mm	-1.11	0.03	-	-0.36	0.58	1.68	0.82
Tuber Number per plant	-1.24	0.08	-0.32	-	0.55	1.71	0.77
Tuber Weight per plant	0.71	0.06	-0.19	-0.96	-	2.09	0.99
Marketable Yield	2.11	0.07	-0.89	-1.01	0.70	-	0.98

Residual Effects = 0.195

Table 5. Path analysis among various characters contributing to tuber yield in potato cultivars at Irrigation after 25 percent discharge of available water

Characters		Indirect Effects					Total
	Direct effect	Tuber Weight between 35-55 mm	Tuber Weight bigger than 55	Tuber Number per plant	Tuber weight per plant	Marke-ting Yield	Correlation coefficients
Tuber Weight between 35-55mm	-0.70	-	-0.15	0.16	-0.32	1.79	0.82
Tuber Weight bigger than 55mm	-0.54	-0.15	-	-0.06	-0.28	1.75	0.72
Tuber Number per plant	0.26	-0.44	0.12	-	-0.15	0.59	0.38
Tuber Weight per plant	-0.39	-0.57	-0.39	0.09	-	-2.25	0.99
Marketable Yield	2.27	-0.55	-0.42	0.07	-0.39	-	0.98

Residual Effects = 0.169

Table 6. Path analysis among various characters contributing to tuber yield in potato cultivars at Irrigation after 35 percent discharge of available water

Characters	Direct effect	Indirect Effects					Total Correlation coefficients
		Tuber Weight between 35-55 mm	Tuber Weight bigger than 55	Tuber Number per plant	Tuber weight per plant	Marke-ting Yield	
Tuber Weight between 35-55mm	-0.07	-	0.001	0.08	0.15	0.62	0.78
Tuber Weight bigger than 55mm	0.02	-0.001	-	-0.06	0.08	0.58	0.62
Tuber Number per plant	0.13	-0.04	-0.008	-	0.05	0.15	0.28
Tuber Weight per plant	0.19	0.06	0.007	0.04	-	0.81	0.99
Marketable Yield	0.82	-0.05	0.012	0.023	0.19	-	0.99

Residual Effects = 0.09

Table 7. Path analysis among various characters contributing to tuber yield in potato cultivars at Irrigation after 50 percent discharge of available water

Characters	Direct effect	Indirect Effects					Total Correlation coefficients
		Tuber Weight between 35-55 mm	Tuber Weight bigger than 55	Tuber Number per plant	Tuber weight per plant	Marke-ting Yield	
Residual Effects = 0.169							
Tuber Weight between 35-55mm	0:60	-	0.04	0.05	-0.07	0.09	0.71
Tuber Weight bigger than 55mm	0.69	0.03	-	-0.001	-0.073	0.11	0.73
Tuber Number per plant	0.13	0.24	-0.59	-	0.023	-0.42	-0.24
Tuber Weight per plant	-0.1	0.43	0.56	-0.03	-	0.13	0.99
Marketable Yield	0.13	0.42	0.59	-0.04	-0.099	-	0.99

Residual Effects =0.094

weight of tuber per plant and tuber yield. The tuber yield was the most important index of selection and its improvement is possible by the way of increasing the weight of each tuber plant. Factor analysis produces only a pattern of effects in correlation to a matrix of characters. However if the above factors be valid (they are dependable to physiological origins and destinations), it seems that breeders can select potato plants with a higher tuber number and weight per plant and longer height.

The path analysis showed that a marketable tuber yield, with the weight of tubers between 35-55mm weight of tubers per plant and tuber weight per plant had positive direct effects on the tuber yield and negative by weight of tubers bigger than 55mm and tuber number per plant (Table 4). A marketable tuber yield had positive direct effects on tuber yield in irrigation after 25 and 35 % discharge of available water (Table 5 and 6), so we observed that the weight of tubers 35-55mm and bigger than 55mm had a positive direct effect on the tuber yield (Table 7).

Discussion

Mean comparisons of attributes among cultivars showed that the yield of Savalan, Ceaser and Kennebec was higher than others in stress and non-stress conditions.

The rate of decreasing in a yield for the cultivars was different. In mild stress as compared with sever stress conditions, Savalan, Ceaser and Kennebec showed less decreasing in a yield and was a drought tolerant cultivar. These cultivars can produce a higher tuber weight per plant in all of the conditions. Yield decrease in mild and severe drought conditions to control was 4.12 and 9.22 ton/ha, respectively.

We observed some susceptible attributes in the weight and number of tubers between 35-55mm and bigger than 55mm. We concluded that, the drought decreased the total tuber yield and marketable tubers. Marfona had the maximum decreasing in both the mild and severe drought stress.

The most affected attributes were marketable tuber yield, tuber weight per plant, number and weight of tubers bigger than 55mm and the tuber number between 35-55mm, but yield had no correlation with the tuber number in plants smaller than 35mm and between 35-55mm, plant height or the main stem number and tuber weight smaller than 35mm.

Conclusion

Savalan and Ceaser had the highest tuber yield. Marketable tuber yield, weight of tubers between 35-55mm, weight of tubers per plant and tuber weight per plant had positive direct effects on the tuber yield and a negative effect by weight of the tubers bigger than 55mm and tuber numbers per plant.

References

1. Agrama, H.A.S. 1996. Sequential path analysis of grain yield and its components in maize. Plant Breed. 115:343-346.

2. Bhagowati, R.R. and Saikia, M. 2003. Character association and path coefficient analysis for yield attributes in open pollinated and hybrid true potato seed populations. Crop Res. 26(2):286–290.

3. Birhman, R.K. and Kang, G.S. 1993. Analysis of variation and interrelationship in potato germplasm. Euphytica. 68:17–26.

4. FAO Statistical Databases. 2005. URL:faostat.fao.org/faostat. Accessed 26 July 2007.

5. Gopal J., Gaur, P.C. and Rana M.S. 1994. Heritability and intra- and inter- generation associations between tuber yield and its components in potato (Solanum tuberosum L.). Plant Breed. 112:80–83.

6. Gunel, E., Oral E. and Karadogan, T. 1991. Relationships between some agronomic and technologic characters in potatoes (In Turkish). J. Ataturk Univ. 22:46–53.

7. Hair, J.R., Anderson, R.E., Tatham, R.L., and Black, W.C. 1995. Multivariate Data Analysis with Readings. Prentice Hall, Englewood, NJ.

8. Li, C.C. 1975. Path Analysis: A Primer. Boxwood Press, Pacific Grove, CA.

9. Maity, S. and Chattarzee, B.N. 1977. Growth attributes of potato and their inter relationship with yield. Potato Res. 20:337–341.

10. Mohammadi, S.A., Prasanna, B.M. and Singh N.N. 2003. Sequential path model for determining interrelationships among grain yield and related characters in maize. Crop Sci. 43:1690–1697.

11. Samonte, S.O.P.B., Wilson, L.T. and McClung, A.M. 1998. Path analyses of yield and yield-related traits of fifteen diverse rice genotypes. Crop Sci. 38:1130–1136.

12. Tuncturk, M. and ÇiftÇi, V. 2005. Selection criteria for potato Breeding. Asian J. Plant Sci. 4:27–30.

13. Yildirim, M.B., Çali°kan, C.F., Çaylak, Ö. And Budak, N. 1997. Multivariate relationships in potatoes (In Turkish). Second Turkish Field Crops Symposium, 22-25 September. Samsun, Turkey.

22

Structural and Functional Characterization of a 4-Chlorosalicyclic Acid Degradation Microbial Community in the Chemostat for the Treatment of Pulp and Paper Mill

V. VENKAT RAMANAN , SHACHI SHAH AND I.S.TAHKUR

Introduction

The pulp and paper industry is one of India's oldest and core industrial sectors. The paper and pulp industry is classified as a highly water intensive one and it also forms a major polluter of soil and water resources (Udayasoorian and Prabu, 2003). Paper mill effluents pollute water, air and soil, causing a major threat to the environment. Approximately 400 pulp and paper mills, which generate significant quantities of solid and liquid waste, carry significant quantities of fibee, fines, filler and other wet end additives that contribute to total suspended solids (TSS), chemical oxygen demand (COD), and biochemical oxygen demand (BOD) exist in India (Murugesan, 2003).The dark brown colour of pulp & paper mill effluent is due to the lignin and lignin derivatives which offer resistance to degradation due to the presence of carbon to carbon diphyhyl linkages (Dence et al.,1980).

Conventional methods of waste water treatment are rather ineffective for the degradation of waste water, whereas indigenous communities seem to be more promising due to the ability to degrade toxic compounds and their intermediary metabolites (Shah and Thakur, 2002). Bacterial isolates obtained from nature have hot been proved effective in the complete degradation of toxic compound present in pulp and paper mill effluent (Chuphal, *et al.*, 2006). Consequently a microbial consortium is usually required to provide all the metabolic capabilities for complete mineralization.

Most of the synthetic chemicals present in the environment are degraded by the microorganisms present in nature. An indigenous community has the potential for bioremediation of a recalcitrant compound. If an indigenous community is enriched in the presence of intermediary metabolites of toxic compounds, significant strains will be evolved in the process of adaptations (Thakur, 1995). This pragmatic approach of biotreatment is analyzed by way of using chemostat for enrichment and bioreactor for testing the efficacy of treatment. Mixed community can overcome the problem faced by monoculture in the environment such as nutrional limitation and toxic substrate. The microbial growth in the bioreactor ascertains the utilization of compounds in the effluent as carbon and energy source. Therefore the present study aims at the enrichment of microbial community for the treatment of effluent. In the light of these facts, it is important to elucidate such biological interaction in a continuously operated system in order to develop a potential microbial consortium. The Study could help in developing strategies for enhancing degradation and large scale removal of chorolo-lignin compounds from the environment.

Materials and Methods

Sampling Site

The study was conducted on the effluent released from the Century Pulp and Paper Mill Ltd., Ghanshyamdham, Lalkuan, Nainital (Uttaranchal), situated at about 7 km from the G.B.Pant University of Agriculture and Technology, Pantnagar, Uttaranchal, India, on the Nainital-Bareilly Highway. This Industry came into operation in June 1984. The industry uses Eucalyptus wood as a raw material. The effluent was collected from inside the premises near the Rayon Grade Paper Unit Laboratory and stored in a refrigerator at 4^0C.

Chemostat culture and enrichment of Microbial community

Sediment samples together with liquid effluent (1:10 w/v) were collected from three sites of the main channel of the effluent released from the Century Pulp and Paper Mill Ltd., Ghanshyamdham, Lalkuan, Nainital (Uttaranchal) and was enriched in a chemostat culture vessel (2lit, effective size 1litre) containing a minimal salt medium (MSM)With 4-chlorosalicyclic acid (4-CSA, 2.5mM) as a carbon source. The media was prepared in a minimal base essel with the following composition (g/l): Na_2HPO_4, $2H_2O$. 7.8; KH_2PO_4, 6.8; $MgSO_4$, 0.2, ammonium ferric citrate, 0.01; Ca $(NO_3)_2$, $4H_2O$, 0.05; $NaNO_3$, 0.085, pentachlorophenol, 0.1 and trace element solution, 1 ml/l(Thakur 1995).

Structural Analysis of Microbial Community

After a continuous enrichment of the microbial community for six months in a chemostat bacterial cells were removed and cultured on LB agar plates for structural and chemo-taxonomical analysis of the colony. The morphologically distinct isolates were identified on the basis of biochemical properties in accordance with Bergy, manual of systematic Bacteriology (Palleroni, 1984). The isolates were also identified by a commercial micro plate test (Biolog, In Corporated, Hayward, CA) based on the utilization of 95 carbon sources (Thakur, 1995). The tests were repeated five times.

Treatment of effluent

Enriched microbial community from the chemostat was applied for the treatment of pulp and paper mill effluent in a bioreactor for pH, colour unit, lignin, COD and phenol fóllowing APHA (1995).

pH

pH of the sample as determined by pH meter

Chemical Oxygen Demand (COD)

The COD was determined by the dichromate reflux method (ref.7). In this method, the sample was refluxed with potassium dichromate and sulphuric acid, and titrated with ferrous ammonium sulphate.

Colour Estimation

The colour content in the effluent was measured. In this method, sample was centrifuged at 10,000 rpm for 30 min and the balance pH was adjusted to 7.6. Absorbance was measured at 465 nm and transformed into the colour unit (APHA, 1995)

Lignin Content

Lignin in the effluent was estimated by a reaction of the effluent with acetic acid and sodium nitrite and ammonium hydroxide and measuring absorbance at 430 nm (APHA,1995)

Phenols estimation

Total phenol was extracted from the effluent by using dichloromethane and assessed by 4-aminoantipyrine method (APHA,1995).

Results and Discussion

There is a great concern about the deleterious effects of aromatic chlorinated phenols in the natural environment. Mixed bacterial community, originating from the sediment core is contaminated with chlorinated lignins in pulp and paper mall effluent and can degrade toxic compounds (Chuophal *et al.* 2006). During the past decade a broad range of microorganisms has been isolated and characterized by the removal of organic compounds from the environment. The lignocellulosic raw materials and chemicals used during the manufacture of the pulp and paper are major contaminants in the effluent of the paper industry. The chemicals in the effluent can be removed by biotreatment and biodegradation. In the present investigation, microbial community was applied, instead of monoculture, because of the possible interactions in the metabolism of the chemical compounds having diversified structural formulation. Chlorolignins, chlorosulphonic acid, chlororesins, chlorophenols, and cholrocatecol are major toxic compounds present in the effluent of the pulp and paper mill (Somenshekhar *et al*, 1984). Chlorinated dibenzo-p-dioxin and dibenzofurans are also formed unintentionally in pulp and paper mill effluent. 4-chlorosalicylic acid is released during co-metabolic turn over of dibenzo-pdioxin and dibenzofuran (Wittch, *et al*, 1992). There are few reports of microorganisms degrading dibenzo-p-dioxin, dibenzofuran and related polychlorinated phenols. Therefore, mixed microbial community was developed in the presence of 4-chorosalicylic acid in the chemostat and applied for the treatment of effluent.

The enrichment of microbial community was performed in a glass vessel containing a mineral salt medium and 4-chlorosalicylic acid (5mM) as the sole carbon and energy source. The growth of the bacterial cells was determined by taking absorbance at 560nm and biomass, and also the carbon source utilization was determined by High Performance Liquid Chromatography (HPLC). It was observed that initially there was a significant increase in the growth of the cells, but after day 11the turbidity of the culture medium went down. However, there was an increase in the turbidity after day 33, which reached to the maximum on day 41. The data indicates fluctuation in the utilization of 4-chlorosalicylic acid initially, but after day 27 the percent utilization of salicylic acid was increased and it reached to 94 per cent on day 41, by the mixed microbial community. During the initial and subsequent enrichments, bacterial strains utilize 4-chlorosalicylicacid as their sole carbon source indicated by the increase in turbidity accompanied by the utilization of the carbon source.

The liquid culture removed from the glass vessel was diluted, in a 10-fold serial dilution. The liquid culture (0.1mL) was spread on nutrient -agar plates. Four different types of colonies were observed which were morphologically characterized, based on diameter colour, opacity, form, elevation, margin, smoothness and texture of microbial colonies. The biochemical tests for all of the four isolates were performed (Table 2). Based on the data of the biochemical test and an earlier report of the Biolog test method, one of the isolate was identified as *Klebsiella* sp. The other three isolates were identified as Pseudomonas sp. The data of the study were similar to earlier studies, which indicated the presence of *Pseudomonas fluorescens, P. mundocina* and *P. cichorii* identified by the Biolog test method, as described earlier.

The pulp and paper mill effluent is colored due to the chlorinin compounds it discharges from the bleaching step of the manufacturing process. Table 1 gives the characteristics of the pulp and paper mill effluent. The effluent treated by microbial community indicates that significant reduction in colour was observed for different periods and is presented in Figure 1. The colour unit was found to decrease for both the control and treated. The per cent colour removal during 15d period for control and treated were 32 and 80 percent. More than 50 per cent of the colour was removed in the treatment reactor after 3 d of incubation period and 71 per cent day 7 of the retention period.

The lignin removal by the mixed community was determined. Data of the study indicated a 25 percent removal of lignin on day 3 which reached to 41 per cent after 15 d of the retention period (Figure 2). The removal of colour by the biological treatment could be correlated for the mingralization of the chlorolignin compounds. The chlorolignins are not only huge in amount in the pulp and paper industry but are also found to be recalcitrant and are the cause of higher chemical oxygen demand. COD, a major indicator of organic and inorganic load of the effluent is presented in Figure 3. The COD removal in the control and treated reactor were found to be 38 and 70 per cent, respectively. This showed the ability of the bacterial consortium to reduce the pollutant load of the effluent by microbial treatment.

The phenol content of the untreated and treated sample at different intervals, was measured. As shown in Figure 4, percent reduction for the phenol content was community. However, only 22.5 per cent phenol was removed in the control reactor having an indigenous population of effluent. Various chlorinated phenols are present in pulp and paper mill effluent which was transformed into highly recalcitrant compounds. The polychlorinated phenols are not degraded easily. However the reduction in the phenol content may be due to the ability of bacterial community to cleave to benazene ring.

This study has shown the successful treatment of Pulp and Paper Mill effluent in a chemostat during enrichment in a continuous culture by a microbial population. The microbial community selected by enrichment was characterized based on the structure and physiology of the individual strains. Further studies on the genetic characterization of these strains and a scale up of the process, are in progress.

Table 1. Characteristics of pulp and paper mill effluent

Characteristics	Value (mean of five replicates)
pH	7.8
COD (mg/l)	5643.00
Colour unity	5422.00
Lignin (mg/l)	
BOD (mg/l)	2563.00

Table 2. Biochemical characterization of bacterial isolates

Biochemical test	Strain 1	Strain 2	Strain 3	Starin 4
Oxidase test	-ve	+ve	-ve	+Ve
Catalase test	+ve	+ve	-ve	-ve
Glucose fermentation test	+ve	+ve	+ve	-ve
Nitrate treduction test	+ve	+ve	+ve	+ve
Indole production test	-ve	-ve	-ve	-ve
Methyl red test test	-ve	-ve	-ve	-ve
Starch Hydrolosis test	-ve	-ve	-ve	-ve
Casein utilization test	-ve	-ve	-ve	-ve
Gealtin liqufiaction test	-ve	-ve	-ve	-ve
Urease test test	+ve	-ve	-ve	-ve
Lactose fermentation test	-ve	-ve	-ve	-ve
Simmon citrate test	+ve	+ve	+ve	+ve
Voges proskauer test	+ve	-ve	-ve	-ve
Hydrogen sulphide test	-ve	-ve	-ve	-ve

23

Optimization of Laccase dose and the reaction time for the detoxification of Saccharum spontaneum hydrolysate for bioethanol production using Pichia stipitis

GUPTA PRITI

Introduction

Energy consumption has increased steadily over the last century as the worlds population has grown and more countries have become industrialized. Crude oil has been the major resource to meet the increased energy demand. Campbell and Laherrere (1998) used several different techniques to estimate the current known crude oil reserves and the reserves as yet undiscovered and concluded that the decline in worldwide crude oil production will begin before 2010 [1]. Therefore, there is a great interest in exploring alternative energy sources. Unlike fossil fuels, ethanol is a renewable energy source produced through the fermentation of sugar.

The cost of many fermentation products depends very much on the cost of the raw materials. Even 'high value' products are not immune from these effects. The future scope and importance of the fermentation technology industry will depend to a large extent on the cost of its carbohydrate containing raw materials.

Lignocellulosic biomass is the most abundant renewable resource that can serve as substrate for the production of alternative fuels because of its availability in large quantities at low cost [2] [3]. Lignocellulose contains lignin, hemicellulose and cellulose as a major component and other components like proteins, vitamins, pectins, soluble sugars, minerals etc. of which only soluble sugars, hemicellulosic and cellulosic fraction are used for producing fermentable sugars to produce ethanol. The production of fuel ethanol from lignocellulosic biomass feedstock has several benefits including domestic availability, pollution reduction and ease of introduction into existing gasoline and diesel distribution networks. The efficient utilization of the hemicellulosic component offers an opportunity to reduce the cost of fuel ethanol production by 25% [4].

Nature of the problem

Ethanol has attracted worldwide interest as an alternative liquid fuel, especially for transportation, for three reasons. Firstly, the oil crisis in the mid 70s stressed the dependence on the supply of petroleum, which can be reduced by the use of alternative fuels, from renewable resources such as ethanol, produced from lignocellulosic materials. Secondly, as the ethanol production process from lignocellulosics uses energy from renewable sources, no net carbon dioxide is added to the atmosphere because the amount of carbon dioxide released during production and combustion of fuel ethanol would be equivalent to the amount of carbon dioxide being absorbed by the replanted biomass for ethanol production. Thirdly, the emission and toxicity of ethanol are lower than those of petroleum [5]. This makes ethanol an environmentally beneficial energy source but production of ethanol is a difficult task as it faces problems of suitable substrate ie.carbohydrate containing raw materials and the development of fermentation technology using lignocellulosic substrate.

The feasibility of *Saccharum spontaneum* as a lignocellulosic substrate for ethanol production has been exploited. This particular plant has been selected because of the following attributes: - it has no economic value, it grows as an extensive weed in India, it is drought and flood tolerant, it can grow even on sandy soils, it is competitive over other weeds as it has an extensive root system that allows its firm establishment in the soil and possesses the lightest of seeds that are of high avail in the seed dispersal to distant places and their easy establishment at the new place [26]. Extensive research has been completed on the conversion of lignocellulosic materials to ethanol in the last two decades [6] [7] [8] [9] [10] [11] [12]. The conversion includes two processes: hydrolysis of cellulose and hemicellulose in the lignocellulosic materials to the fermentable reducing sugars, and fermentation of the sugars to ethanol by yeast or a bacterial species.

Dilute acid hydrolysis has been successfully developed for the hydrolysis of lignocellulosic materials. Dilute sulphuric acid hydrolysis is a favorable method for either the pretreatment before enzymatic hydrolysis or the conversion of lignocellulosics to sugars [13]. Formation of aromatic and polyaromatic compounds with a variety of substituents during the dilute acid hydrolysis has been reported in literature [14] [15] [16] [17] [18] [19] [20]. These compounds are regarded as degradation products of lignin [21]. Phenolic compounds are phenol, vanillin, vanillic acid, vanillyl alcohol, 4-hydroxybenzaldehyde, 4-hydroxybenzoic acid, coumaric acid, syringaldehyde, syringic acid, cinnamaldehyde, dihydroconiferyl alcohol, hydroquinone, catechol, homovanillic acid and Hibbert's ketones. Other substances reported in the literature are formaldehyde, methylglyoxal, 2-hydroxymethylfuran, dihydroxyacetone, glyceraldehydes and maltol [22].

Many of the hydrolysis by-products are considered to be toxic for the growth of microorganisms in order to carry out fermentation. Therefore detoxification is carried out prior to fermentation of hydrolysate to ethanol by microorganisms.

Study Area

The biological process for converting the lignocellulosic material to fuel ethanol requires delignification to liberate cellulose and hemicellulose from their complex with lignin, depolymerization of the carbohydrate polymers to produce free sugars and fermentation of the mixed hexose and pentose sugars to produce ethanol. The present work deals with the fermentation of lignocellulosic hydrolyzates to ethanol by *Pichia stipitis*. Hydrolyzates are not just sugar solutions. They are, in actuality, rather nasty solutions that microorganisms are not necessarily pleased to stay with. They contain a myriad of chemical compounds (acetic acid, furfural, phenolics, etc) most of which inhibit microbial growth, even at very low concentrations.

The productivity of the ethanol production process depends upon the substrate chosen, as substrate should be easily available in plenty throughout the year without adding enough cost for its transportation. It also depends upon the composition of substrate as the higher lignin content poses problems in the release of sugars in the hydrolyzates. Large amounts of inhibitory compounds are produced if the lignin content is high as harsh conditions are used to release the sugars. *Saccharum spontaneum* is the substrate of choice in my work as it has a less lignin content (only 20%) and, hence, dilute - acid hydrolysis releases good amount of pentose and hexose sugars. The biggest advantage of this substrate is that even upon acid hydrolysis very minute quantities of inhibitory compounds (acetic acid, phenolics, and furfural) are liberated which reduces the required steps of detoxification in order to enable the microorganisms to grow but detoxification is necessarily required.

In order to remove the inhibitors and increase the hydrolysate fermentability, several detoxification treatments, including chemical, physical and biological methods have been used. These methods include direct neutralization, over liming, ion exchange resins, absorption onto activated charcoal, and treatments with enzymes such as peroxidase and laccase from fungi [23] [24]. Our focus is on the detoxification by the laccase treatment as being an enzyme it is specific in its action. Since detoxification increases the cost of the process [25], it is important either to overcome detoxification steps or to develop cheap and efficient methods. However, the needs for detoxification must be evaluated in each case since it depends on the chemical composition of the hydrolysate and its specific strain.

To bring about the production of ethanol, fermentation of *Saccharum spontaneum* is carried out using *Pichia stipitis. Pichia stipitis* is preferred because of its industrially required attributes: (1) it has a short generation time. (2) It can also ferment pentose sugars along with hexose sugars present in lignocellulosic substrate. (3) It produces very negligible amounts of xylitol, an inhibitor (4) It ferments a broad range of substrate containing sugars to ethanol in a short period of time [26].

Methodology

Lignocellulosic substrate: *Saccharum spontaneum* was obtained locally from the outskirts of Delhi, India. A dried condensed leaf-sheath of substrate was finely chopped to attain the size of 5 - 6 mm using laboratory scale disintegrator. The substrate was properly washed with water to remove dust and other unwanted particulate matter followed by drying at 60^0C overnight.

Chemical analysis of the substrate was carried out according to the TAPPI test methods (Atlanta, Georgia, U.S.A. 1992). *Saccharum spontaneum* was hydrolyzed according to the sulfuric acid pretreatment method of Sun Y. et al [27].

The chopped *Saccharum spontaneum* (5 – 6 mm) was soaked in dilute sulfuric acid (final concentrations: 0.5% to 3.5%) and pretreated in an autoclave at 100°C and 120°C and Russian autoclave was used to pretreat the substrate at 140°C and 160°C with residual time of 10, 20, 30 and 60 minutes to obtain the optimum acid concentration, optimum temperature and optimum time to obtain the maximum sugar concentration in the hydrolysate.

The contents were filtered through a vacuum filter and both filtrate and solid residue were collected. Filtrate was analyzed for total sugars following the method of Dubois et al [28], Phenolics by following the method of Singleton [29] and reducing sugars by following the method of Miller [30].

Organism: *Pichia stipitis* used in this work was obtained from stocks in the Department of Microbiology, South campus, University of Delhi, India. *Pichia stipitis* was maintained on MGYP slants

Detoxification procedures

Detoxification procedures were carried out according to Carvalheiro et al [31] with some modifications. These were carried out over raw or pH balance-adjusted hydrolyzates as required. In the later, the pH was raised to 5.5 (fermentation pH) by the addition of Ca $(OH)_2$. After 1 h at pH 5.5 the precipitate was removed by centrifugation at 7500×*g* for 25 min. Instead of centrifugation as stated in the above-referred protocol the detoxified hydrolyzates were recovered by vacuum filtration using Whatman no. 1 filter paper.

Activated charcoal treatment

Granular activated charcoal was mixed with the acid treated hydrolyzates of the pH 5.5 in the ratio of 1:100 (w/v) and stirred for ½ an hour at room temperature. The detoxified hydrolyzates were recovered by double vacuum filtration using Whatman no. 1 filter paper. When needed the pH balance of treated hydrolyzates was adjusted to 5.5 with Ca $(OH)_2$ or H_2SO_4.

A pH balance of 5.5 hydrolyzate was obtained through the addition of calcium hydroxide and stirred for ½ hour and then filtered to recover the hydrolyzate. The approach is called direct *neutralization* after which further detoxification is carried out.

Over liming

The pH balance of the raw hydrolyzates was increased to 10.0 by the addition of Ca $(OH)_2$. The beaker containing hydrolyzate was then kept at shaking conditions for ½ hour at pH10. The hydrolyzate was then recovered by double vacuum filtration using Whatman no.1filter paper, acidified with sulfuric acid to a obtain fermentable pH balance (5.5).

Treatment with ion-exchange resins

The hydrolyzates and the resins were mixed in the ratio of 10:1 (v/w) and kept overnight at room temperature. The resins were removed by filtration and the pH balance of the treated hydrolyzates was corrected, when needed, to 5.5 with H_2SO_4.

Treatment with laccase

Laccase can be employed for detoxification of hydrolyzates according to Jonsson et al [21]

Laccase from *Cyathus stercoreus* was used to detoxify the hydrolyzates. The optimum dose and time for laccase action in order to bring about detoxification was found by setting up two experiments: in one experiment different doses of enzyme [2.5ml to 12.5 ml; 10 ml corresponds to 883 units by the Guaiacol method [32]] and in another experiment the optimum dose was added to all tubes and phenolics were checked after every ½ hour by the Singleton method [29] till the phenolics were reduced to zero.

Fermentation

The fed – batch fermentation was carried out using in situ a sterlizable 12.0 l fermenter (Sartorius Pvt. Ltd, Banglore, India) with the operating capacity (10.0 l). The fermenter was initially started as a batch fermenter with a volume of 3.5 l; furthermore the total volume was maintained to 10.0 l by the feeding of hydrolysate intermittently at a pH balance 5.5, temperature of 30^{0}C and the total dissolved oxygen content was kept at 35 % throughout the course of fermentation. A 10 ml of antifoam (Thomas Baker, Mumbai, India) was added to avoid froth formation during the course of fermentation. The sample was taken out, intermittently and analyzed for sugar consumption, biomass production, xylitol and ethanol production.

The fermentation of hydrolyzates was carried out only after detoxification of the hydrolyzates.

Analytical methods

Ethanol was estimated by GC (Perkin-Elmer, Clarus 500) with an elite-wax (crossbond-PEG) column at 120°C, flame ionization detector at 210 °C and an injector at 180 °C using pure ethanol from Bengal chemicals as standard. The carrier gas was nitrogen

Estimation of reducing sugars: pentoses were estimated by the method of Fredrickson and Tsuchia [33] and hexose sugars were estimated using DNS method of Miller [30]. Total sugars were estimated by the phenol sulfuric acid method of Dubois et al. [28].

Phenolics were estimated using spectrophotometric methods of Singleton et al [29]. In this method, 50ìl of test sample was mixed with 250 ìl of folins reagent. The mixture was vortexed well and 750 ìl of 20% Na_2CO_3 was added after 1 minute but before 8 minutes of addition of folins reagent. The volume was made up to 5 ml. The reaction mixture was incubated for 1 hour in the dark at 23^{0}C to develop a blue color and it was read at 760 nm against the blank containing distilled water in the reaction mixture instead of the test sample.

The standard curve was prepared using vanillin 50 – 500 ìg/ml. in order to evaluate the spectrophotometric method given by Singleton et al. [29] and the phenolics were also estimated by gas chromatography.

Major Findings

Among the lignocellulosic biomass are forests, agricultural and agro-industrial residues, which are a widespread and inexpensive source of hexose and pentose sugars, *Saccharum spontaneum* has proven to be an ideal substrate for ethanol production as it is evident from its chemical composition (Table.1). Optimization of chemical hydrolysis of the substrate was also carried out in order to release maximum reducing sugars. Different parameters that were studied for maximum sugar recovery include temperature, acid concentration and time. The substrate released maximum sugars (17.4 g/l corresponding to 85% hydrolysis) at 120°C at 2.5 % of sulfuric acid concentration when the process was carried out for 60 minutes.

As during chemical hydrolysis different concentrations of inhibitors are released in the hydrolyzate (Table2) and several detoxification strategies were carried out and compared for hydrolyzate of *Saccharum spontaneum* (Fig. 1). Different detoxification strategies include over liming, activated charcoal, direct neutralization, ion-exchange resin and laccase treatment. For laccase treatment, dosage (table3) and time (table 4) were optimized to get efficient removal of inhibitors and it was found that the enzyme reduces the phenolics completely in 6 hours and with 7.5ml of enzyme.

After detoxification of the hydrolyzate, it was converted to ethanol using *P. stipitis* in both batch and fed-batch cultivation. Parametric optimization of ethanol production from pentose sugars was also carried out using *Pichia stipitis*. It was found that *P. stipitis* give maximum yield of ethanol at pH balance 5.5, temperature 30°C and agitation rate of 150rpm. In the fed-batch fermentation, ethanol production was much higher as compared with batch fermentation reaching the concentration of 26.2 g/l in 72 hours as against 5.0 g/l in 22 hours in batch fermentation.

Table1. Chemical analysis Of *Saccharum spontaneum* (*TAPPI, AtlantaUSA.1992*)

COMPOUND	%(BY WEIGHT)
LIGNIN	20 ± 0.026
HOLOCELLULOSE	68.8 ± 0.048
PENTOSANS	19.1 ± 0.032
DRY WEIGHT	8.0 ± 0.012
ASH	2.2 ± 0.010

Table 2. Phenolics released (g/l) at different temperatures (°C) at 60 minutes

Acid conc. (%)	Phenolics (g/l) →			
	100°C	120°C	140°C	160°C
0.0	0.056± 0.015	0.058± 0.116	0.147± 0.008	0.159± 0.022
0.5	0.196± 0.010	0.197± 0.016	0.297± 0.011	0.364± 0.010
1.0	0.197± 0.019	0.202± 0.065	0.325± 0.016	0.400± 0.009
1.5	0.206± 0.014	0.209± 0.108	0.352± 0.016	0.483± 0.004
2.0	0.217± 0.016	0.216± 0.025	0.382± 0.012	0.512± 0.007
2.5	0.232± 0.045	0.235± 0.08	0.391± 0.011	0.591±0.010
3.0	0.254± 0.056	0.275± 0.068	0.397± 0.013	0.676± 0.011
3.5	0.228± 0.045	0.286± 0.065	0.505± 0.018	0.791± 0.009

*Fig*1. Comparison of different detoxification strategies on phenolics reduction. (DN stands for direct neutralization)

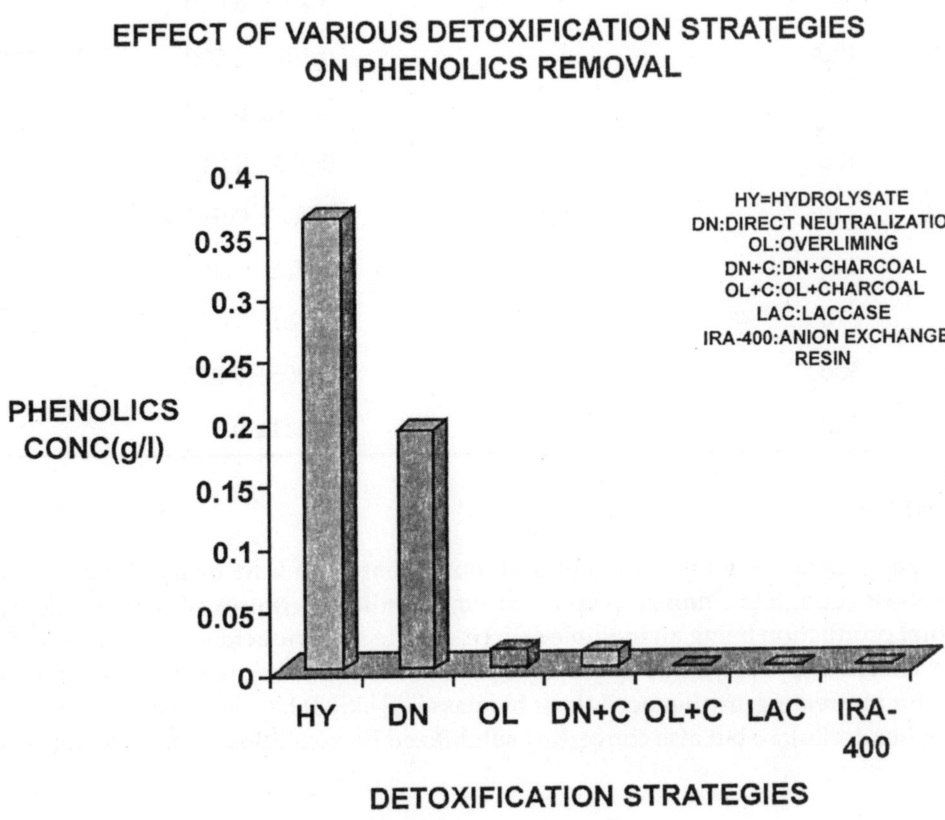

Table 3. Optimization of laccase dose (*C. stercoreus*) for complete removal of phenolics

Laccase (U/ml)	Phenolics (g/l)
2.5ml	0.353± 0.010
5.0ml	0.047± 0.026
7.5ml	**-0.011± 0.022**
10.0ml	-0.010± 0.020
12.5ml	-0.011± 0.028

10 ml of enzyme have 883 units as estimated by the Guaicol method.

Table 4. Optimization of reaction time for Laccase activity for complete removal of phenolics

Time (minutes)	Phenolics (g/l)
30	0.474± 0.018
60	0.401± 0.020
90	0.400± 0.020
120	0.383± 0.020
150	0.350± 0.018
180	0.312± 0.017
210	0.268± 0.017
240	0.053± 0.018
270	0.020± 0.016
300	-0.005± 0.015
360	-0.011± 0.014

Conclusion

Saccharum spontaneum is among those lignocelluloses that have little or no economic value. Also, it is robust enough to colonize even in extreme conditions. It proved to be an ideal substrate for fuel production being an inedible source. Dilute sulfuric acid was employed for the purpose of chemical hydrolysis as dilute acid pretreatment has become a state of the art of technology for pretreating any lignocellulosic biomass [34] [35]. It has the advantage of not only solubilizing hemicellulose but also converting solubilized hemicellulose to fermentable sugars [36].

As inhibitors are released during chemical hydrolysis of substrate that is unfavorable for the growth of microorganisms, detoxification needs to be carried out before proceeding to fermentation. The laccase, from *Cyathus stercoreus*, treatment led to the complete removal of the inhibitors from the lignocellulosic hydrolysate as laccase causes the polymerization of the inhibitory compound and usually the phenolics that settle down forming precipitate are easily filtered out paving the way for fermentation of sugars to ethanol production. Our result is in accordance with the finding of Jonsson et al [21] as they found that treatment with the enzymes peroxidase and laccase, obtained from the ligninolytic fungus *Trametes versicolor*, have been shown to increase the maximum ethanol productivity in a hemicellulosic hydrolysate of willow two to three times.

After the process of detoxification of the hydrolysate by laccase treatment, *Pichia stipitis* was efficiently able to ferment the hydrolysate to ethanol giving a production of 4.6g/l at 20 hours which is a considerable good yield. A fed - batch fermentation of the hydrolyzate on 2% xylose was carried out. A fresh sugar solution was added into the medium after every 24 hours. In the fed-batch fermentation, ethanol production was much higher as compared with a batch fermentation reaching the concentration of 26.2 g/l in 72 hours and it remained constant for 120 hours after which the ethanol production decreased

References

1. Campbell, C.J., Laherrere, J.H. 1998. The end of cheap oil. Sci. Am. **3**, 78–83.

2. LyndLR.1989. Production of ethanol from lignocellulosic materials using thermophilic bacteria: critical evaluation of potential and review. Adv. Biochem. Eng./Biotechnol. 38, 1-52.

3. Parisi F. 1989. Advances in lignocellulosics hydrolysis and in the utilization of hydrolysates. Adv. Biochem. Eng/Biotechnol. 38, 53 - 87.

4. Hinmen ND; Wright JD; Hoagland W; Wyman CE.1989. Xylose fermentation, an economic analysis. Appl. Biochem. Biotechnol. 20/21, 391 - 401

5. Wyman, C.E., Hinman, N.D. 1990. Ethanol-fundamentals of production from renewable feedstocks and use as a transportation fuel. Appl.bioch.biotechnol.24, 735-753

6. Bjerre, A.B., Olesen, A.B., Fernqvis, T. 1996. Pretreatment of wheat straw using combined wet oxidation and alkaline hydrolysis resulting in convertible cellulose and hemicellulose. Biotechnol. Bioeng. **49**, 568–577.

7. Azzam, A.M. 1989. Pretreatment of cane bagasse with alkaline hydrogen peroxide for enzymatic hydrolysis of cellulose and ethanol fermentation. J. Environ. Sci. Health B. **24** (4), 421–433.

8. Dale, B.E., Henk, L.L., Shiang, M. 1984. Fermentation of lignocellulosic materials treated by ammonia freeze-explosion. Dev. Ind. Microbiol. **26**, 223–233.

9. Duff, J.B., Murray, W.D. 1996. Bioconversion of forest products industry waste cellulosics to fuel ethanol: a review. Bioresour. Technol. **55**, 1–33.

24

Antifungal and Antitumor activities of a novel protein from Indian squill Urginea indica Kunth

M.N. SHIVA KAMESHWARI AND K.J. THARASARASWATHI

Introduction

The plant kingdom has helped us in the discovery of new drugs useful for the alleviation of diseases both in plants and humans. Fungal diseases have been one of the principal causes of crop losses. The epidemic spread of fungal diseases is mainly controlled by crop rotation, breeding of fungus resistant cultivars of crops and the application of agrochemicals. While conventional breeding procedures to produce cultivars resistant to fungal diseases are time consuming processes, the use of agrochemicals proves to be costly and also has potentially harmful effects on the environment (l). Therefore the identification of genes that encodes enzymes or proteins, which inhibit the growth of fungi, help to obtain resistance to fungal infection (2). Antimicrobial peptides that have been characterized so far include cecropins and meganins, thionins, defensins, lipid transfer proteins, hevein and knottin type antimicrobial peptides (3).

The search for new drugs to combat cancer for which there is no satisfactory solution as yet, continues relentlessly. Plant based drugs such as bisindole alkaloids, vinblastin and vincristin ftom *Catharanthus roseus,* teniposide and etoposide developed ftom podophyllotoxin, a liginin obtained from *Podophyllum emodi* and *Podophyllum paclitaxal* a diterpenoid from *Taxus wallichiana,* irenotecan and topotecan obtained from quinoline alkaloid, campothecine, a constituent from *Campotheca acuminata* are used for the treatment of various types of cancer (4,5,6).

Urginea is one of the extremely interesting polytypic genus with about 100 species and is represented in India by nine species (7,8). *Urginea indica* is also khown as the Indian squill. The Squill bulb has long been used as a source of a medicinal product with pharmaceutical and biocidal applications. Bufadienolides of the squill are glycosides used as cardio-tonic agents. The toxicity of the squill bulb is attributed to its scilliroside content. (9,10). However there are no reports describing the presence of antifungal or antitumor protein(s) from squill bulbs.

In the present study efforts have been made to identify the biological activity of a protein component from *Urginea indica,* a plant endemic to India, Africa and the Mediterranean regions. We have isolated a basic protein of approximately 26 KDa from *Urginea indica,* which has antifungal and antitumor activity. The antibodies raised against the protein neutralized the antifungal activity of the protein. The antifungal activity was specific towards *Fusarium oxysporum* and *Rhizoctonia solani.* The protein showed an inhibition of proliferation of mouse mammary carcinoma cells *in vivo.* This is the first report describing the isolation and identification of biological activities of a novel protein from *Urginea indica,* the Indian squill

Materials and Methods

Collection of *Urginea indica* bulbs

The squill bulbs grown in the wild were collected from various forest areas in Kamataka, India. About thirty different species were collected.

Preparation of crude extract from *Urginea indica* bulbs

The thirty different species of *Urgineaindica* bulbs were grown in a green house under uniform environmental conditions before the preparation of a crude extract from one of the species. Bulbs (25gm) were chopped into small pieces and were homogenized in a blender using 25mM Tris-HCI (pH 6.8) buffer. The homogenate was passed through cheese cloth prior to centrifugation at 5000 rpm for 15 minutes at 4°C.

Plant pathogens

Pure cultures of plant pathogens *Aspergillus flavus, Colletotrichum dematium, Fusarium oxysporum, Rhizoctonia solani* and *Alternaria alternata* were a kind gift from Dr. K. A. Raveesha, Department of Botany, University of Mysore, Mysore, India.

Assay for antifungal activity against plant pathogens

Partial Purification of the antifungal protein from *Urginea Indica*

Gel filtration chromatography

Gel electrophoresis

Culture and Isolation of Ehrlich ascites tumor (EAT) cells

In vivo growth assay for EAT cells

IMMUNOLOGICAL METHODS

Production of polyclonal antibodies

Immunodiffusion

Determination of neutralization assay

Western Blot Analysis

Results

Antifungal activity of Urginea indica crude extract.

The crude extract from *Urginea indica* when verified for antifungal activity (Figure 1) was used towards five plant fungal pathogens such as A) *Fusarium oxysporum*, B) *Rhizoctonia solani* C) *Alternaria alternata*, D) *Aspergellus flavus* and E) *Colletotrichum dematium*, and showed an inhibition of growth of *Fusarium oxysporum and Rhizoctonia solani*. However at that similar concentration no antifungal activity was detected for *Alternaria alternata, Aspergillus flavus and Colletotrichum dematium*. The antifungal activity could also be detected in an acid extractable fraction of the crude extract (data not shown) indicating that a basic component in the extract is responsible for the antifungal activity.

Fractionation of antifungal activity.

The neutralized acid extractable component was lyophilized and loaded on to a gel filtration column. As is shown in figure 2A, two major protein peaks resolved and when the column fractions were verified for the antifungal activity towards *Fusarium oxysporum* as the test fungus, all of the antifungal activity was associated with the peak II fraction obtained from the gel filtration chromatography (figure 2B). The inset of figure 2A depicts the SDS-PAGE profile of pooled fractions from the gel filtration column chromatography. As shown in the figure, a major protein with a molecular mass of 26 KDa was detected in crude .extract (lane 1), acid extract (lane 2) and peak II from the gel filtration column (lane 5).

Neutralizing antibodies against 26 KDa antifungal protein.

The 26 KDa antifungal protein from the peak II of gel filtration when used for the production of polyclonal antibodies showed to be a potent immunogen. In figure 3A is shown a single precipitin band upon reactivity of rabbit antiserum with the respective antigen present in the peak II of gel filtration chromatography. As is seen in the figure, no cross reactivity was obtained with PBS (vehicle) or BSA or pre-immune serum. In figure 3B is shown the Western blot analysis, which indicates a strong detection of 26 KDa antifungal protein in a crude extract, acid extract and peak II fraction from the gel filtration column. In figure 3C is shown the growth of Fusarium oxysporum (plate I) over loaded with fungal spores), when the plate with an equal spore load was charged with purified fraction of 26 KDa and a protein clear antifungal activity is visible (plate 2), with the antigen-antibody complex no such antifungal activity was detected (plates 3&4).

Antitumor activity of 26 KDa protein.

The result in figure 4A shows the growth profile of mouse mammary carcinoma cells in the peritoneal cavity of mice. The tumor cells grow for a period of 10-12 days with production of some ascities fluid. From day 4th to 8th, is the exponential growth period during which the purified 26 KDa protein (50 ìg) was injected (i.p.). The result indicates that immediately thereafter there is a significant decrease in the body weight of the animals treated with the protein. Further, the result in figure 4A&B clearly indicates the dramatic effect of the inhibition of the proliferation of tumor cells in the peritoneal cavity by the injected 26 KDa protein and a significant decrease in the ascites volume confirms the antitumor effect of the protein.

Discussion

Pathogenic fungi are major problems in agriculture. Genetic engineering provides an opportunity to protect plants from fungal diseases and to reduce the use of synthetic fungicides. The genes for an antifungal protein can be engineered into plants to increase the resistance of crop plants to fungal attack, thus decreasing the use of environmentally unfriendly pesticides. The major factor limiting the application of this technology is the identification and isolation of useful genes. Plants themselves are a potential source of new antifungal activities. For example, two antifungal proteins have been isolated from radish. The genes for these proteins were cloned and constitutively expressed in transgenic tobacco plants, which showed enhanced resistance to pathogenic fungi (12,13).

The main objective in our study was to identify novel biologically active protein (s) from *the Urginea indica,* a plant with components rich in medicinal attributes. Preliminary data in this present paper describes the partial purification of a novel protein with antifungal and antiproliferative properties from the Indian squill *Urginea indica.* This protein is basic in nature because it could be enriched in an acid extractable fraction of a crude extract of *Urginea indica.* Plants protect themselves against microbial pathogens by various induced defense responses such as the modification of plant cell walls and the production of antimicrobial peptides, secondary metabolites, chitinases, glucanases, membrane interacting proteins and peptides (14). However the proteins isolated from *Urginea indica* extract are not a pathogen-induced protein, but a constitutively expressed one. A particular family of antimicrobial peptides that inhibit growth of a broad range of fungi at micromolar concentration are plant defensins, which are small basic peptides (15). Thionins represent a family of peptides with 45 to 47 amino acids in them. The antimicrobial activity spectrum of thionins includes several gram-positive bacteria and about twenty different phytopathogenic fungi. The SDS-PAGE and western blot analysis of the partially purified protein revealed a molecular weight of approximately 26 KDa and exhibited antifungal activity towards Fusarium oxysporum and Rhizoctonia solani. Neutralization of the antifungal activity of the protein by its antibodies confirms the specific biological activity of the protein. Studies have to be conducted using a spectrum of other plant fungal pathogens as targets for the isolated protein. Onion seed lipid transfer protein is reported to have antifungal activity (16). Hevein and Knottin- type peptides from Amaranthus caudatus and Mirabilis jalapa seeds inhibit the whole range of fungi and gram positive bacteria (17). All plant antimicrobial peptides characterized so far contain multiple disulphide bridges (Cystein rich) and are basic peptides (3).

The antifungal protein isolated from the bulbs of *Urginea indica* exerted an antiproliferative activity on mouse mammary carcinoma cells *in-vivo* as evidenced by the decreased body weight and decreased tumor cell number in the animals treated with the 26 KDa antifungal protein. Further, effective inhibition of ascites formation in-vivo clearly suggests that the injected protein influences either the expression or secretion of permeability factors from mammary carcinoma cells in-vivo. The antifungal protein, thionins are shown to exert adverse effects often involving permeabilisation of various cultured plant protoplast (18). The Calvatia caelata protein with a molecular weight of 8 KDa has been reported to have antiproliferative activity of human breast carcinoma cells (18). However the mechanism of antiproliferative action of the the 26 KDa antifungal protein from *Urginea indica* remains to be elucidated. Preliminary data (not shown) indicates that the protein is not toxic to mice. The isolation of 26 KDa protein in the present study and the identification of its antifungal activity to plant pathogens and antiproliferative activity on tumor cells is a unique finding which may have a potential application both in agriculture and in human health care. Currently the studies are being undertaken to exploit the antifungal activity of the *Utginea indica* protein for engineering disease resistant crops.

References

1. A.J. Verbiscar, J. Patel, T.F. Banigan, R.A. Schatz. Scilliroside and other Scilla compounds in Red squill, J.Agric. Food Chern. 34 (1986) 973-979.

2. B.J.C. Cornelissen and L.S. Melchers. Stratergies for control of fungal diseases with transgenic plants, Plant Physiol. 101 (1993) 709-712.

3. B.P.A. Cammue, K. Thevissen, M. Hendriks, K. Eggermont, 1.1. Goderis, P. Proost, J. Van Danme, R.W. Osborn, F. Guerbette, J. C. Kader, and W.F. Broekaert. A potent antimicrobial protein from onion (Allium cepa L) seeds showing sequence homology to plant lipid transfer proteins, Plant Physiol. 109 (1995) 445-455.

4. C. Camido, R.M. Morae, F.E. Dayan and D. Fessriro. Molecules of Interest, Podophyllotoxir. 54 (2000) 11-120.

5. F.R.G. Terras, K. Eggermont, V. Kovaleva, N.V. Raikhel, R.W. Osborn, A. Kester, S.B Rees, 1. Vanderleyden, B.P.A. Cammue, W.F. Broekaert. Small cystein-rich antifungal proteins ffom radish (*Raphanus sativus* L). Their role in host defense, Plant Cell. 7 (1995) 573-588.

6. H.K. Airy shaw. A Dictionary of flowering plants and ferns. 8th edition (Wills, J.C.Ed. Cambridge University Press, Cambridge. (1993).

7. K. Broglie, 1. Chet. M. Holliday, R. Cressman P.H. Biddle, G. Knowlton, C.J. Mauvais, R. Broglie. Transgenic plants with enhanced resistance to the fungal pathogen *Rhizoctonia solani,* Science. 254 (1991) 1194-1197.

8. K. Hemadri. and S. Swahari. Urginea nagarjunae. A new species of liliaceae ITomIndia. Ancient Sci, Life. 2 (1982) 105-110.

9. K. Thevissen, R.W. Osborn, D.P. Acland and W.F. Broekaert. Specific banding sites for an antifungal plant defensin ffom Dahlia *(Dahlia Merckii)* on fungal cells are required for antifungal activity, Mol. Plant Microbe Interac. 13 (2000) 54-61.

10. K.M.M. Swords, J. Liang and D.M. Shah. Novel approaches to engineering disease resistance in crops, Genetic Engineering. 19 (1997) 1-13.

11. M.F.C. De Bolle, R.W. Osborn, 1.J. Goderis, L. Noe, D. Acland, C.A. Hart, S. Torrekens, F.V. Leuven and W.F. Broekaert, Antimicrobial Peptides ftom Mirabilis jalapa and Amaranthus caudatus. Expression, processing localization and biological activity in transgenic tobacco, Plant Mol. Biol. 31 (1996) 993- 1008.

25

Feeding preferences of *Eisenia fetida* Zinia

K.G. SONI AND CHETNA

Introduction

Agriculture was a seasonal profession. It has become a cyclic profession with the chemical fertilizers and irrigation facilities. Modern technologies in agriculture have largely contributed to the green revolution and helped to tide over the food crisis during the last four decades. The green revolution was the result of the combination of high yielding varieties, irrigation, fertilizers and pesticides. India is now paying for the ecological, environmental, social, cultural and economical costs of this revolution. The consumption of chemicals in the form of fertilizers and pesticides has gone up seven times in the last 20 years but production has only increased by a miserable two- fold. Another problem, which has emerged during the past few years, is the decomposition of complex organic waste resources.

Darwin (1881) was the first to show that earthworms affect soil formation and development. Earthworms play a key role in soil biology as versatile bioreactors. They effectively harness the beneficial soil micro flora, destroy soil pathogens and convert organic wastes into antibiotics, biofertilizers, vitamins, growth hormones and worm biomass. Vermitechnology has been proposed globally as a potential tool to stabilize natural and anthropogenic waste, such as sewage sludge, industrial waste, agro-industrial solid waste, plant derived waste and animal dung, etc by the joint action of earthworms and microorganisms. Earthworms are important for soil as they act physically as aerators, crushers and mixers, chemically as degraders and biologically as stimulators, in the decomposition subsystem. Earthworms can be cultured and put to various uses i.e. to improve soil fertility, to convert organic waste into manure, to produce earthworm based protein food for livestock, for a drug and vitamin source as well as natural detoxicant and a bait for the fish market.

Review of Literature

The Involvement of earthworms for the degradation of organic waste and the production of vermicompost is catching up for scientific investigation and for commercialization.

Vermicompost, a product of non-thermophilic biodegradation of organic material through interactions between earthworms and microorganisms is a peat like material with high porosity, aeration, drainage, water holding capacity and microbial activity (Atiyeh et al 2000).

It is an aerobic, stabilization and non thermophilic process of organic waste decomposition (Guandi *et al* 2002). The viability of using earthworms as a treatment or management technique for numerous organic waste streams has been investigated by a numbers of workers. Earthworms feed voraciously on organic waste and while utilizing only a small portion for their body synthesis, they excrete a large part of these consumed waste materials in a half digested form. Considerable work has been carried out on vermicomposting of various organic materials and it has been established that epigeic earthworms hasten the composting process to a significant extent along with the production of a better quality compost.

Since the intestines of earthworms harbour a wide range of enzymes, hormones etc., these half digested substrates decompose rapidly and are transformed into a form of vermicompost within a short time (Lavelle 1998). Different animal organic waste has already been studied for conversion into vermicompost by different species of earthworms including:

- cattle dung (Edwards *et al* 1985; Mitchell 1997) ,
- turkey waste (Edwards *et al* 1985),
- cow slurry (Hand *et al* 1988),
- pig waste (Chan and Griffiths 1988),
- horse waste (Hartenstein *et al* 1979),
- plant litter (Bansal and Kapoor 2000),
- poultry dropping (Ghosh *et al* 1999),
- mango leaves (Talashilkar *et al* 1999),
- water hyacinth (Gajalakshmi *et al* 2001),
- paper waste (Gajalakshmi *et al* 2002),
- municipal sludge (Baker *et al* 2002) and
- agricultural waste (Zinia 2007)

The life cycle of the species has been documented in the literature, but the growth, population spectra and reproductive parameters of *E. fetida* in different crop residues, cattle dung, and vegetable waste has not been studied yet. The survival and fecundity of the species in different waste must be known to ensure a successful vermicomposting. Moreover, the growth and reproduction of the species is an indication of its favored feed. This problem was designed to study the population spectra and maximum biomass achieved in different feeds by both the species.

Methodology

Ecologically, earthworms can be classified as epigeic, endogeic and anecic. The epigeic and anecic have been harnessed for use in the vermicomposting process. The selection of the correct earthworm species for a particular vermiculture application is important. Earthworm stock of species *E. fetida* was kept in cattle dung as culturing material, and specimens were randomly picked from this stock culture. Different feeds used were:

- Cow dung (F1),
- Cattle dung and wheat straw(F2)-1:1 ratio
- Cattle dung and rice straw(F3)-1:1 ratio
- Cattle dung and leaf litter (F4)-1:1 ratio
- Mixed vegetable waste (F5).

The worms were fed with fresh, urine free cattle dung as good results from such feeding were obtained by Neuhauser *et al* (1979). Leaf litter of *Sterculia alata,* commonly known as oplar, was used as substrate. The crop residues used in the experiment were dried, crushed into small pieces and mixed with cattle dung in 1:1 ratio.

The feeding preference was related to the maximum biomass gained, high population and more young ones. For studying the growth and reproduction of the *E.fetida,* earthworms in pairs were kept in each petri dish (14.8 cm x 14.8 cm) and plastic containers with different formulated feeds F1-F5. Three replicates for each feed were maintained. The moisture content was kept to 70-80%. Cocoon production was recorded weekly. The population spectra was studied by introducing 20 young worms in each feed at the start of the experiment. The earthworms were allowed to grow in the feeds and no additional feed was added. The different life forms including cocoons, juveniles, young forms and adult were counted in different feeds.

Statistical analysis was done using an analysis of variance (ANOVA) and a student t-test was done using statistical packages like GSTAT and CPCS 1.

Results and Discussion

Population spectra of *E.fetida* in different feeds

Five earthworms of the epigeic species were introduced in each feed for studying growth. Three replicates were maintained to avoid the experimental error. *E.fetida* gave maximum weight in (F4) Cattle dung + Leaf litter. The maximum weight achieved was 3.543 ± 1.63 gm followed by (F3) Cattle dung +Rice straw which was 3.262±2.54 gm, then (F1) Cow dung which showed 2.732 ± 3.15 gm and then 2.539 ± 2.81 in (F2) Cattle dung +Wheat straw and 2.430 ± 1.39 gm in (F5) in Mixed vegetable waste after seven weeks(**Fig.1**). Thus *E.fetida* favoured F3 the most. The results indicated that agricultural and animal residues can be utilized.

Different life forms viz., adults, sub adults, juveniles and newly emerged were counted in different feeds containing *E. fetida* separately. Initially 20 young worms of both species were

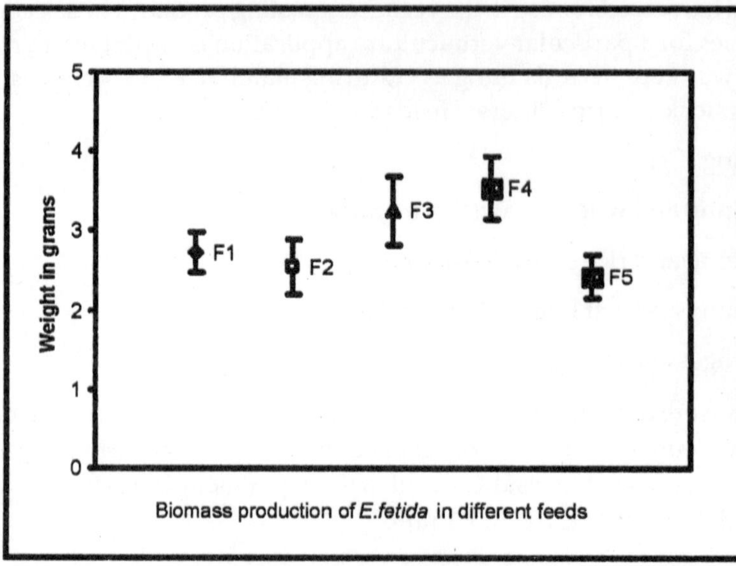

Fig 1. Biomass production of E.fetida in different feeds after 7 weeks

added. The different life forms were counted after 11 weeks. The population spectrum of *E.fetida* is presented in Table 1.

Table 1. Population spectra of *E. fetida* in different feeds after 11 weeks

Feed	Adults	Sub Adults	Juveniles	Newly Emerged
Feed 1	14	08	86	69
Feed 2	18	10	49	88
Feed 3	15	08	59	96
Feed 4	16	12	65	102
Feed 5	14	12	74	58

Reproductive Parameters

E.fetida showed the shortest life cycle of 63±1.24 days and 64±1.69 days in F4 and F3 feed, respectively. F1, F2 and F5 feeds showed a life cycle of 66±1.63, 69±0.47 and 70±2.05 days (Fig.7), respectively. Reinecke *et al* (1992) also reported life duration of 70 days for *E.fetida* which is comparable with our results. Tripathi and Bhardwaj (2004) who reported 4 and 5¼ months for *E.fetida* and *L.mauritti* completed their life cycle in the desert region of Rajasthan. The life cycle varied significantly in different feeds in our study ($p<0.05$). The calculated CD at 5% was 3.667 (F-ratio = 7.15).

E.fetida showed a mean cocoon production of 0.45, 0.41, 0.51, 0.51 and 0.37 per worm per day in different feeds. There existed a significant difference ($p<0.05$). The calculated CD at 5% was 0.169 (F-ratio = 4.33). The mean cocoon production ranged from 0.37-0.51 during our study. This indicated that the quality of the organic material affected the reproductive potential. Feed containing Cattle dung + Rice straw and Cattle dung + leaf litter was favoured by both species of earthworms. Average life cycle duration in days and mean cocoon production per worm per day in *E.fetida* has been presented in Table 2.

Table 2. Average life cycle duration in days and mean cocoon production per worm per day in *E.fetida*

Feed	Life cycle	Mean cocoon production per worm per day
F1	66±1.63	0.45±0.02[a]
F2	69±0.47[b]	0.41±0.07[a]
F3	64±1.69[a]	0.51±0.04[b]
F4	63±1.24[a]	0.51±0.02[b]
F5	70±2.05[b]	0.34±0.08[a]

†Different small letters indicate significant difference between feeds ($p<0.05$) CD (5%) =3.667 CD (5%) = 0.169

Garg *et al* (2005) studied the cumulative cocoon production by *E.fetida* in different animal waste. Sheep waste showed 0.44, 0.39 in cow, 0.19 in buffalo, 0.37 in horse, 0.28 in donkey, 0.36 in goat and 0.32 in camel waste. The number of cocoons produced per worm per day was in the order: sheep > cow H" horse H" goat > camel > donkey > buffalo. The difference between the rate of cocoon production could be related to the biochemical quality of the feed.

E.fetida showed an incubation period of 19-23 days of cocoons to hatch. Reinecke *et al* (1992) showed an incubation period of 21 days for *E.fetida*, 18 days for *P.excavatus* and 17 days for *E.eugeniae*. Dominguez *et al* (2001) reported an incubation period of 14 days in *E.eugeniae*. This confirmed that the substrate had a little effect on the incubation period of the cocoons.

Hatchlings from a single cocoon were observed. In case of *E.fetida*, F1 showed a 3.66±0.47 mean number of hatchlings, F2 had 3.33±0.82, F3 had 3.37±1.24, F4 had 4.66±0.47 and F5 showed 4±0.51 mean number of hatchlings. *E.fetida* produced 1-5 hatchlings from a single cocoon. Evans and Guild (1948) observed 1-4 hatchlings from one cocoon in *E.fetida*. The mean incubation period and mean number of hatchlings from one cocoon in *E.fetida* in different feeds has been presented in Table 3.

Table 3. Mean incubation period and mean number of hatchlings from one cocoon in *E.fetida* in different feeds.

Feed	Incubation period (days)	Mean number of hatchlings from one cocoon
F1	19 ± 0.82	3.66 ± 0.47
F2	23 ± 1.24	3.33 ± 0.82
F3	21 ± 1.70	3.37 ± 1.24
F4	22 ± 1.63	4.66 ± 0.47
F5	20 ± 2.05	4.00 ± 0.51

Values are mean±SD

Conclusions

Thus earthworms can be cultured in different agricultural and animal wastes. Results showed that the epigeic species *Eisenia fetida* is capable of growing in different substrates. *E.fetida* favored agricultural wastes the most. Rice straw and leaf litter was preferred by the species. The reason was the organic nutrients present in these agricultural wastes which resulted in higher biomass production and better reproductive parameters. In a nutshell, a cleaner environment can only be achieved if we change our attitude to consider the waste not a nuisance but an asset and a new source of secondary raw materials for generating gold from garbage and silver from sewage.

Vermicomposting is an economic, easy and eco friendly technology for the disposal of organic wastes. It has considerable potential being economically viable.

References

1. Atiyeh R M, Dominguez J, Subler S and Metzger J D (2000) Earthworm - processed organic wastes as components of horticultural potting media for growing marigold and vegetable seedling, *Compost Sci. Util.* **8**: 215-223.

2. Baker G, Michalk D, Whitby W O and Grady S (2002) Influence of sewage waste on the abundance of earthworms in pastures in South-Eastern Australia. *Eur. J. Soil. Biol.* **38**:233-237.

3. Bansal S and Kapoor K D (2000) Vermicomposting of crop residues and cattle dung with *Eisenia fetida. Bioresour. Technol.* **73**:95-98.

4. Chan L P S and Griffiths D A (1998) The vermicomposting of pretreated pig manure. *Biol. Wastes* **24**: 57-69.

5. Darwin (1881) The formation of vegetable mould through the action of worms with observations on their habits. John Murray, London, 326 pp.

6. Dominguez J, Edwards C A, Ashby J (2001) The biology and population dynamics of *Eudrilus eugeniae* (Kinberg) (Oligochaeta) in cattle waste solids. *Pedobiologia* **45**: 341-353.

7. Evans A C and Guild W J (1948).Morphological studies n relationship between earthworms and soil fertility. *App. Biol.* **35**:471-434.

8. Edwards C A, Burrows I, Fletcter K E and James B A (1985) The use of earthworms for composting food wastes, In: *Composting of agriculture and other wastes*, Gasser J K (Eds.), Elsevier Applied Science, Amsterdam, pp 229-242.

9. Gajalakshmi S, Ramasamy E V and Abbasi S A (2001) Potential of two epigeic and two anecic earthworm in vermicomposting of water hyacinth. *Bioresour. Technol.* **76** (3): 177-181.

10. Gajalakshmi S, Ramasamy E V and Abbasi SA (2002) Vermicomposting of paper waste with the anecic earthworm *Lampito mauritii, Indian. J. Chem. Technol.* **9**:306-311.

11. Garg V K, Chand S, Chillar A and Yadav A (2005). Growth and reproduction of *Eisenia fetida* in various animal wastes during vermicomposting. *App. Eco. Environ. Res.* **3**(2):51-59.

12. Ghosh M, Chattopadhyey G N and Baral K (1999) Transformation of phosphorus during vermicomposting. *Bioresour. Technol.* **69**:149-159.

13. Guandi B, Edwards C A and Arancon Q (2002) Changes in trophic structure of soil arthropods after the application of vermicompost. *Eurp. J. Soil. Biol.* **38**:161-165.

14. Hand P, Hayes W A, Satchell J E, Frankland J G Edwards C A and Neuhauser E F (1988) The vermicomposting of cow slurry. *Pedobiologia*, **31**: 199-209.

15. Hartenstein R, Neuhauser E F and Kaplan D L (1979) Reproductive potential of the earthworm *Eisenia fetida. Oecologia* (Berlin) **43**: 329-340.

16. Lavelle P (1988) Agastrodrilus omodeo and Vaillaud, a genus of carnivorous earthworm from the Ivory Coast. In: *Earthworm Ecology from Darwin to vermiculture*, Satchell J E, (Eds.), Chapman and Hall, New York, pp 425-429.

17. Mitchell A (1997) Production of *Eisenia fetida* and vermicompost from feedlot cattle manure. *Soil Biol. Biochem* . **29**:763-766.

18. Neuhauser E F, Hartenstein R and Kaplan D L (1979) Second progress report of the potential use of earthworms in sludge management, In: *8th National sludge Conference,* Information transfer, Silver Springs, Md, 9pp.

19. Reinecke A J, Vilizoen S A and Saayman R J (1992) The suitability of *Eudrilus eugeniae, Perionyx excavatus* and *Eisenia fetida* (OLIGOCHAETA) for vermicomposting in Southern Africa in terms of their temperature requirements. *Soil. Biol. Biochem.* **24**(12): 1295-1307.

20. Talashilkar S C, Bhangarath P P and Mehta V B (1999) Changes in chemical properties during composting of organic residues as influenced by earthworm activity. *J. Indian. Soc. Soil. Sci.* **47**:50-53.

26

Impact of Tourism on Biodiversity: A Case Study of Sariska Tiger Reserve

NIDHI GANDHI

Introduction

The biodiversity of the Sariska Tiger Reserve is impacted adversely by human activities in many different ways such as habitat destruction, over harvesting, mining, environmental pollution, subsistence use of rare plants and animals, introduction of alien species, and uncontrolled touristic activities. Human population pressure is considered an important underlying cause of species over-exploitation and habitat biodiversity loss. The relationship between humans and biodiversity at the Sariska Tiger Reserve is quite alarming.

There is a phenomenal variation in diversity across the entire range of living systems. No single process or theory can explain a phenomenon as complex as biological diversity. The intellectual challenge and scientific value of the study of diversity lies in the conceptual synthesis required to understand a complex phenomenon that is influenced by many different interacting factors and processes (McIntosh, 1987). The escalating rate of erosion of biodiversity in India, which is one of the top twelve mega-diversity countries, in the world, is a cause for grave concern. With numerically the third largest scientific and technical manpower, India ought to be well placed to face new challenges, and reap benefits from new opportunities. Unfortunately, we are poorly equipped to do so. Evidently, we need a new culture and new institutions to take good care of our human and natural resources. This new culture, these new institutions have to be appropriate to the new age of information. This requires that we embrace a new democratic culture of inform and share, eschewing the current bureaucratic culture of control and demand. At the same time, we must create a stake for the people in taking good care of the natural resource base. We must awaken in people the spirit of innovation and enterprise to marry the rich traditional knowledge base with modern science and to capitalize on our wealth of genetic resources (Gadgil and Rao, 1998).

According to Agenda 21: "The current decline in biodiversity is largely the result of human activity and represents a serious threat to human development." (Agenda 21, Chapter 15, Section 2.)

The problem of managing biological diversity is fairly straightforward conceptually. Biological diversity is generally used to describe the variability among living organisms and the ecological complexes of which they are a part. Without this variability in the living world, ecological systems and functions would breakdown, with detrimental consequences for all life including humans. Consequently, as biological diversity is essential to survey the basic ecological services and resources necessary to maintain human welfare and even existence, then a minimum level of biological diversity is required to sustain the well-being of not only current but also future generations (Forlke *et al.*, 1994).

Indeed, ecologists now believe that the highest levels of diversity in terms of numbers of species present in a locality occur in a community with moderate levels of human intervention rather than in a biological community totally free of human disturbance. This is because disturbance can create more heterogeneous environmental regimes permitting not only species favouring less disturbed conditions to persist, but also making room for other species thriving under more disturbed conditions (Gadgil and Rao, 1998). This does not mean that disturbance is wholly favourble to diversity. Rather it plays a negative role in the protected areas. Drastic human disturbances often depress levels of the diversity of life.

Driving Forces for Biodiversity Loss

The main driving forces behind the biodiversity loss arise from human activities, and can be distinguished in terms of proximate and underlying causes. Proximate causes refer to the direct over-exploitation of a species (e.g.: hunting, fishing, poaching) and the indirect impact of ecosystem degradation or destruction that leads to a species loss (e.g.: habitat alteration).Underlying causes refer to the economical, social and cultural factors that lie behind the economic activities that lead to the direct depletion of species, and the destruction and degradation of their habitat. These underlying causes include the scale and growth of human population, culture and ethics, economic incentives and institutions (Barbier et al., 1994).

India has a rich and varied heritage of biodiversity, encompassing a wide spectrum of habitats from tropical rainforests to alpine vegetation and from temperate forests to coastal wetlands. India figured with two hotspots - the Western Ghats and the Eastern Himalayas - in an identification of eighteen biodiversity hotspots carried out in the eighties (Myers, 1988). Recently, Norman Myers and a team of scientists have brought out an updated list of 25 hotspots (Myers et al., 2000). In the revised classification, the two hotspots that extend into India are The Western Ghats/Sri Lanka and the Indo-Burma region (covering the Eastern Himalayas); and they are included amongst the top eight most important hotspots. In addition, India has 26 recognised endemic centres that are home to nearly a third of all the flowering plants identified and described to date.

However, this rich biodiversity of India is under severe threat owing to habitat destruction, degradation, fragmentation and over-exploitation of resources. According to the Red List of Threatened Animals (IUCN, 2000), 44 plant species are critically endangered, 113 endangered and 87 vulnerable. Amongst animals, 18 are critically endangered, 54 endangered and 143 are vulnerable (Table II and III). Ten species are Lower Risk conservation dependent, while 99 are Lower Risk near threatened. India ranks second in terms of the number of threatened mammals, while India is sixth in terms of countries with the most threatened birds (IUCN, 2000).

Table I Threatened Animals of India by Status Category

Ex	EW	CR	EN	VU	LR/cd	LR/nt	DD
0	0	18	54	143	10	99	31

Ex-extinct; EW-Extinct in the Wild; CR- Critically Endangered; VU-Vulnerable; LR/cd-Lower Risk conservation dependent; LR/nT- Lower Risk near threatened; DD-Data Deficient. Source: IUCN, 2000

Table II Threatened Plants of India by Status Category

Ex	EW	CR	EN	VU	LR/cd	LR/nt	DD
7	2	44	113	87	1	72	14

Ex-extinct; EW-Extinct in the Wild; CR- Critically Endangered; VU-Vulnerable; LR/cd-Lower Risk conservation dependent; LR/nT- Lower Risk near threatened; DD-Data Deficient. Source: IUCN, 2000.

The Nature of the Problem

The exponential growth of the human population, making humans the dominant species on the planet, is having a grave impact on biodiversity. This destruction of species by humans will eventually lead to a destruction of the human species through natural selection. Man has proved to be the worst enemy of not only mankind but of the animal and plant kingdom also. The entire history is a mute witness to what man can do to the ecosystem that supports him. Several conflicts have engendered between local communities - heavily dependant on the resources of these areas and Protected Areas (PAs) managers. The human activities are putting a lot of pressure on the resource of the Sariska Tiger Reserve like the destruction of the habitat due to the grazing of the cattle; withdrawal of water below a critical limit that leads to a water-crisis in the tiger reserve; Protected Area personnel and people conflict; increasing population of people and the livestock; unsustainable tourism; increasing traffic on the two highways that criss-cross Sariska; the poaching of animals for various purposes and fuel wood and fodder collection by villagers.

The destruction of the habitat for all kinds of purposes, including tourtic activities, agriculture, mining, construction of roads, clearing of forest for agriculture and pasturing is probably the most important threat to biodiversity in Sariska. Biodiversity will continue to decline as long as we continue to remove and constrict the natural habitats in which wild species live. This loss is bound to be costly, because natural ecosystems provide vital services to human societies. Recreational, aesthetic, and commercial losses will be inevitable. The loss of wild species, therefore, will bring certain and unwelcome consequences, because it is linked directly to the degradation or disappearance of ecosystems.

The emergence of wildlife tourism is the most significant change that has taken place in the national parks after the enforcement of the Wildlife (Protection) Act, 1972. The wildlife tourism industry has contributed significantly to the national economy through the increase in the arrival of domestic and international tourists to wildlife tourism destinations, but due to a lack of effective tourism managerial policies suited to wilderness areas, the wildlife resources and their habitat is subjected to negative impacts associated with growing influx of tourists and related infrastructure development. The unregulated tourism also affects the capacity of the national parks to withstand the tourist pressure. The problem also becomes acute when the wildlife habitat is misused by destructive land uses. The situation further deteriorates with lack of active participation by stakeholders in eco-development activities.

Barring a few exceptions, the rapidly expanding tourism industry is proving highly destructive to wildlife and local people. Also there is an adverse and serious impact of increased and commercialised pilgrimage activities on protected areas and the Sariska Tiger Reserve is no exception. Wildlife tourism is a contemporary form of early aristocracy recreation in wilderness areas. Earlier the recreation was confined to the wealthy kings, emperors and rulers but the contemporary wildlife tourism is open for all income classes of people and any one who has leisure time at his disposal can participate in the wildlife tourism. The creation of the tourist infrastructure inside the national parks attracts the regional, national and international tourism, thus, the wildlife has become transformed into a recreational resource.

There is a diversity of the wildlife species surviving in a few of the remaining natural wildlife habitats and people are interested in watching dangerous animals- carnivores, herbivores and poisonous creatures in natural open settings. The motivational factors associated with wildlife tourism are primarily to see wildlife in the natural jungle habitats, to explore the mystique of nature, to encounter dangerous animals and to experience the thrill and risk involved in visiting the home of the wild animals. Scenery, culture, geology, and history of the wilderness areas complement the wildlife tourism and enhance the recreational experience for wildlife tourists. The impact of tourism on wildlife can be both positive and negative. However, if the intensity of the touristic development is high and continuous, then, the impact of tourism on wildlife will be negative.

The human use of habitats often simplifies them. We might, for example, remove fallen logs and dead trees from woodlands for firewood, thus diminishing an important microhabitat on which several species depend. When a forest is managed for the production of a few or one species of tree, tree diversity obviously declines, and with it so does the diversity of a cluster of plant and animal species dependent on the less favoured trees. Streams are sometimes "channelized" – their beds are cleared of fallen trees and riffles, and sometimes the stream is straightened out by dredging. Such alterations inevitably bring on a loss of diversity of fish and invertebrates that live in the stream.

The unregulated tourism is affecting the reserve's capacity to withstand the tourist pressure. There are a number of temples in the area. Bharathari temple near Bharathari, and the Hanuman temple at Pandupole in the core area are visited by several hundred thousand people every year. Tuesdays and Saturdays are days of free entry for the pilgrims. Pilgrim pressure increases particularly during the monsoon season and virtually becomes unmanageable on festival occasions. Also, in the buffer zones of Nahar Sati, Sati Mata Narayani Ji, Talvriksh and Garbaji are the religious places visited by several hundred thousand people every year. A temple within the park's core area attracts pilgrims, particularly on Tuesdays and Saturdays, causing disturbance to wildlife. They come in groups by vehicle or on foot, and are never refused entry.

The local people have been visiting road checkpoints into the Core Area. There is an enforced ban on commercial exploitation. The effectiveness of the checkpoints appears a little weak. There are a number of reasons for the weak monitoring and enforcement of rules. One reason is the inadequate pay, education and equipment of forest guards, and the fact that single unarmed guards often live in villages they are supposed to control.

A second reason is the failure to formally settle the rights of local villages. This prevents the declaration of Sariska Tiger Reserve as a National Park and in that case, utilisation would be much more restricted. At present, not all villagers appear to be aware of the rules.

The third reason is the slow legal system in India. Court cases are delayed for years. This makes the sanctioning mechanism available to the Reserve Management ineffective. The fourth reason is that the reserve employs a fairly large but low-paid staff

Development of Infrastructure and Environmental Pollution

The two major causes are the location of mines in the vicinity and the heavy traffic due to tourists, passengers and roadway buses on the state highways that criss-cross Sariska. Roads

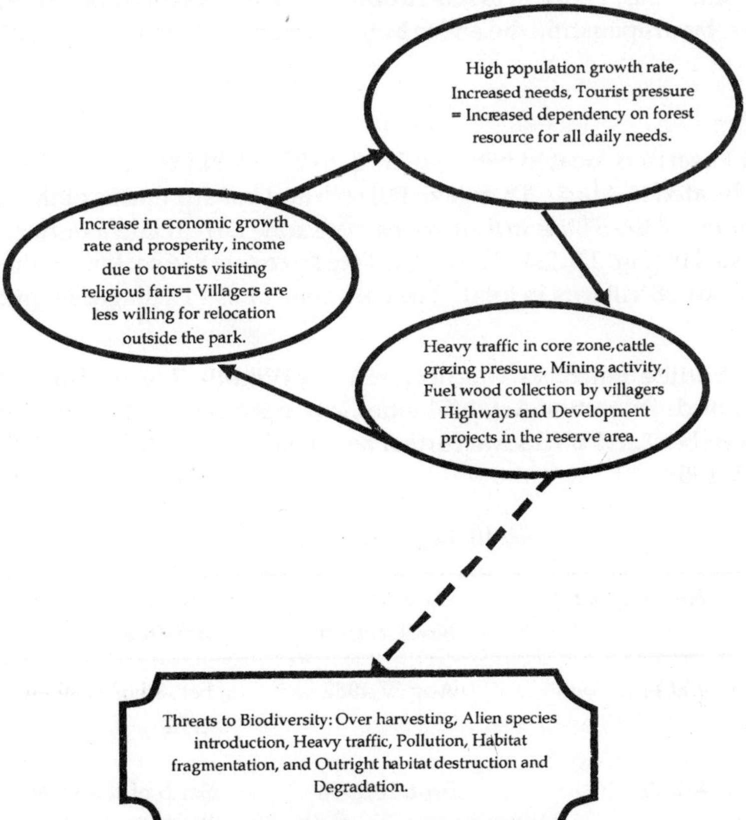

Flowchart Depicting the Relationship between Human Activity and Biodiversity in the Sariska Tiger Reserve

and other infrastructure also impact on the wildlife by modifying animal behaviour and species distribution in areas with infrastructure. Wildlife is impacted directly by infrastructure through substantial noise, disruption of the physical environment, alteration of the chemical environment, and the introduction of exotic species, but most of all, by accelerating processes like hunting, logging, slash and burn and tourism.

Animals avoid areas near infrastructure, their breeding success decreases in developed areas, and habitats become fragmented. The ecological impacts of losses of habitats and redistribution of animals away from the development may again affect the foraging success or survival substantially in areas beyond these initial zones of disturbance, and, hence, result in overgrazing, erosion, changes in the predation pressure and breeding success. Avoidance of developed areas therefore affects a much larger area than that of the physically altered footprint of development. The extent of the zones within where the wildlife will become affected by the infrastructure varies according to species, season and type of disturbance, habitat and other environment factors.

The cumulative effects of the construction of roads, tourist pressure, changes in the vegetation composition may result in the disruption in the ecosystem function. The outcome is that infrastructure, by propagating the entire human associated activity, is leading to a loss of biodiversity.

Study Area

The Sariska Tiger Reserve is located between 27° 5′ to 27° 33′ N Latitude and 79° 17′ to 76° 34′ E Longitude. It is located in Alwar, Thangazi, Rajgarh and Bansur tehsils of the Alwar district, Rajasthan. The zoning of the STR is in four zones: core zone I, II, III and a buffer zone covering a total area of 866 sq. km (Fig. 2.3, 2.4). The present study covers 866 sq. km. of the Sariska Tiger Reserve. This area has 28 villages in total. The core zone I has 11 villages include the grazing settlements.

The Sariska Wildlife Sanctuary was declared as a Wildlife Reserve under the Rajasthan Wild Animals and Birds Protection Act, 1951 vide Govt. notification Miscellaneous No.F.39 (2) For/ 55 dated November 7, 1955. This was further amended vide notification No.F.39 (2) Rev.A/ 54 dated August 5, 1958.

Table III Legal Status of STR

NP/WLS	Area (sq. km.)	Date of Notification	Sec. Under which notified
National Park (proposed)	400.14	Aug.27,1982	Sec. 35 of WLPA, 1972
Wildlife Sanctuary	492.29	Sept. 18,1958	Sec.5 of the Wild Animals & Birds Protection Act, 1951 and Sec 66(4) of WLPA,1972

Source: Status Report 2 003-04, Sariska Tiger Reserve.

Human habitation is denser in wider valleys, which has streams and a comparatively higher water table. According to the 1991 census, there was a population of 2,54,011 in total in the area. The human population inside and around the reserve is increasing rapidly because of illiteracy. The increasing population and the consequent biotic pressures have resulted in the degradation of the forest areas, particularly on the fringes. It is important to shift those villages from the National Park, which have been paid some compensation and have them allocated to alternate land. Cattle rearing is the main profession of the local people and they depend on the forest area for grazing. The population of livestock is also increasing steadily. Domestic livestock entering the reserve spreads diseases.

Sariska is situated on the Aravalli Range, which extends for about 700 km from the southwest of Palanpur in Gujarat, northeastward and up to the Delhi Ridge. The Aravallis appears somewhat like two fans, joined handle-to-handle, meeting near Beawar. They fan out in the form of the Alwar hills in the north and the Mewar hills in the South. The hills vary in elevation, ranging on an average between 300 and 900m above mean sea level (msl), the highest point being 1290m. Local relief ranges from 20 to 300m and dominant slopes range from 5 to 80%. The width of the mountain is 8 to 30km in the central part. It widens to 160km in the southern fan and 172km in the north.

The climate of this tract is sub-tropical characterized by distinct winter, spring, summer and Monsoon. The summer season commences from middle of March and the heat soon becomes very intensive in April. Hot westerly winds known as "Loo" are common during April, May and part of June. This period is extremely hot. The nights are generally pleasant even during summer. The rainy season commences from late June and continues till the middle of October. The winter season commences from November. It becomes cold in December – January and the temp, drops down to sub zero during the nights.

Plate: Main Gate of Sariska Tiger Reserve

Sariska, which used to be a hunting place *(Shikargah)* of the princely state of Alwar, in the past, is nowadays a "Tiger Reserve" of international reputation. Sariska is very rich in biodiversity; it is equally rich in flora and fauna. The vegetation of Sariska constitutes an ideal habitat for Tiger and other varieties of animals. Broadly, the forests and vegetation of Sariska fall under the category of a dry deciduous forest.

The Sariska Tiger Reserve, with its awe inspiring craggy canyons and the tropical dry deciduous scrub jungles and limited water supply is one of the extreme limits of the distribution of the tiger in India - perhaps in the world. In this seemingly hostile habitat the ecological tolerance of the tiger is on test. Sariska therefore is a unique tiger reserve in India. Also, the natural history of the ecosystem combines with the rich history of the country.

Methodology & Findings

The objectives of the present work were fulfilled through secondary and primary sources of information. The secondary information source includes statistical information, reports, articles, toposheets, land use records and maps of STR. It also includes a District Census Handbook-District Alwar (1991 & 2001). The study is supplemented by personal observation exercises. For primary data, questionnaires were prepared for the villagers, tourists and forest officials of STR. The questionnaire is an integrated part of the primary data collection and it is included in the following aspects for villagers: human awareness of the problem, the life style and the socioeconomic structure of the villagers, the land use pattern of the village, the level of dependency of the local population on the resources from the reserve, NGOs and their efficiency in the area, probable measures suggested by the villagers and their practical applicability. Apart from the questionnaire survey, interviews were conducted in person and also in groups to gather information regarding the general perception of the residents regarding various aspects of problems like the depletion of natural resources, and biodiversity and how they react to tourism.

The present study is based on a multiple measurement technique. The study has combined a personal survey for villagers, observations, tourist survey and a forest official survey with vital secondary statistical information obtained from the Sariska Tiger Reserve Headquarters, several governmental bodies and non-governmental organizations. Secondary data collected was analysed both qualitatively and quantitatively. Data on forest degradation was obtained in the village survey through the visual inspection of the extent the of forest degradation while interviewing (comparing the cover in core zone I, II, III and buffer zone); by asking villagers to compare the condition of the forest today with earlier times (25 years back) and by a visual determination of forest use penetration, that is, the depth into the forest from the village boundary where user pressure was evident. The recurrent conflict with the villagers on the matter of grazing in the STR has alienated the villagers from the forest authorities.

A net impact scoring matrix is developed to scale the human impact indicators as per the environmental items in Sariska quantitatively. The indicators were rated between the scales of magnitude to check the severity of the impact. A net score was computed by adding the score obtained by the indicators. A scoring key was prepared and the magnitude of the indicators were classified into: negligible impact, slight, moderate, high and severe human impact on the STR.

In discussion with the Field Director and the Conservator of Forests, STR, the following details were noted about the tourists visiting the park. The local people have been visiting Pandupole Temple located inside the Core-I area of the Sariska National Park since the State times. A toll tax of Rs. 1.00 was charged from people going on foot in the Park. Later motorized vehicles changed the whole scenario in a negative way. Due to an old lower court order, the entry inside the Park was made free on Tuesdays, Saturdays and Mela days. However, it is being experienced that many people from Delhi and the surrounding towns of Jaipur and Alwar are coming in their private cars on these two special days with the main objective of enjoying the nature and picnicking. Unfortunately, incidences of alcohol consumption in the Park by some of these 'devotees' have also been reported. The vehicles entering the Park are a major source of disturbance to the wild animals and it has been observed that the sighting of herbivores and predators has significantly reduced. The sighting of langoors and monkeys increases as they come near the road for the food thrown by the ignorant people for them to eat. This is changing the behavioural pattern of these animals. Apart from disturbing the wild animals, the free entry into the Park causes a loss of revenue to the State exchequer.

The following table indicates the number of tourists visiting the Park and the revenue generated (On days other than Tuesdays and Saturdays) Fig. A shows the trend of tourists' visiting Sariska (1993-2004).A Munsif Court had passed orders that entry inside the Park for people visiting Pandupole on Tuesdays and Saturdays would be kept free. This order has resulted in a drastic loss of Government Revenue and has certainly affected the development of the Park.

SARISKA TIGER RESERVE - Administrative Map

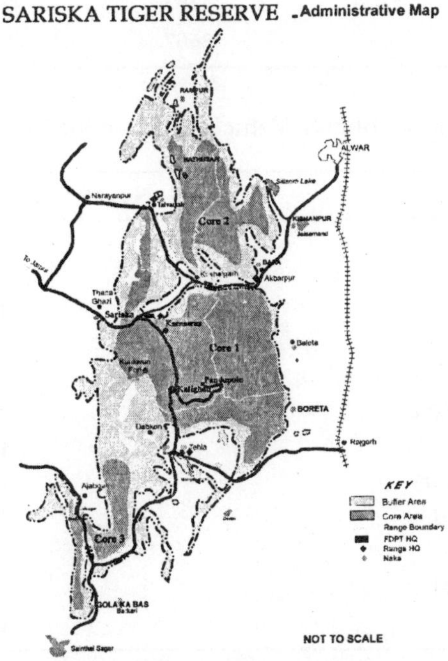

Fig. Administrative Map of STR

Source: Office Records, Sariska

Since there is no limit of vehicle entry into the Park on these special days, there is lot of disturbance to the wild animals as the pucca road to Pandu Pole runs through the heart of the core area. Data was collected for the last financial year (2003-2004) pertaining to the number of free vehicles allowed in the Park on Tuesdays and Saturdays. From the Table V it is clear that in one year about 15709 vehicles have been allowed free entrance in the Park on Tuesdays and Saturdays.

Table IV Number of Tourists Visiting Sariska and the Revenue Generated

Years	Indians	Foreigners	Students	Total	Entry Fee Collected (Rs.)
1993-94	43246	6127	6937	56310	11,44,282.00
1994-95	48070	4883	5457	58410	11,73,310.00
1995-96	45911	6799	5732	58442	12,27,621.00
1996-97	60020	7502	5450	72972	15,11,082.00
1997-98	49746	9138	5078	63962	30,95,754.00
1998-99	41300	11558	3333	56191	35,93,602.00
1999-20	38134	8194	8659	54987	37,95,423.00
2000-01	41139	19181	4580	64900	57,43,070.00
2001-02	40716	15289	2705	58710	56,06,771.00
2002-03	36097	6824	4163	47084	33,61,705.00
2003-04	32079	6897	3657	42633	34,42,251.00

Table V Number of Free Vehicle Entry in the Year 2003-04

Month	No. of Vehicles Allowed Free
April, 2003	989
May, 2003	999
Jun-03	1071
July, 2003	2049
August, 2003	1884
September, 2003	2758
October, 2003	1059
November, 2003	1109
December, 2003	1121
January, 2004	718
February, 2004	880
March, 2004	1072
TOTAL	**15709**

Source: Office Records, Sariska

Fig. A Trend of Tourists Visiting Sariska (1993 to 2004)

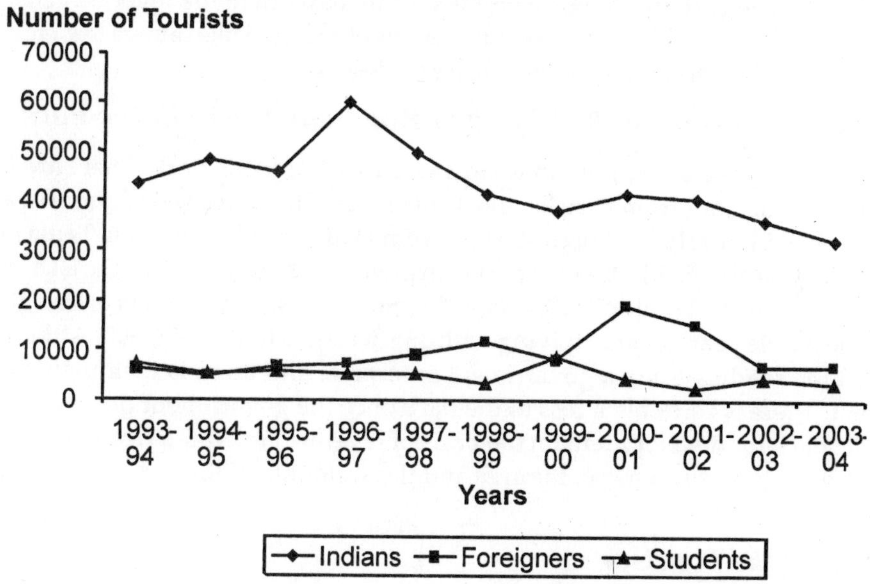

Source: Office Records, Sariska.

The entry fee of the vehicle in the Park is Rs. 125.00 per vehicle. Thus the loss of revenue can be estimated as:

Entry fee per vehicle	*Total free vehicle entry*	*Loss of revenue (Rupees)*
(Rupees) 125	*(Numbers)* 15709	19, 63, 625

The entry fee for Indians is Rs. 25.00 per individual. If the average occupancy per vehicle is taken to be 4 then the loss of government revenue can be estimated as:

Average occupancy per vehicle	*Entry fee per Indian individual (Rupees.)*	*Loss of revenue (Rupees)*
04	25	15,70,900

Thus the estimated total loss to the government revenue is Rs. 35,34,525. Thus we find that the loss of revenue to the Government is slightly more than the revenue generated in the Park in 2003-2004 i.e. Rs. 34,42,251. Also, other items of revenue like camera fees etc. has not been taken into account, which could further increase the losses.

Loss of Park Revenue by the Plying of Roadway Buses in the Park

There is a free entry for the Rajasthan Roadway buses inside the park. There are twelve such buses that enter the Park from Sariska and Tehla gates. The buses start from Alwar and go to Dausa via Tehla. Similarly the buses starting from Dausa reach Alwar via Tehla and Sariska. As discussed with the Field Director and Conservator of the Forest of Sariska, if we take a general estimate, then 12 multiplied by Rs. 125 is equal to Rs. 1500 per day is straight loss to the Park. For the whole year, on multiplying with 365, it comes to Rs. 5,47, 500. Although it can be argued that the roadways are a government enterprise and there is seemingly no loss to the government, there is certainly a loss to the park once the government decides to allocate the funds to the park for development on the basis of revenue generated from the entry fees. The buses are causing a disturbance in the area and the accidental killing of wild animals has been reported in the park.

Perception of the Tourists

The perception of tourists visiting the STR was collected by a structured questionnaire. Tourists visit Sariska for different reasons like adventure, holidays and visiting religious places. The mode of travel inside the park is varied like two-wheelers, jeeps, cars, and buses. The number of questionnaire distributed was 200. There was a good response from tourists. The response percentage was 60% (120). The following two graphs depict their responses. The responses were collected about their views on restricted tourist entry, ban on the development of the commercial projects, charge fee from the Indian tourists on Tuesdays and Saturdays, the use of CNG vehicles in the park and their level of satisfaction about the biodiversity of the Sariska Tiger Reserve.

68% respondents were in favour of restricted tourist entry; 82% supported the ban on the commercial developments in and around the reserve, 68% of tourists supported the idea not to charging entry fee on all days from the Indian tourists, 87% support the use of CNG vehicles in the park and 100% support the strict implementation of the anti-poaching laws. From this we can conclude that all responses received are quite positive for the STR except the charge of entry fee from the Indian tourists on Tuesdays and Saturdays. This was a little expected, as the number of Indian tourists responses were more than foreign tourists.

The level of satisfaction on the biodiversity of the park was averagely good as 45% of the tourists were satisfied. It is not a very good percentage as the frequency of spotting a tiger in this Tiger Reserve is decreasing everyday.

The Impact on the Spatial Distribution of the Tiger in STR

After making a detail survey of all possible issues and after a discussion with the Field Director and Conservator of the Forest, in the Sariska reserve it can be said that earlier the tiger population was distributed throughout the Tiger Reserve. In olden times when the disturbance was due to the increasing population, agriculture and mining was the least and the habitat was secure, and the tigers were reported from Tehla, Akbarpur Sariska and the Talvriksh Ranges. The Ramgarh-Sariska Corridor was intact and the movement of tigers from the nearby Jamva Ramgarh Sanctuary was reported earlier, but due to the encroachment on the revenue lands situated in the corridor area, an increased mining and heavy biotic pressure, the movement of wild animals specially the tigers have been adversely affected. Then census has shown that the distribution of the tigers is restricted to the Sariska Range only. The movement of tigers is there in Akbarpur and the Tehla Ranges but the tigers have come back to the Sariska Range. This is due to the heavy disturbance of mining and biotic pressure in the two Ranges. The movement of tigers has been reported in the Talvriksh Range also but instead of residing in the area, the animal has preferred to come back to Sariska. The population of tiger is static and no change in its population had been observed. The birth of cubs has also not been reported for the last three years. This is quite disturbing and requires immediate managerial intervention.

As per the latest information, a stage has come that the tigers are being reintroduced in the STR to maintain the status of it being a Tiger Project.

Net Impact Scoring Matrix for the Human Impact on the Biodiversity in the Sariska Tiger Reserve

A net impact scoring matrix was developed to scale the human impact indicators as per the environmental elements in Sariska quantitatively. This paper is a part of a larger study that helped in analyzing other parameters also. The indicators were rated between the scales of magnitude to note the severity of the impact. Scale used for scoring is:

- No impact : 0
- Slight impact : 1
- Moderate impact : 2
- Severe impact : 3

The matrix is based on several environmental elements viz. habitat quality, migratory movement, local movement, animal behaviour, feeding & breeding, predator-prey relation, nutrition, water availability, tree cover, endangered sp.(flora & fauna), eco-sensitive & corridor areas of the STR. The human impact threat indicators are agriculture, dairying, deforestation, developmental activity, encroachment, human habitation, livestock grazing, mining (waste discharge), NTFP collection, poaching / forest mafia, recreational structures, religious fairs, roads, timber collection, tourism, transportation/ traffic. A net score was computed by adding the score obtained by the indicators. A scoring key is prepared and the magnitude of the indicators were classified into:

Negligible impact (below 4)	:	dairying, NTFP collection
Slight impact (4-8)	:	religious fairs
Moderate impact (8-12)	:	poaching, agriculture, timber collection
High impact (12-16)	:	roads, recreational structures, mining, development activity, encroachment, tourism, livestock grazing
Severe impact (above 16)	:	deforestation, human habitation

Table VI : Net Impact Scoring Matrix for Human Impact on Biodiversity in Sariska Tiger Reserve

Human Impact Action	Habitat quality	Migra-tory move-ment	Local quality	Animal beha-viour	Feed-ing & breed-ing	Predator prey relation	Nutri-tion	Water availi-bility	Tree cover	Endan-gered sp. (flora) & fauna	Econsi-tive & Corridor areas	Total Score
Tourism	2	0	1	2	3	3	0	0	0	2	2	15
Poaching/ forest mafia	0	0	2	2	1	1	0	0	0	3	2	10
Deforestation	3	1	0	1	2	1	2	3	3	3	2	21
Agriculture	1	0	0	0	0	0	1	2	3	1	2	10
Livestock grazing	3	0	0	2	1	1	1	2	1	0	2	13
Transportation/ traffic	1	0	1	2	0	0	0	0	0	2	1	9
Development activity	2	1	1	1	0	0	0	2	2	1	3	13
Encroachment	3	0	1	1	0	0	0	2	2	1	3	13
Timber collection discharge)	3	0	1	1	0	0	1	2	1	1	2	12
Recreational structures	2	1	1	2	0	0	0	1	2	2	3	14
Roads	3	2	2	1	0	0	0	0	1	1	2	12
Dairying	1	0	0	1	0	0	0	0	0	0	1	3
Religious fairs	1	0	1	1	0	0	0	0	0	1	3	7
Human habitation	3	1	2	2	2	1	1	3	3	2	2	22

Conclusion

Once the private hunting grounds of Alwar's former royal family, today only 20 % of this vast expanse of jungle is tiger habitat. Biotic pressure from humans and cattle is not the only cause for the degradation of this forest. One must not underestimate the mining lobby, which remains a potential threat to wildlife for they can twist the verdict of the courts in their favour at any time. Sariska awaits National Park status that will restrict activities such as grazing and mining. Also there are other recommendations that need to be worked on relating to the state of non-sustainable tourism in STR. Tourism comes in the "high impact" category of the net impact scoring matrix, so there is a need to promote the ecotourism activities.

The growth of the human and livestock population is one of the most important root causes in the decline of the biodiversity. Apart from the growth in numbers, the patterns and levels of resource consumption have changed. The increasing population has diminished the extent of forest and other uncultivated lands, depleted lands of native vegetation and spoil the remaining lands and ecosystems by the disposal of waste and other forms of pollution. Increasing the pressure of livestock for grazing has left the land bereft of any vegetation for the wildlife at Sariska.

There is a direct link between poverty and biological resources at Sariska Tiger Reserve. That poor people are forced to hunt rare and endangered animals and sell in the illicit market is still to be established. Options for living for the people here are limited like the cultivation on a patch of the forest, rearing livestock on the basis of migration into the forest, selling forest produce at a local market causing direct loss to the biodiversity in the STR.

Knowledge about the extent, status, and role of species, natural habitats, and ecosystems in supporting human life is lacking especially among the villagers. On one hand, there is a shortage of knowledge and trained scientists in the domain of biological diversity, and on the other hand, the traditional knowledge has not been recognized in the mainstream public domain. The process of policy development and decision-making in the public domain in the villages of Sariska remains highly insular.

The causes described above are responsible for triggering various threats to the biodiversity. However, the root causes do not operate in isolation. They tend to act with and exacerbate one another (Rajasthan Biodiversity Strategy and Action Plan, 2002). Though a well-managed park, the future of Sariska hangs in the balance because its future depends on the awareness and action to maximize wildlife in a setting where nature is in harmony with wildlife, where there is water and where an enhanced tiger and leopard population can roam in a much bigger wilderness. The need of the hour is to support a less disturbed environment for the wildlife and a better biodiversity in Sariska, so that there are no wildlife-cattle and villagers-foresters conflicts anymore.

References

1. Baskin, Y. (1997) The Work of Nature: How the Diversity of Life Sustains Us; A Project of SCOPE, Island Press, Washington, D.C.

2. Frenkel, R.E. (1970) Ruderal Vegetation along Some California Roadsides, University of California Publications, 20.
3. Gadgil, M. and Rao Seshagiri, P.R. (1998) Nurturing Biodiversity: An Indian Agenda, Centre for Environment Education, Ahmedabad.
4. Goudie, A. (2000) The Human Impact on the Natural Environment, Blackwell Publishers, Oxford, U.K.
5. Government of Rajasthan (2002) Forest Department, Management Plan, Sariska Tiger Reserve (2004-2014).
6. Hannah, L., Lohse, D., Hutchinson, C., Carr, J.L., Lankerani, A. (1994) A Preliminary Inventory of Human Disturbance of World Ecosystems, Ambio 23:246-250.
7. Huston Michael, A. (1994) Biological Diversity: The Coexistence of Species on Changing Landscapes, Cambridge University Press, UK.
8. IUCN (2000) **Red List of Threatened Species,** Gland, The World Conservation Union
9. Jodha, N.S. (1985) Population Growth and the Decline of Common Property Resources in Rajasthan, India, Population and Development Review,11(2), 247-264.
10. Joshi, P.C. and Joshi, N. (2004) Biodiversity and Conservation, A.P.H. Publishing Corporation, New Delhi.
11. McIntosh, R.P. (1987) Pluralism in Ecology, Annual Review of Ecology and Systematics, 18, 321-41.
12. McNeely, J.A., Gadgil, M., Leveque, C., Padoch, C., Redford, K., Arden Clarke C. (1995) Human Influence on Biodiversity, in Heywood, V.H., Watson, R.T., (eds), Global Biodiversity Assessment, Cambridge University Press, Cambridge, U.K. 711-821,
13. Ministry of Environment and Forests (2000) Annual Report 2000-2001, Government of India, New Delhi.
14. Myers, N. (1988) Threatened Biotas: 'Hotspots' in Tropical Forests, Environmentalist 8: 187-208.
15. Myers, N., Mittermeier, R.A., Mittermeier, C. G., da Fonseca, G A B, Kents, J. (2000) Biodiversity Hotspots for Conservation Priorities, Nature 403: 853-858.
16. Project Tiger Status Report (2001) Ministry of Environment and Forests, Government of India, New Delhi.
17. Rajasthan Biodiversity Strategy and Action Plan (2002) Centre for Management Studies, HCM, Rajasthan Institute of Pubic Administration, Jaipur.
18. Shanmugaratnam, N. (1996) Nationalisation, Privatisation and the Dilemmas of Common Property Management in Western Rajasthan, Journal of Development Studies 33, 163-187.

27

Ground water analysis in the rural area of district Gurdaspur (Punjab)

CHETNA MAHAJAN AND ZINIA

Introduction

Ground water acts as a reservoir and source of water for well, spring, bore wells and hand pumps. The increase in human population and fast development led to the scarcity of drinking water. In India, more than 80 per cent populations live in rural areas. In these rural areas, the main source of drinking water is ground water. As the ground water is used for drinking purpose by most of the population, it is very essential to test the quality of water for drinking purpose. Therefore, in present study an attempt was made to characterize the ground water quality of district Gurdaspur with respect to physico-chemical characters and to find the suitabilities of these waters for drinking and domestic purpose.

Methodology

For the analysis of underground water in rural areas of District Gurdaspur, 48 random water samples were taken. All samples were collected in plastic bottle of one litre capacity. The bottles used for sample collection were thoroughly cleaned before sampling. They were rinsed thrice with the sample water before collection. The analysis of the samples was done on 7 days after the collection of samples. The various parameters analyzed (Table 1) for physical for physical and chemical examination of samples by following the procedure given by Goel and Trivedy (1986).

Table 1. Different parameters and their methods of analysis

S.No.	Parameter	Method of Analysis
1	pH	Water and soil analysis kit
2	TDS	Water and soil analysis kit
3	Conductivity	Water and soil analysis kit
4	Hardness	Complexometric titration
5	Chloride	Argentometric titration
6	Sodium	Flame photometry
7	Potassium	Flame photometry
8	Alkalinity	Acid-base titration

Results and Discussion

The pH value in the district Gurdaspur ranged from 6.90 to 7.30 (Table 2). The highest values (7.3) was found in Dinanagar block and lowest value (6.9) was found in Dharkalan block. The prescribed WHO and BIS range for pH of potable water is 6.5 -8.5. None of the samples found to be having pH more than 8.0. So, from pH point of view, water is good for drinking purpose in the district Gurdaspur.The average TDS value ranged from 220 mg/l to 959 mg/l (Table 2). The highest value was found in Kalanaur block, while lowest value was found in Dharkalan block.

The TDS in water includes inorganic salts and a small amount of organic matter and main ions contributing to TDS are carbonates, bicarbonates, sulphate, nitrate, sodium, potassium, calcium and magnesium. The TDS value in district Gurdaspur found in desirable limit as the the permissible limit for TDS is 1000 mg/l as prescribed by WHO. The hardness value in the district Gurdaspur ranged from 17 mg/l to 73.3 mg/l (Table 2).

The highest value of hardness was found in Kalanaur block and the lowest value was found in Dharkalan Block. So, water is considered as soft in the district Gurdaspur. Hard water is nuisance for domestic purpose because it does not form froth with soap. The drinking of hard water might lead to increased incidence of urolithiasis.The chloride value of drinking water in the district Gurdapur ranged from 75-22.3 mg/l (Table 2). The lowest value was found in the Dinanagar block and a highest value was found in Kalanaur block.

WHO and BIS limit for chloride is 250 mg/l. So chloride value is much below the desirable limit prescribed by WHO. The average value of sodium was highest (87 mg/l) in Fategarh Churian block and lowest value (5.75 mg/l) was found in Khanuwan district. It was seen that average value of potassium in ground-water was highest in Batala district (20.7 mg/l). Potassium was almost absent in Dharkalan block. Alkalinity was found highest in Dhariwal block and lowest (261.25 mg/l) in Dharkalan block (Table 2). The alkalinity of water is the measure of hydroxide, normal carbonate or bicarbonate content. In natural water, these are usually present

~~as compounds~~ of K, Na, Ca or Mg with Ca and Mg bicarbonates, the alkalinity usually equals the carbonate or temporary hardness, but where Na and K carbonate are present, the alkalinity will lie above the hardness level carbonate alkalinity should not exceed 120 ppm. So, it was conclude that in different blocks of district Gurdaspur, pH value ranges from 6.5-8.0 and TDS value ranges from 300-900 mg/l.

These parameters are well with the range of standard laid down by WHO. Sodium level in water varies widely but normally ranges between 6-130 mg/l (Table 2). Potassium is found almost absent in these samples except in one location. Alkalinity was found more in all the samples and it was due to the presence of sodium and potassium carbonates. The hardness value and chloride was found low in all the samples. Therefore, the people living in these areas are advised to take more sdalts in their diet to avoid the deficiency of these salts.

Table:2 Concentration of different parameters of ground water in various blocks of district Gurdaspur

Block	pH	TDS (mg/l)	Hard-ness (mg/l)	Chloride (mg/l)	Na (mg/l)	K (mg/l)	Alkal-inity (mg/l)
Dinanagar	7.305	301.25	22.25	7.5	18.5	0.25	288
Kahnuwan	7.2	282.5	26.5	8.25	5.75	0.5	272
Narot Jaimal Singh	7.205	518.75	43.75	9.25	13.5	1.5	292
Bamial	7.18	524.6667	50.33333	10	11.33333	1.333333	285.3333
Pathankot	7.01	372.7	28.75	9.5	18.75	0.75	263
Dhankalana	6.8975	220.25	17	9.25	11.5	0	261.25
Gurdaspur	7.1475	306.25	30.75	10.75	7.75	0.75	284.75
Dhariwal	7.64	510.3333	53	9	33.33333	2.333333	301.6667
Shri Hargobindpur	7.3825	408.75	40.5	9.25	13.5	0.5	299
Batala	7.0975	656.25	43.5	13.25	43	20.75	287
Dear Baba Nanak	7.1925	637.66	53	15	45	8.66	288
Kalanaur	7.08	959	73.33	22.33	85.66	1	280
Fategarh Churian	7.18	798.6667	54.33333	16.66667	87	7	289.6667

Conclusion

Ground water acts as reservoir of water for wells, springs, bore wells and hand pumps. The increase in human population and development led to the scarcity of drinking water and addition of certain harmful substances. Therefore the present study made an attempt to characterize the ground water analysis of Gurdaspur district where most of population lives in rural areas.

Reference

1. Goel, P.K. and Trivedy, S.K. (1986). Chemical and biological methods for water pollution studies. Environmental Publications, Karad (India), pp. 35-96

28

Ecotourism: A Modern Approach to Improving Biodiversity, Environment and Sustainability

DR. PREETI RAJPAL SINGH AND DR. ANTONIA AGMATA PALIWAL

Introduction

Tourism-related activities make up the world's largest economic sector, contributing directly and indirectly. Tourism has also been identified as one of the major high growth industries of the new century. The World Tourism Organization (UNWTO) estimates that income from tourism and related activities will amount to over $1.5 trillion in the year 2010. According to the World Tourism Organization, about 694 million tourists traveled internationally in 2003 and spent about US$ 514 billion. Global tourism continued to move upward during 2006 with the number of international tourist arrivals worldwide reaching about 846 million.[1]

The UNWTO forecasts that international tourism will continue growing at the average annual rate of 4 %.[2] and that by 2020 there will be 1.6 billion international tourist arrivals, worldwide who will be spending about US$ 2000billion.

Tourism and the developing nations

The pro-tourism literature for developing countries highlights tourism as a primary source of income and job employment, making it indispensable to the economy of the country. It also illustrates other positive externalities, an important one advocating tourism as being a source of foreign currency and wealth creation in these countries as a result of tourist spending in these destinations and the enhancement of international relations, through global cultural exchange. For developing countries, tourism is a major platform used to address and fight poverty in a sustainable way. Some countries follow the trajectory of generic mass tourism for economical reasons with a risk of entering the tourist trap – the sheer number of visitors is endangering the future of these destinations. They are sacrificing the long-term interests of the local population and the natural environment of their area in order to maximize the short-term opportunities that tourists can bring.

Also, the concerns and/or protests over displaced local communities, environmental degradation like the one along Kerela state's coastline, drug trafficking, prostitution and various cultural and social problems that accompany this boom have long been subdued and at times even legitimized.

The main challenge, therefore, that the tourism industry faces in these countries is to sustain economical growth while ensuring the long-term protection of the social and natural environment.

What is Ecotourism?

Ecotourism is a solution to this challenge. It allows for touristic activities to take place, while keeping in mind ways to overcome the negative impacts of mass tourism Ecotourism means ecological tourism, where ecological has both environmental and social connotations.

According to the International Ecotourism Society(TIES) ecotourism is:

'responsible travel to natural areas that conserve the environment and sustain or even improve the well-being and welfare of local people'.

Baobab suggests the following as good ingredients for ecotourism (i) it should be nature-based — with the main motivation for traveling being the observation and appreciation of nature as well as the traditional cultures prevailing in natural areas (ii) it should generate economical benefits for host communities and provide alternative employment and income opportunities for local communities and increase awareness towards the conservation of natural and cultural assets (iii) it should minimize the negative impacts upon the natural and socio-cultural environment.[3] Besides it should provide a positive experience for tourists and local people and enhance environmental awareness.

As Vail and Hultkrantz point out, there are three interconnected dimensions of tourism sustainability: ecological, economic and social.[4] Ecological sustainability refers to the need for limiting the touristic demand pressure in accordance with the carrying capacity of the local ecosystem. Economic sustainability points to the viability of touristic activities evaluated in terms of profitability and capacity to generate quality employment as opposed to menial and informal jobs. Vail, Lapping, Richard & Coate, refer to the quality of employment as a socially constructed notion. As a benchmark, they suggest that a quality touristic job should reflect a cluster of attributes, such as compensation, benefits, job security, opportunities to acquire and use skills, career development potential and work place congeniality.[5]

Social sustainability emphasizes the necessity for a net positive contribution from tourism to the civic and cultural life in the host region with co-ordination required between the environmentalist and the economist.

Ecotourism is one of the current buzzwords in promoting biodiversity, environment and sustainability.

Background

The United Nations designated 2002 as the International Year of Ecotourism at The Commission for Sustainable Development 7 meeting held in New York in April 1998. With the support of the WTO and UNEP, the aim was to hold a number of preparatory meetings leading up to the World Ecotourism Summit (WES) convened in Quebec City in May 2002. TIES received funding from the Ford Foundation and UNEP to support six regional meetings around the globe.

The four themes discussion framework for each regional meeting agreed upon by the UNWTO & U NEP were:

- Tourism Planning: The Sustainability Challenge
- Regulation of Ecotourism: Institutional Responsibilities and Frameworks
- Product Development, Marketing and Promotion of Ecotourism: Fostering sustainable Products and Consumers
- Costs And Benefits of Ecotourism: Ensuring Equitable Distribution among all Stakeholders.

The South Asia regional meeting was jointly convened by the Ecotourism and Conservation Society of Sikkim (ECOSS) and The Mountain Institute in Gangtok, Sikkim, India from 21-24 January 2002. The countries attending included Bhutan, India, Nepal and Sri Lanka. Pakistani and Bangladeshi representatives were unable to attend due to political turmoil resulting from the attacks on the Indian Parliament in December 2001, and Maldivian representatives had their own WTO meeting in early February. The main themes discussed were community participation and benefits from ecotourism, access to credit and financing and establishing effective regional partnerships.

The International Year of Ecotourism was sending a message that tourism should be considered an instrument of poverty alleviation. Putting it on the agenda of the Johannesburg Summit on Sustainable Development would convey the same message. It was possible to reconcile economy with ecology and the environment with development. The process leading up the Summit in Quebec City resulted in the **Quebec Declaration on Ecotourism.** [6]

The economical impacts of ecotourism, (or any economical activity) refers to the change in sales, income, jobs, or other parameter generated by ecotourism. The impact can be grouped into three categories: direct, indirect, and induced and these may be both tangible and intangible.[7]

Direct impacts are those arising from the immediate tourist spending, such as money spent at a tea stall or *dhaba*. The indirect impacts result from the demand generated for goods and services (inputs) from other businesses by this tea stall; while the induced impacts, arise when the tea stall owner and his employees spend their earnings to buy hitherto unbought goods and services. Of course, if the entrepreneur purchases the goods and services from outside the region of interest, then the money provides no indirect impact to the region — it leaks away.

By identifying the leakages, or conversely the linkages within the economy, the indirect and induced impacts of tourism can be estimated. In addition, this information can be used to identify what goods are needed but are not being produced in the region, how much demand there is for such goods, and what the likely benefits of local production would be. This enables policy makers to determine priorities for developing inputs for use by the tourism or other industries.

This is one reason why ecotourism has been embraced as a means for enhancing income and jobs of local communities along with the conservation of natural resources.

The Indian Scenario

The development and promotion of tourism is primarily undertaken by the State Governments/ Union territories themselves. However, the Ministry of Tourism provides financial assistance for the development of a touristic infrastructure on the basis of project proposals received from the State Governments/ Union Territories.

The foreign exchange earnings from tourism in INDIA during 2003 were Rs.16,429 crore (US$ 3,533 million). As per the estimates of the Department of Tourism, total direct employment in the tourism sector in India was about 20 million during 2003-04, while the indirect employment multiplier in tourism is fairly high and is estimated at 1.36. During the year 2003-04, the tourism industry registered a growth of 17.3 per cent in foreign tourist arrivals. Tourism has been India's strength for a long time. However, we have succumbed to the tourist trap in many instances. The local infrastructure at many tourist destinations is inadequate for the number of tourists that visit it. The local taxes are not enough for financing and rebuilding it.

The Government of India announced a new **National Tourism Policy** in 2002. It envisages to encourage people's participation in tourism development including Panchayati Raj institutions, local bodies, Co-operatives, non-governmental organisations and enterprising local youth to create public awareness and to achieve a wider spread of tourist facilities. Among other things, it also talks of a greater focus on ecotourism.

However, ecotourism is still mostly hype and inaction in India. Himachal Pradesh and Madhya Pradesh have ecotourism policies, but Governments fail to be proactive and supportive and these remain plans in government files. The Department of Tourism in Punjab is planning a major eco touristic initiative and will implement the proposed project by creating a special Directorate or Society under the Forest and Wild Life Department or Punjab State Forest Development Corporation (PSFDC). This development comes in the wake of the Eco Tourism Policy for Punjab drafted recently by the state Forest Department. The proposed mega project with an outlay of an investment of Rs 128 crore and time frame of about eight years will begin from the next financial year . How effectively and efficiently this project will translate from file to field, only time will tell.

The chief obstacles in the rapid development of Ecotourism in India are the lack of communication, collaboration and cooperation among stakeholders. Everybody's talking, but few are listening. In India, the true potential of ecotourism has yet to be realized due to the very absence of partnership between the forest and tourism departments. In Kerala, the creation of a separate Department for Eco-tourism is an entity administratively and structurally defined as different from the Department of Tourism and this is an interesting instance exemplifying this trend. 'Tourism' and 'Eco-tourism' are projected as separate while the mechanisms and practices followed under them are fine-tuned for mass tourism's aggressiveness. An indicator of the potential harmful effect of tourism, particularly beach tourism, in Kerala, is the fragility of its coastal zone. An important work on documenting Coastal Regulation Zone (CRZ) violations along the coast in Kerala pointed to the possibility of tourism as being a major source of such violations, which threaten the coastal ecosystem as well as the livelihood of fish workers.[8]

Another impediment has been the powerful nexus between bureaucracy, politicians and the highly influential hospitality industry. The government has refused to devolve power to local bodies in cases involving decisions on touristic projects. More often than not in such cases the final decisions of the government run counter to the position taken by many local bodies. The case of the Kumarakam Bird Sanctuary provides a good example. Massive construction works were carried out in the vicinity of the sanctuary in order to build tourist resorts, ignoring the objections raised by the local Panchayat that the construction activities, which adversely affected the bird sanctuary, were also leading to irreversible destruction of the mangrove ecosystem in the region. Similarly, in the case of another major tourism project in the state, Bakel, the 'Development Report' of the local body said that all steps must be taken to ensure that large hotels would not be permitted in the locality. Nonetheless, within months the Department of Tourism announced its decision to hand over the land acquired by the government from local people to large hotel monopolies for the construction of luxury hotels in Bakel.

Inspite of all this, there are some ecotourism success stories in India. Ecotourism programmes in the Periyar Tiger Reserve, Chinnar Wildlife Sanctuary and Chimmony Wildlife Sanctuary are the examples of the best practices, because of the participation of the local community. The Periyar Tiger trail programme in the Periyar National Park, Kerala, has employed 23 former poachers as guides. The forests have been their second home and their long enduring relationship with the plant and animal life becomes their USP. They provide tourists with the very information that they are looking for. The guides are also very effective in preventing the plunder of forest resources by outsiders, since they have a direct stake in it.

With many of the western tourists now disenchanted by the generic trips to over-populated and "happening" tourist locations, there is a wide scope for ecotourism in India. The tourist is beginning to search for simplicity, a natural environment and authentic cultures that India has in abundance. However, we do need to provide quality infrastructures.

Allowing multi-stakeholder partnership in forest and other protected areas would facilitate private investments in eco-tourism. Suppose there exist patches of green cover areas that are separated by non-green, inaccessible patches of land. Then corporates, NGO's or others may be invited to develop the non-green patch and to invest in infrastructure, on a benefit-sharing basis with other stakeholders like the Forest Department and the local community. This would not only provide access to the green covers but also provide jobs to the local inhabitants in the area and therefore bring with it the direct, indirect and induced benefits mentioned earlier.

Another obstacle Ecotourism faces is that it is not producing enough money to be successful. Whereas comforts and luxuries once played the major role in a tourist's decision-making process, today the environment has become an important criterion. They are willing to pay for the preservation of the natural and social environments they seek to explore. To cover costs and get returns on investment, the firm can collect access charges for a certain period of time but the locals should have free access in the area. Spending per visitor can be increased through provision of local handicrafts and cuisine. Backward linkages can be increased through a greater use of local agricultural and other products.

Recommendation

Now, for a word of caution. Applying the label "eco" or "green" does not automatically mean that all is ecologically good and environmentally fair. Ecotourism must guarantee that certain standards are met; which gives rise to an urgent need for an **Ecotourism and Environment Accreditation Programme.** For this, we recommend that **University of Delhi** set up a **Centre for Ecotourism** in association with an industry confederation or chamber and The International Ecotourism Society (TIES). The recommended constitution of the body responsible for awarding the Accreditation could be as follows:

Permanent members – nine

- V.C of Delhi University and his nominee
- Nominee of The International Ecotourism Society (TIES).
- Confederation/chamber of industry – 2members
- Secretary, Ministry of Environment & Forests and his nominee
- Secretary, Ministry of Tourism and his nominee

Co-opted members – nine (with each case under consideration)

- Two representatives of the applicant party
- Nominee of the local (government) university
- Nominee of State Ministry of Environment & Forests
- Nominee of State Ministry of tourism
- Nominee of the regional hospitality industry association
- Three nominees of the affected local community, including at least one woman, from say the local panchayat, local SEWA (self employed women's association, members of the local micro-credit association, etc

For a start, the university has requested one room and minimal funds to cover administrative operations. Once functional, the Centre will approach the **Asian Development Bank (ADB)** for a grant since the ADB does have a provision for giving on operation and research grant for this purpose.

This **Accreditation Programme** will ensure a "win-win" experience for all stakeholders. The Centre will help identify genuine ecotourism products in India and its stamp will give the multi-stakeholders an assurance that the accredited product is backed by a commitment of best practice environmental management to ensure the conservation of the flora and fauna; safeguarding and enhancing the economical, social and cultural interests of the local communities and the provision of quality experiences to the tourist.

The economical and environmental viability of an Ecotourism project depends primarily on the characteristics of the natural resources such as flora and fauna systems in an area. Hence, proper benchmarking, monitoring and the evaluation of these systems is critical to their implementation and sustainability. A highly trained individual in the field of ecotourism, say a **Certified Ecotourism Analyst (CEA),** must be involved with every ecotourism project right from its stage of conception. He/she will be authorized by the government to look into the design and the implementation of the project, particularly the ecosystem assessment and monitoring of the project and will be responsible for submitting this information to the respective authorities in the state. He will initiate the development of manuals for the particular ecotourism project for the state. He/she will develop training modules on ecotourism for the participants of the project, organize workshops, forge linkages in hospitality sectors across the states and abroad in direct consultations with the Ministry/Departments of Environment and Forest or other stakeholders on ecotourism and represent Ecotourism in India in the country and abroad. The **Centre for Ecotourism** will, after the requisite training and evaluation, **award this certification** to interested applicants.

The rapid development of mass tourism in the past has led to a mushrooming of touristic managerial institutes offering various diploma and certificate courses with no emphasis or even mention of ecotourism. Avenues for imparting knowledge and skills on ecotourism are limited. A Centre for Ecotourism at the University of Delhi will **develop and conduct courses on ecotourism** such as Ecotuorism Appreciation Course, Concepts of Ecotourism, Ecotourism Planning and Design, Development and Management of Ecotourism, Ecotourism as a Business, Ecotourism and Conservation tools, Human Dimensions in Ecotourism, Biodiversity Assessment and Monitoring/Evaluation. It is proposed that the **Centre for Ecotourism** will offer these Courses as part-time, weekend or through its School of Open Learning.

Ecotourism needs proper scientific support and there is a general lack of analysis and research with respect to the opportunities that the ecotourism market represents. Policies and programmes should be based on the expertise, studies and research carried out by this Centre. It is strongly recommended that the **current conference be made an annual event** under the aegis of the proposed Centre.

The success of Ecotourism in Improving Biodiversity, Environment and Sustainability depends largely on awareness, education and people's participation, and it is here that the Centre will play a pivotal role and enlarge its sphere of activities. It is recommended that the Centre will initiate the formation of **E.T. (Ecotourism societies)** in all the constituent colleges of Delhi University with the objective of mobilizing the youth and involving them in ecotourism. Interaction with fresh college entrants has revealed that most of them are aware and even enthusiastic about ecotourism, having been exposed to it during the SUPW (socially useful productive work) sessions of their class 12 curriculum. Membership to this society would be voluntary and the colleges would be asked to make an annual budget provision to it while allocating funds to other societies. The activities of the ET societies will include: promoting awareness among the people at all levels by organizing *Nukkad natak* (street theatre), seminars, workshops, excursions and camps, exhibitions, quiz/essay/debate/painting/poster competitions. The societies will be encouraged to submit regular activity reports to the Centre and the most outstanding efforts be rewarded every year at the International Conference on Biodiversity, Environment and Sustainability.

The Centre would also initiate a proposal to the government authorities to start observing an **Ecotourism Day** and lead the way by celebrating a similar day or week at the university. The celebrations would include a documentary/short film (on ecotourism) competition – to be shot by amateur students using only a hand camera or mobile phone. It will encourage students to move away from their regular hip-hop excursions towards those based on ecotourism. Partial subsidy for these trips can be easily achieved by industry partnership/sponsorship.

The Centre would also explore the possibility of developing linkages of ecotourism projects to **Carbon Offset Programmes** and also an **incentive programme of credit points to tourists** patronizing ecotourism projects.

We also wish to propose to the University that the subject of Ecotourism and its Appreciation as one of the **Qualifying subject options at the honours level** of its ongoing under-graduate degree programme, be introduced.

In conclusion, even though ecotourism may not appear as glamorous as traditional tourism and short-term economic considerations may conflict with ecotourism goals, today, six years after the International Year of Ecotourism, ECOTOURISM IS A NECESSITY AND NOT JUST ANOTHER OPTION FOR INDIA.

Annexure

LINKS TO THE ORGANISATIONS WORKING FOR ECOTOURISM IN INDIA:

The Mountain Institute is an international non-profit organization dedicated to the conservation, community development, and cultural preservation in the Andes, Appalachian, Himalayan, and other mountain ranges of the world.*www.mountain.org*

ECOSS: Eotourism Conservation Society of Sikkim has a mission to see that there is a thorough and correct understanding of the concept of ecotourism. *www.sikkiminfo.net/ecoss/*

ATREE: Ashoka Trust for Research in Ecology and the Environment (ATREE) was founded in an effort to address the environmental challenges facing India. It strives to conserve biodiversity and promote sustainable development while seeking to advance the protection of the environment.*www.atree.org*

HELP TOURISM: Help Tourism is an organization working to bridge tourism with conservation in the East and North East India. Established in 1991,theyhave been able to identify and promote destinations based on ecotourism guidelines. They have pioneered sustainable development of communities through eco-tourism, preservation of culture and nature. *www.helptourism.com*

References

1. UNTWO, UNTWO World Tourism Barometer, Vol.5 No.2, June 2007.
2. UNWTO, Long-term Prospects: Tourism 2020 Vision.
3. www.baobabtravel.com/baobab_ecotourism.
4. D Vail & L Hultkrantz, 'Property rights and sustainable nature tourism: institutional adaptation and maladaptation in Dalarma (Sweden) and Maine (USA)', Beijer Discussion Paper Series 124, Sweden: Beijer Institute of Ecological Economics, 1999.
5. D Vail, M Lapping, W Richard & M Coate, Tourism and Maine's Future, Augusta, ME: Maine Center for Economic Policy, 1998.
6. www.ecotourism2002.org/anglais/declaration.html
7. Kreg Lindberg, The Economic Impacts of Ecotourism, Charles Sturt University, November 1996.
8. Equations, Dossier on CRZ Violations in Kerala 1997: Survey on Kasargod, Kannoor, Kozhikode, Vypeen and Kovalam, Bangalore: Equations, Equitable Tourism Options, 1997.

29

Environment and sustainable development – challenges today with a vision for future

DEWAN ANJALI

Introduction

Sustainable development is a socio-ecological process characterized by the fulfilment of human needs while maintaining the quality of the natural environment indefinitely. The linkage between environment and development was globally recognized in 1980, when the International Union for the Conservation of Nature published the World Conservation Strategy and used the term 'sustainable development'. The concept came into general usage following the publication of the 1987 report of the Brundtland Commission — formally, the World Commission on Environment and Development set up by the United Nations General Assembly. This became the most often quoted definition of sustainable development as development that 'meets the needs of the present generation without compromising the ability of future generations to meet their own needs'. It refers to the wise use of resources within a framework in which environmental, economic and social factors are integrated. It is about maintaining and improving the quality of life while safeguarding the quality of life of generations to come. The components of biological diversity have to be utilized in such a manner that does not interfere with the natural functioning of the ecological process and life-support systems. It is a collection of methods to create and sustain development which seeks to relieve poverty, create equitable standards of living and satisfy the basic needs of people.

Sustainable development does not focus solely on environmental issues. The 2005 World Summit Outcome Document, refers to the interdependent and mutually reinforcing pillars of sustainable development as economic development, social development and environmental protection. Green development is generally differentiated from sustainable development in that green development prioritizes what its proponents consider to be environmental sustainability over economic and cultural considerations. Proponents of Sustainable Development argue that it provides a context in which to improve overall sustainability where cutting edge Green development is unattainable. For example, a cutting edge treatment plant with extremely high maintenance costs may not be sustainable in regions of the world with fewer financial resources. An environmentally ideal plant that is shut down due to bankruptcy is obviously less sustainable than one that is maintainable by the indigenous community, even if it is somewhat less effective from an environmental standpoint. Some research activities start from this definition to argue that the environment is a combination of nature and culture.

The three components of sustainable development

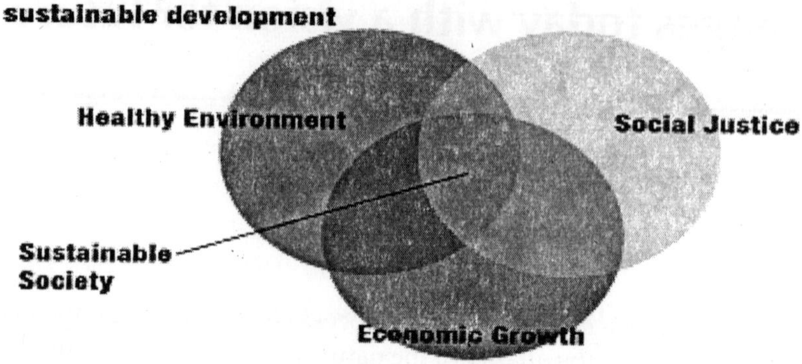

Environmental sustainability is defined as the ability of the environment to continue to function properly indefinitely. This involves meeting the present needs of humans without endangering the welfare of future generations. The goal of environmental sustainability is to minimize environmental degradation, and to halt and reverse the processes to which they lead. An unsustainable situation occurs when the sum total of nature's resources is used up faster than it can be replaced. Sustainability requires that human activity uses nature's resources only at a rate at which they can be replenished naturally. The long term result of environmental degradation would be local environments that are no longer able to sustain human populations to any degree. Such degradation on a global scale could imply the extinction of humanity.

Consumption of renewable resources	State of environment	Sustainability
More than nature's ability to replenish	Environmental degradation	Not sustainable
Equal to nature's ability to replenish	Environmental equilibrium	Steady-state Sustainability
Less than nature's ability to replenish	Environmental renewal	Sustainable development

The gross mismanagement of our planet has much to do with an inequitable distribution of the benefits of development. Perpetuating this inequity can mean only a continuing draw-down on the world's natural resources and the environment. After a century of unprecedented growth, marked by scientific and technological triumphs that would have been unthinkable a century ago, there have never been so many poor, illiterate and unemployed people in the world, and their number is growing. Close to a billion people live in poverty and squalor, a situation that leaves them little choice but to go on undermining the conditions of life itself, the environment and the natural resource base.

The Need for a New Developmental Paradigm

The need for a new developmental paradigm was widely recognized by the mid-1980s. By then, the United Nations had already declared the second and third of its Developmental Decades a failure, especially in their inability to halt the cycle of poverty that existed in the world's poorest and slowest developing countries. Further, the conditioned loan policies of the World Bank and the International Monetary Fund in combination with patterns of development assistance, provided through various bilateral and other arrangements, succeeded in plunging already impoverished countries into even deeper debt. Thus, the combination of high debt, slow economic growth, and rapidly increasing poverty enhanced the profound levels of human suffering that already existed in many of these countries. Groundbreaking events, such as the United Nations-backed Rio Earth Summit on Environment and Development, in 1992, and the Johannesburg Summit on Sustainable Development a decade later, show that policy-makers recognized the grave challenges that lie ahead in preserving the planet's fragile environment. One of today's environmental challenges is to find ways of strengthening the scientific and socio-economic perspectives to help authorities make decisions and produce sustainable developmental strategies.

Issues & Challenges

It is widely accepted that meeting the needs of the future depends on how well India and the rest of the world balances their social, economic and environmental goals when making decisions today. Research carried out under consecutive research Framework Programmes has addressed sustainable development, in particular, and environmental issues, in general, in a number of different contexts. Broadly speaking, the challenges facing us today are as follows:

- The ways in which greenhouse gas emissions and atmospheric pollutants affect climate, ozone depletion, soil and forests
- The water cycle and soil-related aspects, especially looking at the impact of climate change on the hydro-cycle, interactions between land, ocean and atmosphere, distribution of groundwater and surface water, freshwater and wetland ecosystems, management strategies for water (availability, demand, vulnerability, quality, etc.)
- Understanding marine and terrestrial biodiversity to minimise the negative impact of human activities ,relationships between society, the economy and biodiversity habitats,

forecasting potential changes, studying the structures and mechanisms while assessing and managing the risks to biodiversity and ecosystems, as well as looking into conservation, mitigation and rehabilitation options for marine and terrestrial biodiversity

- Mechanisms of Natural Disasters and Desertification such as seismic and volcanic activity – and how these tie in with climate change, land and soil degradation – and studying what science can do to help long-term forecasting of hydrogeological hazards, natural hazard monitoring, mapping and management, and to improve disaster preparedness and mitigation strategies
- Strategies for sustainable land management, including integrated coastal zone management and assessment tools for measuring the impacts of agriculture and forestry on the environment, while looking at new technologies and integrated approaches to make these sectors more environmentally friendly
- Systematic operational forecasting and modelling of climate change (atmospheric, terrestrial, marine-based) to improve understanding of extreme events, and to use data coming from the Global Monitoring for the Environment and Security (GMES) programme and other Global Observation Systems for the climate, oceans and terrestrial more effectively
- Cross-cutting issues, such as putting together sustainable development concepts and tools for measuring the sustainability dimension on relevant policies wards sustainable development in a very thorough, practical and meaningful manner.

Key Documents

The Rio Declaration: A statement of 27 principles upon which nations agreed to base their actions in dealing with environmental and development issues. The Rio Declaration built on the previous Declaration of the United Nations Conference on the Human Environment which was adopted in Stockholm in 1972. The Stockholm conference was the first global environmental meeting of governments, which stated that long-term economic progress needs to be linked with environmental protection.

Agenda 21: An action blueprint on specific issues relating to sustainable development that emerged from the Rio Summit. Agenda 21 explained that population, consumption and technology were the primary driving forces of environmental change and for the first time, at an international level, explicitly linked the need for development and poverty eradication with progress towards sustainable development

The Bruntland Report: 'Our Common Future' which was produced in 1987 by the World Commission on Environment and Development, also laid the foundations for the Environmental Principles. This landmark document highlighted that people needed to change the way they lived and did business or face unacceptable levels of human suffering and environmental damage.

Global Compact: The environmental principles provide an entry point for business to address the key environmental challenges. In particular, the principles direct activity to areas such as research, innovation, co-operation, education, and self-regulation that can positively address the significant environmental degradation, and damage to the planet's life support systems, brought by human activity.

Sustainable Development Strategies

The Government has emphasized the need for a coordinated approach to sustainable development. Its efforts to advance sustainable development in India and around the world are also evidenced by its action on climate change, human health and the environment, its support for the building and maintenance of green infrastructure in and its recognition of, the unique relationship tribal people have with the natural environment. In addition, the Government has also recently made a commitment to regularly update strategies, which will ensure progress. Some of the common themes that have emerged in this regard include:

- Enhancing the capacity of Indians to act and make decisions in support of sustainable development; strengthening and improving access to knowledge in support of integrated decision-making
- Raising the level of understanding/awareness of sustainable development issues; providing the necessary tools to policy makers to a better advanced sustainable development
- Building on existing partnerships and creating new ones; providing leadership in areas such as greening of operations; fostering innovation and technology and promoting the adoption of sustainable development policies and practices in India and abroad
- A coordinated approach, which outlines best practices for departments to adopt in the procurement of waste management, water conservation, energy efficiency, vehicle fleet management, land use management and human resource management, needs to be prepared.
- Eco-efficient business practices that conserve energy, eliminate use of toxic materials, and reduce or reuse resources, innovative eco-efficient practices, tools, technologies and products can increase productivity while improving environmental performance.

The quality of life of the Indian people is intrinsically linked to the health and well-being of their communities. A sustainable community is resilient to change and has a vision for the future. However, to become sustainable, communities face complex environmental, economic and social challenges and require a coordinated approach to information, support and expertise.

Sustainable Development and Healthy Indians

Human health is dependent on a wide range of factors including the natural and built environments, and social, cultural and economic conditions. Certain populations like children, pregnant women and lactating mothers, the elderly and tribal people may be particularly vulnerable. A meeting with representatives from a range of sectors needs to be initiated to discuss the implications of human health when making decisions for sustainable development. The intention is to continue this dialogue whereby departments will explore information needs, strategies for action and approaches to engage the public in order to protect the health and overall quality of life of our people.

Social and Cultural Aspects of Sustainable Development

There is a growing recognition that the concept of sustainable development is multi-faceted with links to social and cultural aspects, such as poverty, distribution of wealth, diversity, equity, gender and health. These may not only affect the environment, but also influence and shape economic behaviour, consumption patterns, and the longer term viability and sustainability of our society. We are at an early stage of understanding the nature and implications of the social and cultural dimensions of sustainable development. Accordingly, various governmental and private agencies will be working together to better understand this area and the relationships involved by undertaking exploratory research and providing social policy perspectives.

Role of Centre for Environmental Education (CEE)

The primary objective of CEE is to improve public awareness and understanding of the environment with a view to promoting the conservation and sustainable use of nature and natural resources, leading to a better environment and a better quality of life. It develops innovative programmes and educational material, builds capacity in the field of education for sustainable development. It undertakes demonstration projects in education, communication and development that endorses attitudes, strategies and is committed to ensuring that due recognition be given to the role of education in the promotion of sustainable development. The activities of CEE and like programmes have been guided by certain strategies for the maximization of quality, effectiveness and impact. These are as follows:

- **Adaptability** to different geographical, cultural, social and economical contexts.
- **Partnerships** utilizing complementary strengths of other organizations to avoid duplication of effort and the network to effectively for the synergy converge of ideas and goals.
- **Building synergies** among Government, NGO's and CEE's for comprehensive impact.
- **Identifying key entry points** for different thrust areas and key targets for initiating and consolidating gains to achieve a multiplied effect.
- **Facilitating networks** to local, national and regional levels through a number of tools such as dialogues, directories, newsletters etc.
- Working to **develop a cadre of professionals** in order to improve and strengthen professionalism in the field of Education for sustainable development (ESD) by capacity build-

ing individuals who in turn would infuse this professionalism into the organizations where they work to the mutual advantage of both those organizations and CEE.

- Ensuring **quality control and excellence** in the production of all material through in-house infrastructural support like the studios, workshops, printing facilities.

International Aspects of Sustainable Development

Ten years ago at the Rio Earth Summit, more than one hundred states proclaimed sustainable development to be fundamental to social and economic progress for all nations. The Indian Government is addressing key sustainable development issues in the international arena, among them: globalization and competitiveness, environmental concerns such as persistent organic pollutants, regional conflicts and their impact on the environment, and the trade and environment interface. Our country is also working with developing countries to help improve their sustainability. Finding ways to improve the capacity of developing countries to negotiate and implement multilateral trade and environmental agreements is of particular interest at this time. The Centre for Environment Education has developed international linkages with numerous agencies, both governmental and non-governmental, that are working in related fields. The linkages are in the form of collaborative projects, exchange programmes , training, symposia and seminars. CEE has international offices in Australia and Sri Lanka.

Fourth International Conference on Environmental Education

The first Intergovernmental Conference on Environmental Education (EE) was held under the aegis of UNESCO-UNEP in 1977 at Tbilisi, Georgia. Conferences have been held every decade thereafter at Moscow in 1987 and Thessaloniki in 1997. The fourth ICEE, co-sponsored by UNESCO and UNEP, was organized by the Government of India. CEE was the nodal agency for the conference which was held in the month of November at Ahmedabad. Being held during the United Nations Decade of Education for Sustainable Development (DESD 2005-2014), the conference looked at building a sustainable future. The recognition that education is a critical agent of transformation in changing lifestyles, attitudes and behaviour, in increasing participation, in visioning and realizing a sustainable world and in facilitating the use of communication, education, participation and awareness (CEPA) to foster the changing needs in different sectors needs to be further strengthened. Reflection, future vision, sharing are the crucial elements of ESD.

The Path Forward

As the sustainable development journey continues, the various departments will continue to learn and improve. There will be further efforts to coordinate between departments as we move to the implementation of the updated strategies. This will include efforts relating to communication and awareness building amongst our countrymen, scientific and technological personnel development and staff training. The Government will also move forward on the current priority areas and identify opportunities for additional collaboration. The transition to a more sustainable society in the 21st Century will not be achieved by one department or on one level of government alone. Moving towards more sustainable development will require the commitment and attention of decision makers at all levels of society. The Sustainable Development Strategies are an important element of the Government of India's response to this significant challenge. The strategies will identify common goals and actions towards ensuring healthy human and natural environments, supporting emerging governing political systems, advancing the appropriate use of natural resources, building vibrant communities, and influencing international activities. Sustainable Development is about a more inclusive society, which provides for better protection of the environment and use of natural resources and shares the benefits of economic growth as widely as possible. It is about achieving a better quality of life.

References:

1. Alger, C. F. (1990). Grass-roots perspectives on global policies for development. *Journal of Peace Research*, Vol.27(2), 155-168.
2. Brandt Commission. (1980). North-South: A programme for survival. Pan books. London.
3. Burek, D. (Ed.). (1991). Encyclopedia of associations: International organizations. New York: Gale Research, Inc.
4. Campfens, H. (1990). Issues in organizing impoverished women in Latin America. Social Development Issues, 13(1), pp.20-43.
5. Ester, R.J. (1993). Towards Sustainable Development-from theory to praxis. Social Development Issues, Vol.15 (3), pp.1-29.
6. Midgley,J. (1990). International Social Work: Learning from the Third World. *Social Work*, Vol.35 (4), pp.295-301.
7. Nayak, R.K. & Siddiqui, H.Y. (Eds.) (1989). Social Work and Social Development. Geetanjali Publishing House, New Delhi.
8. World Commission on Environment and Development (1987). Our Common Future. Oxford University Press. New York.
9. World Bank (1990).World development Report 1990: Global poverty. Oxford University Press, New York.
10. World Bank (1993).World development Report 1993: Investing in Health. Oxford University Press, New York.

30

Environment and Consumer: Global Quest for Sustainable Consumption

Dr. Sheetal Kapoor

Introduction

Rapid changes in the environment are the cause of a number of unknown disasters. Environmental problems have become so severe that they have attracted global attention. The United Nations has identified some key global natural environmental problems such as[1]:

- Global Warming
- Acid Rain
- Ozone depletion
- Water pollution
- Deforestation

Global warming is the underlying factor to many natural disasters, such as, droughts, earthquakes, floods and shifts in the typical monsoon weather pattern. India has recently faced two disasters in recent times. It was the earthquake that rocked Gujarat in the Year 2001and later on it was the Tsunami which took a major toll of around 10,000 official figures in India in 2005. In both of these instances it is the poor people who had to bear the brunt of the disaster.

Following the adoption of the UN declaration on sustainable development, sustainable consumption has become an important part of preserving the environment.

Consumers should protect the environment by:

1. Promoting the use of products which are environmentally sustainable.
2. Encouraging recycling

3. Requiring environmentally dangerous products to carry appropriate warnings and instructions for safe use and disposal.

The Government should further promote the use of non-toxic products by:

1. Raising consumer awareness of alternatives to toxic products
2. Establishing procedures to ensure that products banned overseas do not enter our national markets.
3. Ensuring that the adverse social impacts of pollution are minimised.
4. Promoting ethical, socially and environmentally responsible practices by the producers and suppliers of goods and services.

Sustainable development

The focus today is to achieve a balance between economical development and the protection of the environment in a way that both can co-exist. This approach is known as achieving a sustainable development.

Sustainability is the concept of conducting life in a resourceful, conservative and efficient manner such that actions do not compromise the ability of future generations to meet their own needs. The essential elements of this manner of living are the promotion and maintenance of industrial, commercial, and residential developmental strategies that lead to a better environment in the future; one sustained by stable, healthful communities within a clean, safe environment[2].

The term "sustainable development" was perhaps used for the first time at the Cocoyoc declaration in the early 1970's. In 1980 and 1983, the Independent Commission on International Developmental Issues, also known as the Brandt Commission, published two significant reports that sought to establish a three way link between development, environment and poverty. In 1983, the United Nations (UN) general Assembly set up the World Commission on Environment and Development (WCED) for the purpose of[3]

(a) examining the environmental and developmental crises and

(b) proposing a framework for sustainable development.

The report stated that the global environmental problems were the result of

(i) non-sustainable patterns of consumption and production in the developed countries; and

(ii) massive poverty in the developing countries.

---

[1] www.un.org

[2] Khanna Sri Ram, Kapoor Sheetal, et al *Consumer Affairs* ((Universities Press (India) Private Ltd, 2007)

[3]Banerji Pranab Environment and Consumer, *Monograpgh* (IIPA, 2005)

In 1992, the United Nations Conference on Environment and Development (UNCED) was held in Rio De Janeiro. It was the largest Environmental Conference ever held and is also known as the Earth Summit. This conference provided international recognition to sustainable consumption ideas and production by adopting the Agenda 21 action plan.

The UN Commission on Sustainable Development (CSD) was set up under the Economic and Social Council (ECOSOC) of the UN in December 1992 to oversee the implementation of Agenda 21. The Commission was created to ensure an effective follow-up of the UNCED, to monitor and report on the implementation of the agreements at the local, national, regional and international levels. It was agreed that a five-year review of the Earth Summit progress would be made in 1997 by the United Nations General Assembly meeting in a special session. The full implementation of the Agenda 21, the Programme for Further Implementation of Agenda 21 and the Commitments to the Rio principles, were strongly reaffirmed at the World Summit on Sustainable Development (WSSD) held in Johannesburg, South Africa from 26th August to 4th September 2002. The United Nations has also framed a Global Compact (UNGC) Programme based on the world's largest corporate citizenship initiative aiming for a more sustainable and inclusive world economy. It is supported by over 4000 participants in over 100 countries. It includes many of the world's most influential companies, such as Coca Cola, Levi Strauss, Nestlé, Microsoft and many more. Companies from all over the world report their progress on implementing the 10 Global Compact principles. The United Nations Secretary-General Kofi Annan first proposed the Global Compact in an address to the World Economic Forum on 31 January 1999. The Global Compact's operational phase was launched at the UN Headquarters in New York on 26 July 2000. The Secretary- General invited business leaders to join an international initiative — the Global Compact — that would bring companies together with UN agencies, labour and civil society to advance universal social and environmental principles.

Special attention should be paid to the demand for natural resources generated by unsustainable consumption and to the efficient use of those resources consistent with the goal of minimizing depletion and reducing pollution. Although consumption patterns are very high in certain parts of the world, the basic consumer needs of a large section of humanity are not being met. This results in excessive demands and unsustainable lifestyles among the richer segments, which place immense stress on the environment. The poorer segments, meanwhile, are unable to meet food, health care, shelter and educational needs. Changing consumption patterns will require a multi-pronged strategy focussing on demand, meeting the basic needs of the poor, and reducing wastage and the use of finite resources in the production process.

In view of the above findings, it is advocated that a new strategy is needed that combines environment and development and which involves producing products which are eco-friendly. Further the role of ethical consumption in today's scenario looks highly relevant.

Eco-Friendly production

Green marketing, also known as ecological marketing and environmental marketing, is not a recent phenomenon but it became a worldwide phenomenon during the nineties[4]. On April 22,

[4]Jain, Sanjay K. and Gurmeet Kaur, *Review of Commerce Studies*, Delhi School of Economics, University of Delhi, Vol. 22 (Jan-June 2003).

1990, more than 100 million people took part in Earth Day, also known as 'decade of the environment'. Green marketing provides consumers with better products and a healthier environment. Producing eco-efficient products creates less waste, uses fewer raw materials, and saves energy. Creating products that are more in tune with nature allows each individual consumer to personally contribute to the environmental clean-up and helps ensure a more secure future for their children. Proctor and Gamble's Tide laundry detergent is phosphate-free, contains ``biodegradable cleaning agents'' and is packaged in a ``recycled-content bottle'' making it highly green.

To have a successful green marketing programme, a company must have the commitment of the top management, as well as an internal environment that encourages employees to act as environmentalists. In the long run companies that demonstrate concern for the environment, offers their consumer the prospect of healthier and more fulfilled lives.

Green consumer

The evolution of a "green" consumer means that a consumer exercises his vote not only in favor of a market response to his needs but also to the individuals responsibility towards the environment. A "green" consumer can convert the power of the consumer into a force for positive action to protect the environment. Eco labelling networks that monitor and evaluate green products have been developed in many countries. These networks have done life cycle analyses to understand the impact of products. Governments have also taken several measures that have supported and facilitated such moves by businesses.

Green marketing is a term used to identify concern with environmental consequences of a variety of marketing activities[5]. It is a concept that can be applied to consumer goods, industrial goods and even services.

Green marketing incorporates a broad range of activities, including product modification, changes to the production process, packaging changes, as well as modifying advertising. It is also known as Environmental Marketing and Ecological Marketing. While green marketing came into prominence in the late1980s and early 1990s, The American Marketing Association (AMA) held the first workshop on "Ecological Marketing" in 1975. This workshop defined ecological marketing as the study of the positive and negative aspects of marketing activities on pollution, energy depletion and non energy resource depletion.

The following possible reasons explain why firms are adopting green marketing:

1. Organizations believe they have a moral obligation to be more socially responsible;

2. Governmental bodies are forcing firms to become more responsible and

3. Competitor's environmental activities pressure firms to change their environmental marketing activities;

---

[5]Cateora, Philip R. and John L Graham, *International Marketing* (New Delhi: Prentice-Hall, 2005), pp347-348

4. High costs of disposing of industrial waste.

It appears that all types of consumers, both individual and industrial are becoming more concerned and aware about the natural environment. As demands change, many firms see these changes as an opportunity to be exploited. Firms marketing goods with environmental characteristics will have a competitive advantage over firms marketing non-environmentally responsible alternatives.

Ethical Consumption[6]

Ethical consumption stresses the role of the consumer in preventing labour exploitation, environmental conservation and exercising human rights while buying products. By consuming consciously and ethically we can realistically create change. Before buying anything one should ask: Who makes it? Who needs it? And who profits from it?

Ethical consumption is a way to help us feel that we have power as consumers, that we can vote with our rupees. By using this consumer power we can have impact on the larger economy and help create a world where the economy benefits all people in more equal ways. Some suggestions regarding ethical consumption are:

- Contact retailers and manufacturers for more information about their practices.
- Make a point to ask about the people who make the things you buy.
- Consider the environmental cost of producing the product.
- Think about how you would dispose of the product once you are finished with it.
- Support fair trade.
- Buy local products and services.
- Reuse things. Fix things that are broken.
- Don't get swayed by advertisements.
- Invest your money locally and ethically.
- Boycott companies that exploit workers and harm the environment.

An old saying goes that "Charity begins at home". While shopping we need to apply this practically in today's fast-paced life style. The strategy of "caveat emptor" i.e. "the buyer needs to marketeer beware" of products has now shifted to "caveat vendor" i.e. the vendor or the marketeer needs to be aware of the needs of the consumer. The marketeer can survive only if the consumer purchases his products.

Gandhiji also talked about ethical consumption in terms of swadeshi (local production) and khadi (hand-spun, hand-woven cloth). Gandhi's economical ideals were not about the destruction of all machinery, but a regulation of their excesses. Gandhi's notion of

[6]Kapoor Sheetal, Ethical Consumption: An Emerging Reality, Consumer Voice, March-April, 2004.

revitalizing village India through the spinning wheel struck many as anachronistic, but the logic of his arguments took on greater force after his death.

Gandhi's economic ideals were not about the destruction of all machinery, but a regulation of their excesses. *Khadi* requires a decentralization of production and consumption, which in turn should take place as near as possible to the source of production. Such localization would do away with the temptation to speed up production regardless of the costs and would alleviate the problems of an inappropriately structured economical system. *Swadeshi* is that spirit in us which requires us to serve our immediate neighbours before others, and to use things produced in our neighbourhood in preference to those more remote. So doing, we serve humanity to the best of our capacity. We cannot serve humanity by neglecting our neighbours. Local, short-distance transportation should receive every encouragement but long hauls should be discouraged because they would promote urbanisation and specialisation beyond the point of human integrity.

In such a scenario, the significance of khadi, its local production and employment of local labour for the process, attains even more importance in terms of environmental sustainability as a protection for consumer health. Radically restructuring the usage of the environment is not an empty slogan. Self-imposed curbs on one's wants and production for one's needs can be achieved by discipline and selflessness. Instead of throwing something away, one should take time to fix it or take it to a tradesperson to fix it. The cost may be cheaper than buying a new one and we would be reducing waste.

By limiting our purchases we can use the money we save, to buy something more appropriate. People are increasingly experimenting with living simply but more fulfillingly with 'an elegant sufficiency'. Living with fewer "things" and assuring that all resources, including labour, are used wisely and fairly can help in creating a more equitable and ecological world. It is time to revisit Gandhiji's Principle of trusteeship, "The earth has enough for everyone's need, but not for everyone's greed."

31

Folk Biomedicines and traditional practices to cure various veterinary diseases in Kangra District of Himachal Pradesh, Western Himalaya

DR. SHEETAL KAPOOR

Introduction

The role of indigenous knowledge in sustainable agricultural production in developing countries is beginning to gain recognition within scientific circles. However, the usefulness of supernatural practices for the treatment of disease is still controversial. A careful analysis of supernatural healing practices used by the Fulani pastoralists reveals that some of these practices are routinely applied in animal health management (**Toyang *et al.*, 1995**). Recognition of the potential contribution of these practices could permit a more objective scientific assessment of the effectiveness of many indigenous systems. One of the most important elements of indigenous knowledge is its system and practice in human and animal health care.

In many developing countries, farmers and herders rely on ethno veterinary medicine to treat their livestock because the western-based veterinary health care system is inefficient due to poor staffing and moreover these veterinary drugs are expensive. Therefore, people still use ethno veterinary medicines. According to **Mathius-Mundy and McCorkle (1989)**, ethno veterinary medicines deal with the folk beliefs, knowledge, skills, methods and practices pertaining to the health care of the animals.

In an Indian context, the art of herbal healing has a very deep root in Indian culture and folklore. In India, more than 70 prcent of the population resides in remote villages (**Kumar *et al.*, 2003**) and even today, in most of the rural areas, people are depending on local traditional healing systems for the primary healthcare of their livestock. Since ancient times, animals have played a very significant role in human life for food, milk, leather, transport, drought, warfare, game and recreation. But still today, in remote areas, no system of organized health care facilities for livestock are available, where the rural people are dependent upon their local livestock for most of their needs especially for agricultural production. The ever-declining provision of animal health care services has resulted in the resurgence of a number of epizootic diseases. Under such circumstances, the villagers depend mainly on local herbal medicine for the treatment of their domestic animals.

The Himalayas, are considered as the home of many indigenous communities (**Berreman, 1999; Sharma *et al.*, 2004**). Its favourable and diverse climate supports various traditional farming systems, which involve agriculture, cattle, goats and sheep rearing. Like other parts of the Himalayan region, this traditional practice of farming is considered as the basic factor to raise the economy of rural and tribal people in the Kangra valley of Himachal Pradesh.

Kangra, lying in the lap of the western Himalayas, is located between 31°47¢ & 32°28¢54"N latitude and 76° 3′ 0′′ & 77°6¢45" E longitude. It covers an area of about 5739 sq km. The terrain is hilly and dominated by the Dhauladhar mountain range in the north, which is an offshoot of the Himalayas. The area is covered with thick and luxurious vegetation comprised of tropical sal forest, subtropical chir pine forest, temperate oak forest, conifer forests, sub alpine forests and alpine pastures that provide a variety of plant resources to the local people. The rainfall varies from 290 cm - 380 cm annually. Kangra is inhabited by a number of castes and ethnic groups viz., Rajputs, Brahmins, Mahajans, Kumhars, Kirths, Lohars, Muhasa, Chamars, Gaddis and Gujjars, each with its own diverse religion, cultural and social traditions. The economy of these villagers is predominantly agricultural. Out of the total geographical area nearly 27.30 % is cultivable and ca 52.51% of the total 1,174,072 population is engaged in agriculture (**District Census Handbook, Census of India, 1991**).

Due to the hilly terrain and small land holding, locals of the area depend on livestock for agricultural operations. They also generate their daily livelihood by selling milk from buffaloes and cows and other dairy products. Goats and sheep are preferred to rear because of the perpetual demand for their wool. Therefore to fulfil the objective of maximizing the production from livestock, villagers put in a great effort to keep their animals in a good state of health.

Inspite of the modern medical facilities in the plain areas, villagers still practice herbal therapy for the treatment of animal diseases and disorders but in the villagesm where people sometimes take 1-2 days to reach their farm from the road, head livestock is totally dependent upon herbal medicines to cure even serious chronic diseases. In certain areas of the valley, people have deep indigenous knowledge and well established ethno veterinary practices to cure and control livestock diseases.

At the present time, unfortunately, both the plant species and the traditional knowledge possesed by the people is threatened in various ways. Medicinal plant resources are declining due to their habitat modification and the unsustainable rate of exploitation while traditional knowledge is threatened by the loss of plant diversity, urbanization, modernization and the low incomes of the traditional medical practioners. Therefore, for the conservation of traditional medical knowledge, it is necessary to document such vital information in a systematic manner so that coming generations can implement this knowledge for the well being of livestock in the Himalayan region. The present work highlights some of the ethno- veterinary medicinal plants and their less known uses for the treatment of animal ailments. Some of these uses, which do not seem to be well known or recorded in the pertinent literature, are presented in this paper with a view to drawing the attention of the physiochemists and the pharmacologists to these reported plants for further critical scientific study.

Methodology

Extensive field surveys were conducted during 2003-2004 for the documentation of traditional knowledge on plant resources used to cure various veterinary problems by the inhabitants of the Kangra valley (Himachal Pradesh), India. Preliminary surveys were conducted in 62 important localities (Annexure-1) dominated by the rural and tribal communities in all the tehsils (administrative unit) of the district. Friendly relations were developed with the local people to obtain information. Attempts were made to collect authentic information through interviews and discussions by approaching the local people in their agricultural fields, houses, grazing land and in pastures. Knowledgeable elderly people belonging to each of the traditional communities, who were convinced regarding the objective of this study, formed the source of healthy information.

Detailed information on the plant species, its local name, medicinal use, part used of the plant along with other ingredients added, method of preparation, mode of administration and dosage were recorded for each claim. Plant specimens collected during the surveys were identified by consulting the herbarium in a Botanical survey of India (northern circle), Dehra Dun. These specimens are deposited in the herbarium (PLP) of the Institute of Himalayan Bioresource Technology (CSIR), Palampur, Himachal Pradesh.

Results

In this article, Plants used to cure various livestock diseases are enumerated according to their families, which are arranged alphabetically. Scientific names of the plants along with the collection number of their specimen (in brackets), is followed by their local names and life form. A detailed account is given on the mode of preparation of the traditional medicines and the administration and dosage prescribed by the local healers to cure various veterinary diseases is also given.

ANACARDIACEAE

Mangifera indica Linn. (PLP.6156)
Local name: Aamb,
Life form: Tree

Mode of preparation, administration and cure: About 100 gm of bark of *Mangifera indica* is grinded and mixed with 50 gm of Jaggery, which is locally called as "gudd". Dose of 50 gm is given to cows for 5-6 days, twice a day, to cure the stoppage of chewing the cud. It is very effective to improve the digestion of cattle.

APIACEAE

Foeniculum vulgare Gaertn. (PLP.6187)
Local name: Sounf
Life form: Herb

Mode of preparation, administration and cure: 100 gm dried seeds of Sounf (*Foeniculum vulgare)* is grinded and finely powdered. It is thoroughly mixed with 50 gm Jaggery (gudd). Prepare small balls, which are locally called "laddu" from this mixture and give a dose of one laddu, twice a day, for 4-5 days for **indigestion.**

Cuminum cyminum Linn. (PLP.6193)
Local name: Jeera
Life form: Herb

Mode of preparation, administration and cure: Seeds, about 150 gm are warmed on mild heat, which are grinded and then powdered. This powder is then mixed with half-a litre of water and is given to livestock, twice a day, for 2-3 days for **indigestion.**

Coriandrum sativum Linn. (PLP.6184)
Local name: Daniya
Life form: Herb

Mode of preparation, administration and cure: Seeds of this plant are grinded to prepare the powdered. This is mixed with jaggery in the ratio of 2:1. Small balls (laddu) are prepared and are given to the animals twice a day, for at least 2 days to cure **constipation.**

APOCYNACEAE

Carissa opaca Stapf ex Haines. (PLP.6169)
Local name: Gharnaa
Life form: Shrub

Mode of preparation, administration and cure: Roots (150 gm) are crushed, finely powdered and then mixed with 50 ml of Mustard oil. Paste is prepared by thoroughly mixing these contents and is applied on **mouth and foot parts** of the cattle. It is applied continuously twice a day till required to cure foot and mouth diseases completely.

ASTERACEAE

Bidens biternata (Lour.) Merill & Sherff. (PLP.5116)
Local name: Kathori
Life form: Shrub

Mode of preparation, administration and cure: Decoction is prepared by boiling 200 gm of seeds in one litre of water. 100 ml of this decoction is given orally to the livestock, twice a day as an **anthelmentic.**

BRASSICACEAE

Brassica campestris Linn. (PLP.6185)
Local name: Saroon
Life form: Herb

Mode of preparation, administration and cure: 400- 500 ml oil of *Brassica campestris* and 100 gm of grinded lahsun (*Allium sativum*) are thoroughly mixed. It is given orally, twice a day, for 4- 5 days to cure **gastric troubles.**

CACTACEAE

Opuntia stricta (Haw.) Haw. (PLP.6186)
Local name: Chuu
Life form: Shrub

Mode of preparation, administration and cure: Latex extracted from the leaves and stem of *Opuntia stricta* is applied to cure **skin infection**. It also cures wounds and burned skin. It is applied for 3-4 days for effective results.

CANNABACEAE

Cannabis sativa Linn. (PLP. 6183)
Local name: Bhang
Life form: Shrub

Mode of preparation, administration and cure: About 150 gm seeds of Bhang (*Cannabis sativa)* are crushed and decoction is prepared by boiling in water. This decoction is given to the livestock in a dose of 50 ml, twice a day, for 2-3 days to cure dysentery

CHENOPODIACEAE

Chenopodium album Linn. (PLP. 6163)
Local name: Palak
Life form: Herb

Mode of preparation, administration and cure: Leaves of *Chenopodium album are* grinded to prepare a paste. It is applied on **wounds** for a week and also is found very effective to cure sores on the skin of cattle. It is applied twice a day.

COMBRETACEAE

Terminalia arjuna Bedd. (PLP. 6178)
Local name: Arjuna
Life form: Tree

Mode of preparation, administration and cure: From the bark of this tree, about 50 gm in weight is boiled in one litre of water till it remains one third. This decoction is given in a dose of 150 ml daily for 20 days to oxen and other cattle for general **weakness**

EQUISETACEAE

Equisetum arvense L. (PLP.6189)
Local name: Trutnali
Life form: Herb

Mode of preparation, administration and cure: A whole plant of Trutnali (*Equisetum arvense*) is ground with 50 gm of wheat flour. This mixture is given orally to the livestock, thrice a day for about one month. It is believed to improve the displaced **fractured bone** and also ensures their proper joining.

EUPHORBIACEAE

Mallotus philippensis (Lamk.) Muell- Arg. (PLP. 6168)
Local name: Kamla
Life form: Tree

Mode of preparation, administration and cure: The fruits of *Mallotus philippensis* are crushed, ground and then finally powdered. The powder is given orally twice a day for 2-3 days to cure diarrhoea and dysentery. It is also used as an **anthelmentic.**

Emblica officinallis Gaertn. (PLP.6192)
Local name: Ambla
Life form: Tree

Mode of preparation, administration and cure: The fruits of this tree are dipped in water overnight in a copper/ iron utensil. The pulp is thoroughly crushed and finally filtered to separate the residue. 'Haldi', powder of *Curcuma longa* is also added in the filtered solution and then applied to cure an infection on the **foot,** locally called as 'kharaidu'. The residue of the fruit pulp is given orally with salt to increase the appetite, which also acts as a **cooling agent.**

FABACEAE

Acacia catechu Willd. (PLP.6158)
Local name: Khair
Life form: Tree

Mode of preparation, administration and cure: Young shoots and leaves of khair, are ground and an aqueous extract is prepared from them. Add 10gm powder of Zinger *(Zingiber officinale)* to it and then given to the livestock in a dose of 20- 30 ml, twice a day, for 3-4 days to cure diarrhoea and dysentery.

Acacia nilotica (L.) Willd. (PLP.1296)
Local name: Babul
Life form: Tree

Mode of preparation, administration and cure: About 50 gm pods of *Acacia nilotica* are burnt and the ash collected from the burnt material which is made into a paste by adding 5-6 drops of water. 10 ml of mustard oil is also added. This paste is applied on boils and **wounds,** for 2-3 times a day till it cures completely.

Bauhinia variegata Linn. (PLP.6162)
Local name: Karal
Life form: Tree

Mode of preparation, administration and cure: Immediately after the mating, female cattle are given the leaves of this plant to ensure the conception. Sometimes the leaves are also crushed with the stem of *Musa paradica* (Kaila), which are fed to act as a cooling agent that ensures the successful **conception.**

Cassia fistula Linn. (PLP 6167)
Local name: Kaniar
Life form: Tree

Mode of preparation, administration and cure: Mature/ ripened pods of this plant are collected and are broken into small pieces. These pieces are boiled in water to prepare the decoction which is finally concentrated at a low temperature. Paste like concentrated decoction is given orally to the animals to cure severe pain in their **stomach due to cramps**, locally known as 'butt'. Decoction is also given orally with salt to increase the **urine flow.**

Dalbergia sisoo Roxb. (PLP. 6173)
Local name: Tali
Life form: Tree

Mode of preparation, administration and cure: A paste of the leaves is prepared and is thoroughly mixed with 'crude sugar' (Khand). This mixture is given orally to the animals to cure chronic diarrhoea.

Mimosa pudica Linn. (PLP.6190)
Local name: Lajwanti
Life form: Herb

Mode of preparation, administration and cure: The leaves of the lajwanti plant (*Mimosa pudica)*, about 150 gm and Lahsun (*Allium sativum*) about 50 gm in weight, both are mixed and are ground. Small balls locally called "laddu" are prepared. One "laddu" is given per day to cows and buffaloes, for a month to promote **fertility.**

LAMIACEAE

Mentha longifolia (L.) Huds. (PLP.6175)
Local name: Pudina
Life form: Herb

Mode of preparation, administration and cure: Leaves of *Mentha longifolia* and *Solanum erianthum* are ground in equal proportions with small quantities each of *Piper nigrum* (Kali mirch), *Allium ceipa* (Onion) and Jaggery (gudd). It is given orally, 3-4 times a day to cure severe **pain and cramps in the stomach**.

Ocimum sanctum Linn. (PLP.6177)
Local name: Tulsi
Life form: Herb

Mode of preparation, administration and cure: Leaves and fruits of tulsi (*Ocimum sanctum*) are mixed in equal quantities and then ground. A sufficient quantity of water is added to prepare the extract. This water extract (100- 150 ml) is given twice a day, for about 3 days to livestock for the proper **regulation of urine**. It is also very effective to cure coughs and colds.

LILIACAEA

Allium sativum Linn. (PLP. 6181)
Local name: Lahsun
Life form: Herb

Mode of preparation, administration and cure: Juice of lahsun (*Allium sativum)* and pyaz (*Allium cepa*) is extracted and is given to livestock for 4-5 days to cure cold and cough.

MALVACEAE

Bombax ceiba Burm. (PLP. 6180)
Local name: Semal
Life form: Tree

Mode of preparation, administration and cure: 150 gm of stem bark of the *Bombax ceiba,* tree is crushed, and dried under shade. This plant material is finely powdered and mixed with 100 gm powder of turmeric. The paste prepared from this mixture is applied for one month on the **dislocated joints** to cure the swelling and pain. It also helps to rejoin the fractured bone.

MENISPERMACEAE

Cissampelos pareira Linn. (PLP.6191).
Local name: Petundu / Batindu
Life form: Herb

Mode of preparation, administration and cure: Leaves and roots of this plant are mixed with the roots of the *Costus speciosum* (ghor), Kali mirch (*Piper nigrum*), ajwan and tubers of *Curcuma officinalis* (Kachi halder). It is then powdered and thoroughly mixed to prepare the paste. This paste is given orally to the livestock, twice a day for about 20 days to cure **ulcers and painful swelling in the stomach**. Leaves of this plant are also given directly to cure dysentery.

Tinospora cordifolia Miers. (PLP.6171)
Local name: Galon
Life form: Shrub

Mode of preparation, administration and cure: Stem of *Tinospora cordifolia* is cut into small pieces and decoction is prepared from it by boiling in water. This decoction is given in a dose of about 20 ml, 2-3 times a day, for about 5 days to cure fever.

MIMOSACEAE

Albizzia lebbek Benth. (PLP. 6170)
Local name: Sarin
Life form: Tree

Mode of preparation, administration and cure: Stem bark of *Albizzia lebbek* (200 gm) is crushed and paste is prepared .It is applied on the skin around the **wounds,** 2-3 times a day till the wounds are cured completely.

MORACEAE

Ficus recemosa Wall. (PLP. 6188)
Local name: Gular, Rumbal
Life form: Tree

Mode of preparation, administration and cure: Stem bark of *Ficus recemosa* (150 gm) is grinded and paste is prepared by adding 50 ml of mustard oil (*Brassica campestris*). This paste is applied on skin, twice or thrice a day for healing the wounds. It also cures other types of **skin diseases** and infections, till required.

MYRTACEAE

Syzygium cumini (Linn.) Skeels. (PLP.6151)
Local name: Jaman
Life form: Tree

Mode of preparation, administration and cure: Leaves of this plant are given to the cattle especially to cure **worms in their stomach** and therefore is believed to act as wormicide. It is given with empty stomach after feeding sweet Jaggery 'Gudd' to attract all the worms to be affected by the wormicidal action of the extract of this plant. Bark is also found effective to cure dysentery.

ORCHIDACEAE

Rynchostylis retusa, (PLP. 5104)
Local name: Phangru, Skamman
Life form: Herb

Mode of preparation, administration and cure: Whole plant is grinded with water to prepare paste, which is applied on the broken bones. It is also applied to cure **joint pains and swelling.** Sometimes it is also given orally for the same purposes.

PIPERACEAE

Piper nigrum Linn. (PLP.6194)
Local name: Kali mirch
Life form: Herb

Mode of preparation, administration and cure: Dry seeds of *Piper nigrum* (50 gm) are grinded to prepare the powder. Aqueous extract is prepared from by adding 100ml water in this fine powder. This solution is given to cattle, sheep and goats for **stomach and intestinal troubles** which is locally known as "Marod" in a dose of 20-30 ml, for 2-3 times a day, for about 2 days to cure completely.

POACEAE

Cynodon dactylon Pers. (PLP.6176)
Local name: Druub
Life form: Herb

Mode of preparation, administration and cure: Paste of whole plant is applied on the infected part of the **skin and to cure boils,** twice or thrice a day, till required.

Dendrocalamus strictus Nees. (PLP.6149)
Local name: *Banj* / Baans
Life form: Tree

Mode of preparation, administration and cure: To cure dysentery, leaves are crushed with wheat grains and residue (husk) of paddy. This mixture is given directly to the animals for twice a day, till required.

Oryza sativa Linn. (PLP.5112)
Local name: Dhan
Life form: Herb

Mode of preparation, administration and cure: Rice are boiled in water to prepare a thin paste. 'Haldi', powder prepared from the tubers of *Cucurma longa* is also added in it to increase its healing properties. This paste is applied twice a day, for about 10-15 days, especially to cure the hair fall, **wounds** and scratches on the tail. This disease is locally called as 'Manailli'

RANUNCULACEAE

Thalictrum foliolosum DC. (PLP.5115)
Local name: Bhamrol
Life form: Herb

Mode of preparation, administration and cure: Paste is prepared by grinding the roots of this plant and the whole tubers of Allium *cepa* (Pyaj). It is given orally to livestock as a treatment against fever and shuffle (khurmuri) in cattle.

ROSACEAE

Pyrus pashia Buch.-Ham.ex D.Don. (PLP.6148)
Local name: Kainth
Life form: Tree

Mode of preparation, administration and cure: Ripened fruits of *Pyrus pashia* and *Psidium guajava* (Amrood) are grinded with the seeds of *Zanthozylum armatum* (Tira mira) in equal proportion. The powder prepared after grinding is thoroughly mixed with wheat flour. This mixture is given to livestock twice a day for 2-3 days to cure dysentery.

RUTACEAE

Zanthoxylum armatum DC. (PLP.6166)
Local name: Tirmira
Life form: Shrub

Mode of preparation, administration and cure: Leaves and fruits of this plant are powdered and are mixed with the seeds of *Brassica nigra* (Kali sarson). It is given early morning in empty stomach to cure **stomach poison**, locally called 'jahar'.

Aegle marmelos (L.) Corr. (PLP.6179)
Local name: Billpatri
Life form: Tree

Mode of preparation, administration and cure: Leaves of *Aegle marmelos* about 200 gm in quantity are cut into small pieces, mixed with jaggery. It is given to livestock twice a day, for 10-15 days to kill the worms **inside the stomach**

SOLANACEAE

Capsicum annum Linn. (PLP.6182)
Local name: Lal mirch
Life form: Herb

Mode of preparation, administration and cure: Dried fruits (about 100 gm) of *Capsicum annum* are grinded, finely powdered. This powder is applied twice a day, for a week, on the portion of **the skin of the cattle** bitted by dog to cure it completely.

Solanum erianthum D.Don. (PLP.6160)
Local name: Olla
Life form: Shrub

Mode of preparation, administration and cure: Decoction of the leaves of this plant is given orally to cure livestock against contagious diseases. The leaves are also crushed with the coal of *Pinus roxburghii* (Chir) to cure cough, locally called "khang".

Solanum surattense Burm.f. (PLP. 7271)
Local name: Doda
Life form: Herb

Mode of preparation, administration and cure: Ripened fruits are grinded with ajwain in the proportion of 2:1 and are mixed thoroughly. This mixture is given to livestock for three days with hot water in a dose of 20 gm to cure fever.

VERBENACEAE

Lantana camara Linn. (PLP. 6164)
Local name: Panzfulli
Life form: Shrub

Mode of preparation, administration and cure: About 100 gm leaves of *Lantana camara* are crushed to prepare the paste. This paste is wrapped inside a cloth and then applied on skin, twice or thrice a day, for a week to cure **cuts and wounds.**

Vitex nigundu L. (PLP.6165)
Local name: Vana
Life form: Shrub

Mode of preparation, administration and cure: Sarson oil (*Brassica campestris*) is applied on the fresh leaves, which are warmed and then applied on the wounds to control swelling and **pain.** It is also applied as dressing material to control the pain and swelling due to broken bones.

ZINGIBERACEAE

Curcuma officinallis (PLP.5110)
Local name: Kachi Halder
Life form: Herb

Mode of preparation, administration and cure: Powder prepared from the dried tubers of this plant is used as an ingredient with *Allium cepa* (Pyaz), *Allium sativum* (Lasun), leaves of 'Jaman Khumb' and seeds and leaves of *Zanthozylum armatum* (Tara mira). It is finally mixed with jaggery 'Gudd' and tablets of 150gms, locally called 'Mathuni' are given orally to cure cold. This disease is locally called 'Mailla' in the kangra valley.

Discussion and Conclusions

During the field trips to different remote and rural villages, it was observed that there is much traditional knowledge conceiving livestock diseases and their treatment within the rural and tribal/ ethnic communities of Kangra valley. During the documentation and interviewing the elderly people during ethno botanical surveys, it was revealed that a total 44 plant species belonging to 28 families are used in the folk biomedicines to cure various livestock diseases. Among different families of the plants used in local ethno veterinary practices, maximum of the species (6) used in the folk biomedicines belong to the family Fabaceae followed by Apiaceae (3), Poaceae (3) and Solanaceae (3) respectively. Of 44 plant species recorded, about 44 % is the herbaceous flora which is used traditionally in ethno veterinary practices followed by trees and shrubs (Fig. 1). Most of the plant parts are used as ingredients with the raw material of other plant species and have more than one therapeutic uses. Among the parts of the plants used in

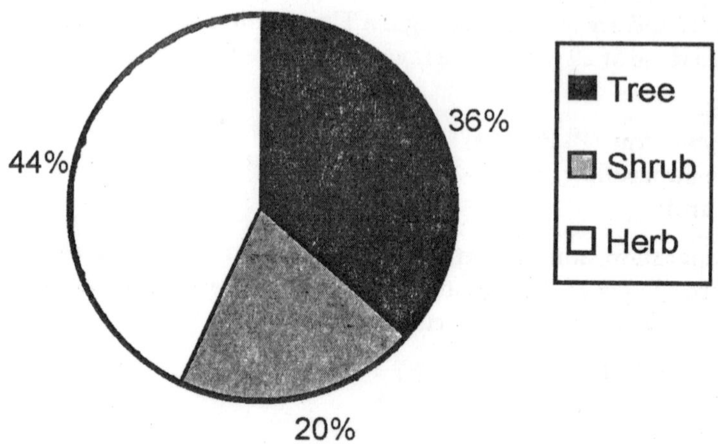

Fig: 1. Habit of plant resources used in veterinary practices

the recipes, leaves are used in 15 preparations whereas seeds (including seed oil) (9), fruits (6), bark (5), underground parts (4), whole plant (3), fresh shoots (2), Pods (2) and Juice (1) are used in descending order. It indicates that local people of the area are quite aware regarding the strong healing properties of leaves which probably is due to the occurrence of effective healing or curing compounds. It was observed that local animals of the area are prone to stomach problems and maximum of the plant species have been found to cure these problems. A total of 10 species are used to cure stomach disorders which includes stomachache, ulcer, painful swelling, poisonous infections, worms in the stomach, gastric troubles and anthelmentic. Dysentery and indigestion is another major problem of the livestock during the rainy season. Maximum intake of polluted water during this season is the major factor responsible for these disorders. 8 plants are used to cure dysentery and 4 have been found effective to cure indigestion. To cure other type of miscellaneous diseases, which include cuts, wounds, burned skin, dog bite, hair fall, wounds and scratches on the tail, a total 9 species are used as ingredients in different medical remedies. Cold and cough is cured by 4 spp., fever by 3, foot and mouth disease by 2, general weakness by 1, fractured bones are cured by 2, sexual dysfunction

(Conception) is ensured by giving 2, joint pains and urinary troubles by 2 species each and 1 species is used as cooling agent (Fig. 2).

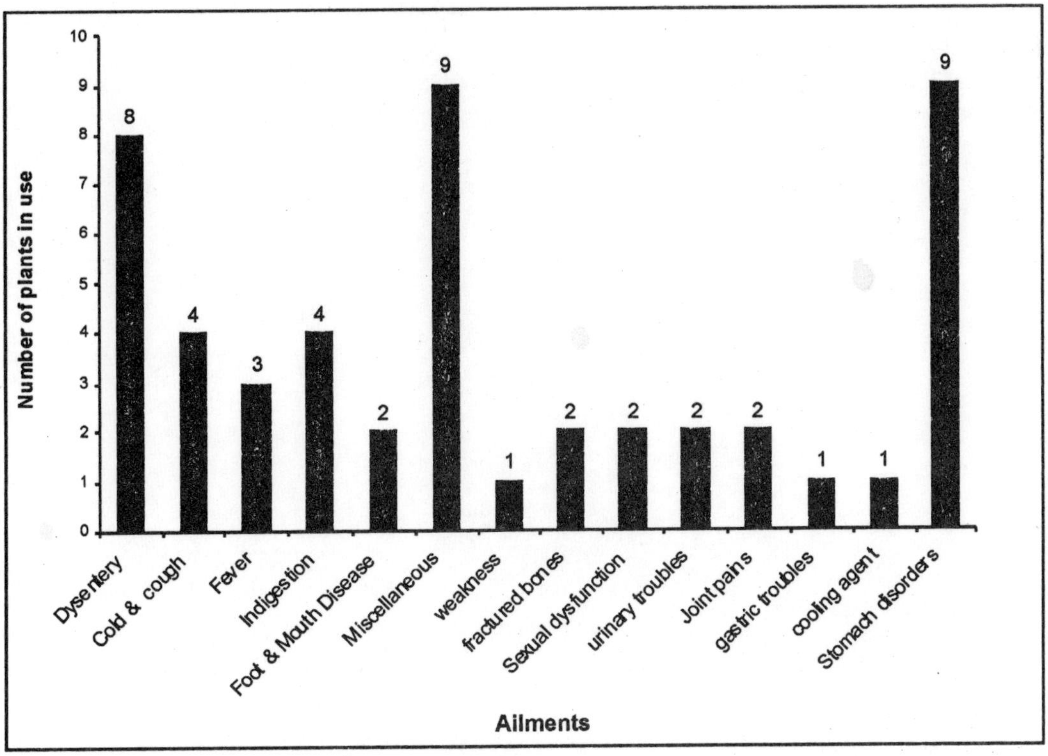

Fig: 2. Extent of curable livestock diseases

It was further observed that traditional information on folk biomedicines and herbal practices to cure various veterinary diseases is restricted with only elderly rural and tribal people of the area. Most of the knowledgeable persons were observed in Gaddi community who are mainly pastoralists in nature. They are having substantial traditional knowledge as they move from various climatic zones of the Kangra district along with their lkivestock. Due to the wide range of harsh climatic conditions and poor medical facilities at the times when they move with their livestock from subtropical zone like Shahpur and Kangra to the subalpine and alpine cold region of Dhauladhar mountains and Bharabhangal area, they are totally dependent upon the plant species present in the wild to cure the above described disease of their livestock. But With the exposure in modern era, ethno veterinary knowledge is decreasing day by day which needs immediate systematic documentation. The plants used for these purpose are also under high threat because of the increase in social demands and should be conserved on priority basis. Further, ethno- surveys are needed to document similar knowledge in the unexplored zones of the Himalayan region before the uses of wild flora to cure the livestock through traditional herbal practices are forgotten.

32

Media Coverage of Climate Change : A Case Study of Indian English Language Dailies

BHARVI DUTT, K.C.GARG AND PRAMILA KUMARI

Introduction

Mass communication is a process in which a message is created by a person or a group of people and sent through a transmitting device (a medium) to a large audience. Here media acts as a connecting link between information generators/providers and the information seekers. If we consider the medium, then in this highly scientific advanced age of audio-visual devices, print media still holds an important place in disseminating information. In print media, newspapers are one of the cost effective primary sources of information for the general public. Newspapers have a wide circulation in national as well as regional languages besides the English language.

Besides disseminating information, media also provides science news to the public. The process of science communication thus involves the flow of information from the scientists to the journalists and then to the general public. Often media uses authentic scientific informational sources to build journalistic stories for the consumption by the general public. Science communication may also involve functional aspects of science and technology relevant to the public at large. Farago said that the presentation of science in the newspapers represents the outcome of a compromise between personal, economic and political pressures (Peter Farrago 1976).

What influences the science coverage in the media and how much does it cover? What role does the media play in influencing public awareness? How do science stories of importance like climate change come to be reported? Who gets cited as a legitimate source in those stories? All of these aspects have been studied to some extent. Studies have been done to analyze the media communication and uncertainty (Allen Bell 1994, Kris M Wilson 2000). Researchers have also tried to analyze the factors that affect the selection of scientific stories to be presented to the public (Meadows 1991, Scientometrics) and what is the public understanding of issues such as of climate change.

Studies have revealed that there public learns a lot about environmental issues through mass media news. In order that correct social action may be taken in matters involving local, national or international issues, awareness and understanding of the effects is essential (Peter Farrago 1976). The issue of climate change has assumed global dimensions encompassing a range of factors including social, political, economic, scientific, and many more. The present study attempts to look at a media representation of climate change.

Media and climate change

United States of America

Recently a number of studies have been conducted to study the role and behaviour of media with respect to the dissemination of anthropogenic climate change information in American newspapers. (Boykoff & Boykoff 2004) in their study of the US prestige-press (the *New York Times*, the *Washington Post*, the *Los Angeles Times*, and the *Wall Street Journal*) have concluded that the journalists have adhered to 'balanced' reporting which has led to a popular discourse being significantly diverged from the scientific discourse.

Another study observed that the increase in the media coverage of climate change is event based. This could be attributed to the release of a first assessment report by IPCC (1990), by the Rio de Janerio summit (1992) and the unveiling of the UN Framework Convention on Climate Change, and by the Kyoto a Climate Summit (1997). The year 2001-2002 witnessed an increased coverage that may be due to a third assessment report by IPCC, climate talks in Bonn, Germany and Genoa, Italy. During 2001, President Bush also declined the signing of the Kyoto protocol (Boykoff 2007).

United Kingdom

A recent study has observed that the UK tabloid press has significantly diverged from the scientific consensus regarding the anthropogenic contribution to climate change. A Sudden increase in the newspaper coverage in a certain year was also observed (Env Res Letter Boykoff 2008). The increase was mainly event based like a natural calamity, intense flooding in England (November 2000) and the G8 summit in Gleneagles, Scotland (July 2005). The year 2006 (September - December) saw a phenomenal increase in newspaper coverage in the UK tabloid attributable mainly to three reasons – release of the Al Gore film 'An Inconvenient Truth', a much publicized multi-billion dollar donation to renewable energy initiatives and bio-fuel research, and the release of a report by Lord Stern which discusses the effect of climate change and global warming on the world economy.

Another study observed that a third of the tabloid press coverage carried the message that the human contribution to climate change was negligible. Content analysis has revealed that in reporting, the views of scientists were used to give legitimacy (Maxwell and Maria 2008). In both the tabloids and broadsheets, little evidence was provided, in the form of data, to substantiate the claims being made. British media has not explored effectively the uncertainties about global warming (Neil Taylor 2002).

Indian scenario

A little effort has been made to study the media dissemination of information regarding the anthropogenic contribution to climate change in India. A study by Dutt and Garg (2000) revealed that the English-language newspapers in India, allocated to environmental science approximately 3% percent of the total space devoted to science and technology. However, this study did not elaborate on the sub-disciplines in the entire gamut of environment. The year for the study was 1996, which was a relatively stable period with no major catastrophes having occurred in that year. The present study was undertaken with the objective to investigate the newspaper coverage of the crucial issues of climate change.

Methodology

English language dailie's readership (18.59%) is second only to Hindi (27.54%) (INS). In some states like Chandigarh, Delhi, Maharashtra and West Bengal, the circulation of English-dailies exceeds that of Hindi-dailies. The present study is based on items related to climate change reported in Indian English-language dailies. The study was carried out for a period of 45 days (1st March 2008 to 15th April 2008), which is a limitation of the present study. The authors are working to collect the data for a period of six months. Table 1 gives the list of 31 newspapers that were used to gather the data. The items retrieved formed the subject matter for the content analysis of the present study.

Table 1: Distribution of items and space covered by different newspapers

S. No	Newspaper	No of Items (%)	Space Cm^2 (%)
1	*The Times of India*	26 (16.05)	4662 (11.78)
2	*The Hindu Business Line*	13 (8.02)	3529 (8.92)
3	*The Hindu*	15 (9.26)	3433 (8.67)
4	*The Pioneer*	8 (4.94)	2819 (7.12)
6	*The Financial Express*	10 (6.17)	2738 (6.92)
7	*The Asian Age*	11 (6.79)	2708 (6.84)
5	*Hindustan Times*	12 (7.41)	2644 (6.68)
8	*Economic Times*	5 (3.09)	2295 (5.80)
9	*Mint*	5 (3.09)	1885 (4.76)
11	*Deccan Herald*	4 (2.47)	1173 (2.96)
12	*Mail Today*	4 (2.47)	1085 (2.74)
13	*The New Indian Express*	3 (1.85)	1076 (2.72)
14	*The Political & Business Daily*	6 (3.70)	1052 (2.66)
15	The Indian Express	5 (3.09)	940 (2.37)

S. No	Newspaper	No of Items (%)	Space Cm^2 (%)
10	*The Statesman*	5 (3.09)	918 (2.32)
16	*Business Standard*	4 (2.47)	876 (2.21)
17	*The Assam Tribune*	3 (1.85)	807 (2.04)
18	*The Financial World*	4 (2.47)	756 (1.91)
19	*National Herald*	4 (2.47)	652 (1.65)
20	*DNA*	2 (1.23)	598 (1.51)
21	*Mid Day*	2 (1.23)	458 (1.16)
22	*Metro Now*	1 (0.62)	400 (1.01)
23	*The Free Press Journal*	2 (1.23)	386 (0.98)
24	*Greater Kashmir*	1 (0.62)	340 (0.86)
25	*Sahara Time*	1 (0.62)	280 (0.71)
26	*Sentinel*	1 (0.62)	210 (0.53)
27	*The Telegraph*	1(0.62)	200 (0.51)
28	*Daily Excelsior*	1 (0.62)	180 (0.45)
29	*Hitavada*	1 (0.62)	168 (0.42)
30	*Tribune*	1 (0.62)	160 (0.40)
31	*Hindustan Times Next*	1 (0.62)	153 (0.39)
	Total	162	39581

The items were subjected to both the qualitative and quantitative analysis. The main issue covered in the items was identified first from the title but in case the title was ambiguous, the contents were used for identification. Each item was also categorized on the basis of the type of work it represented like research, meeting, report, etc. The data was analyzed in terms of the space covered and the source of the items the sources of the items were in terms of country and the illustrations shown along with items and citations, were also analyzed.

Result

Space covered and newspapers: A total of 1,168 items related to science & technology were retrieved, from 31 newspapers published from different parts of the country. Of these 193 (16.52%) items were on different aspects of environmental science, of which 162 (83.94%) items were on different issues related to climate change. These 162 items occupied 39,581 cm^2 (Table 1). The Times of India allocated the maximum space 11.78% (4,662 cm^2) followed by *The Hindu Business Line* 8.92% (3,529cm^2) and *The Hindu* 8.67% (3,433 cm^2). These three newspapers together occupied almost a third of the total space. The top ten newspapers covered almost 70% (27,886 cm^2) of the space and the remaining 21 newspapers covered 30% (11,695 cm^2) of space. Nine newspapers published only one article during the study period. The Hindustan Times and next, *The Hitavada* and *Tribune allocated the* minimum space.

Table 2: Distribution of items in terms of different categories

S. No	Issues	No of items (%)	Space Cm² (%)
1	Impact	58 (35.80)*	12603 (31.76)*
2	Policy	34 (20.99)	9091 (22.91)
3	CO_2 control	24 (14.81)	7047 (17.76)
4	Public awareness	19 (11.73)	4598 (11.59)
5	Natural phenomenon	8 (4.94)	2092 (5.27)
6	Causes	7 (4.32)	989 (2.49)
7	Finance	6 (3.70)	1566 (3.95)
8	CO_2 emission	4 (2.47)	1130 (2.85)
9	Infrastructure	1 (0.62)	204 (0.51)
10	New technology	1 (0.62)	161 (0.41)

Issues covered: The issues covered in the items were identified from both the title and the contents and were classified into ten different categories (Table 2). The Majority of these items 58 (35.80%) related to the impact of climate change and global warming and these items occupied the maximum space 12,603 cm² (31.76%), followed by policy related items 9091 cm² (22.91%) and carbon-dioxide control related items 7,047 cm2 (17.76%). Policy related items followed by items dealing with carbon dioxide control occupied 23% and 18% space respectively. These three issues together occupied about 72% both in terms of the number of items published as well as the space occupied. The remaining 28% items were related to seven other issues. Of all the issues the least number of items were about the development of new technologies and infrastructure, both having a single item only. Though seven items were published on the causes of climate change, space occupied was a meagre 989 cm² that was far less than that of finance (1,566 cm²) and carbon-dioxide emission (1,130 cm²) related items each had only 6 and 4 items respectively.

Geographic Origin: The information regarding the country of origin of the event or activity of an item could be traced in respect of only 150 items (Table 3). The origin was identified from the content of the items. These 150 items had their geographical origin from sixteen different countries, of which 68 (45%) items had their origin in India followed by those from the US 22% (32). About 7% (11) and 4% (6) items were from Australia and the UK respectively. The remaining 22% came from other countries. A major proportion of items from other countries originated from Thailand and were all about international policy agreements/deliberations on climate change in the form of meeting reports held in Bangkok, to generate global cooperation for the reduction in greenhouse gas emission to slow down climate change.

Table 3: Geographic distribution of the items

S. No	Issues	India	USA	UK	Australia	Others	Total
1	Impact	23	15	1	4	10	53
2	Policy	16	1	1	-	14	32
3	CO_2 control	10	4	2	1	6	23
4	Public awareness	14	1	-	3	-	18
5	Natural phenomenon	2	3	-	3	-	8
6	Causes	-	4	2	-	1	7
7	Finance	1	1	-	-	2	4
8	CO_2 emission		3	-	-	-	3
9	Infrastructure	1	-	-	-	-	1
10	New technology	1	-	-	-	-	1
	Total	68	32	6	11	33	150

Citation: Of the 162 items, almost 76% (123) had quoted someone. For the analysis, only items with actual quotation of a source were included. The individuals being quoted included scientists, experts, politicians, environmentalists, etc. More than half the items had quoted a single source. Almost 40% of the items had quoted either two or three sources. A small percent (5 percent) of items had quoted more than four sources.

Illustrations: Out of the 162 items, almost half of the stories carried illustrations (Figure 1). A majority of the items 77% (63) carried visuals in the form of photographs. The remaining 23% included all other categories like animation, graphs, satellite images, etc. and of these maps and tables were present in only about 1% of items were present. Animation was the second most dominating visual medium present in about 6% of the items. Graphs and satellite images were

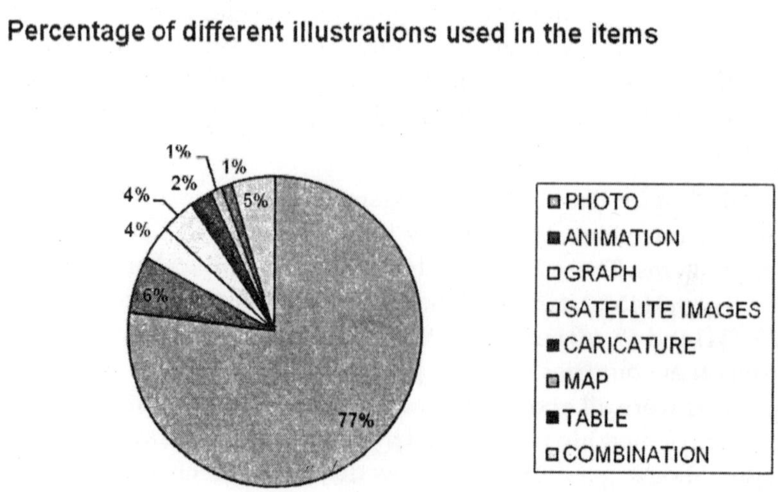

Figure 1

both present in 4% of each item. Data supporting the text in the form of tables and graphs, was present in just 5% of the total items. Impact related items had the maximum number (29) illustrations followed by policy related items (15) and by carbon-dioxide control related items (14).

Sources of news: Out of 162 items, only 7 (4%) items were written either by scientists or experts. Out of these seven items, four were by Indian authors while the remaining three were by foreign authors. Of the remaining 155 items, only 54 items had mentioned the name of the reporter as well as the correspondent of the particular newspapers who had written 11 items.

Of the 162, 100 items had been sourced from different news agencies/services (Figure 2). Of these, *Press Trust of India (PTI)* contributed 17%, followed by *Agence France-Presse (AFP)* and *Reuters each* 14%, and 12% from *Associated Press (AP)* respectively. Of these 100 items, 47% were from various Indian news agencies, whereas the remaining 53% were from different foreign

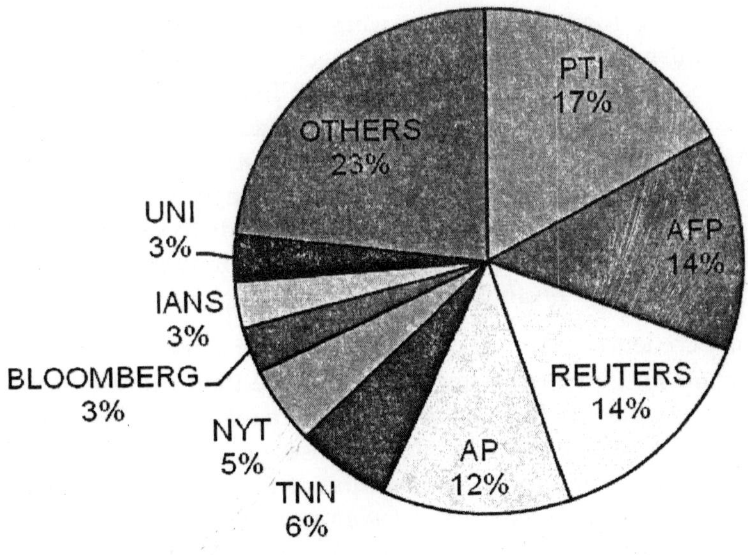

Figure 2: Distribution of items with respect to the news agencies

news agencies. Items were sourced from American, British and French news agencies. The major Indian sources were from PTI and the Times News Network.

Prominence: The placement of a news item in a newspaper indicates the importance accorded to the item. News is categorized and placed in specific sections like national, international, science and technology, sports, entertainment, etc. The study indicated that almost 30% of these items were published in the first five pages, whereas another 27% were published between page numbers six to ten. Six items that appeared on the front page covering a total space of 1,586 cm^2 were basically messages / statements addressed to India by political dignitaries (Tony Blair, Former British Prime Minister). Of these, two items were published in *The Times of India*, and the remaining, one in each *The Hindu*, *The Pioneer*, *The Financial Express* and *The Financial World*. The maximum coverage in the first five pages was given by *The Pioneer*, followed by *The Financial Express*. *The Statesman* and *The Times of India* both published four items in the first five pages. Only 22 items were published in the page number 21 to 40.

Discussion

The entire media coverage seems couched in a larger context of the political hype being raised all over the world over the issue of climate change. As a result of the international activism, concerns of the eminent members of the public including the scientific community, with the involvement of the United Nations and governments of the developed and developing countries, the issue of climate change has come to occupy a persisting and dominant concern of the media in English-language dailies in India. Not only has there been a proliferation in the items relating to climate change in the newspapers but also the media appears to have used all the art of journalistic imagery to arrest the attention of its readers. The media in order to create sensationalism and to attract the readers uses powerful and strong words in the titles. A few examples of this dramatism would highlight this aspect. The *Times of India*, wrote "Climate change will wreak havoc on health" whereas *The Economic Times* used "Global warming wreak havoc with monsoons, bring droughts". The *Statesman* had one item with the heading "Earth in crisis, warns NASA scientist" whereas *The Hindu Business Line* and *DNA* used titles "Glacier melting at alarming rate" and "World's glaciers melting alarmingly: UN" respectively. The use of such forceful words like havoc, fears, crisis, alarming tends to cause apprehension and anxiety in the public mind. The coverage in general appears alarmist in nature with a major emphasis on the negative impacts of climate change.

In order to confer credence and an air of authority to their stories the newspapers tended to quote eminent people and organizations to authenticate their representation of climate change. The *Economic Times* quoted the director of the Indian Institute of Tropical Meteorology in the item "Global Warming wreak havoc with monsoons, bring droughts". In the item entitled "Global warning fuels malaria, dengue outbreak fears", The *Hindustan times* wrote "*The damage, WHO experts say will be huge in developing countries because of lack of preparation to handle disasters and lack of infrastructure to prevent any kind of outbreak*". DNA wrote "*UNEP warned that further ice loss could have dramatic consequences particularly in India, whose rivers are fed by Himalayan glaciers*".

It appears that the agenda of the issue is being set somewhere else and various actors, including the media, are orchestrating to the tune of that agenda. Political activists in the western world too seem to be setting the tone of the agenda. It is known that the USA had earlier declined to ratify the Kyoto protocol on the ground that it would affect the American economy. Now with the Kyoto deadline approaching fast and no future international pact being evolved for slowing down global warming and climate change, the US (despite being the biggest carbon dioxide emitter) is once again dilly-dallying to make a definite effort to deal with this global problem. Now it refuses to make any concrete effort till India and China also agree to definite emission reduction goals. *The Asian Age* in the item "US: deep emission cuts could rattle economies" said *"American negotiators at a UN climate conference warned that calls for a steep emission cuts could further rattle economies, especially in the developing world. But delegates from poor nations and environmentalists said on Thursday that such fears were just a ploy to avoid tough action"*. The media portrayal regarding climate change initiatives and international policy formation in the Indian press coverage also points towards the fact that international leaders are trying to influence India and China to agree to climate goals. Instead of focusing on the US to agree to the goals, they are trying to rope in India and China which then might make the US change its mind too.

In the story, India is the benign, beneficial growth of the future, The *Pioneer quoted* former British Prime Minister Tony Blair *"if America, Europe and Japan prepare for a really strong cut in emissions, with the right incentive, then they have the right to ask Indian and China to play fair and then we could structure a global deal. India and China cannot have the same obligations as the west, but need to meet the obligations that climate change imposes on us all"*. In another story, *The Times of India* (India can take a lead on climate change: Gore) quoted *Al Gore, a former US Presidential contender and Nobel Prize winner, "the correct response to the climate crisis is not a comparison between some level of pollution that has been achieved by other countries a long time ago with dirty technologies but rather what can be achieved in the 21st century with efficient technologies. India is highly vulnerable to the climate change crisis. In league with other nations, India can help be a part of the solution."* Similarly stories in other newspapers with the headings like Be committed, Blair tells India, China (Deccan Herald), India can lead the battle against climate change: Blair (The Hindu) and Climate change: Blair for India in leading role (Hindustan Times) talks about the participation of both the developed and the developing countries in combating climate change rather than arguing and blaming anyone for the present situation.

References

1. Peter Farago. Science and the media. Sir Harrie Massey & Sir Frederick Dainton (Eds). Oxford University Press, 1976.
2. Allen Bell. Media (mis)communication on the science of climate change. *Public understanding of Science*, Vol. 3 (1994), pp. 259-275.
3. Kris M Wilson. Drought, debate, and uncertainity: measuring reporters' knowledge and ignorance about climate change. *Public understanding of Science*, vol 9 (2000), pp 1-13.
4. Julia B Corbett and Jessica L Durfee. Testing public (un)certainity of science:media representation of global warming. *Science Communication*, Vol. 26 (2004), pp. 129-151.
5. Keith R Stamm, Fiona Clark, and Paula Reynolds Eblacas. Mass communication and public understanding of environmental problems: the case of global warming. *Public understanding of Science*, vol 9 (2000), pp 219-237.
6. A J Meadows. Quantitative study of factors affecting the selection and presentation of scientific material to the general public. *Scientometrics*, Vol. 20 (1991), pp 113-119.

7. Maxwell T. Boykoff and Jules M. Boykoff. Balance as bias: global warming and the US prestige press. *Global Environmental Change*. Vol 14 (2004), pp 125–136

8. Maxwell T. Boykoff and Jules M. Boykoff. Climate change and journalistic norms: A case-study of US mass-media coverage. *Geoforum*, Vol. 38 (2007), pp. 1190–1204

9. Maxwell T Boykoff and Maria Mansfield. 'Ye old hot aire': reporting on human contributions to climate change in the UK tabloid press. *Environmental Research Letters*, Vol. 3 (2008), doi: 10.1088/1748-9326/3/2/024002

10. Neil Taylor and Subhasini Nathan. How science contributes to environmental reporting in British newspapers: a case study of the reporting of global warming and climate change, *The Environmentalist*, Vol. 22 (2002), pp. 325–331.

11. Bharvi Dutt and K C Garg. An overview of science and technology coverage in Indian English-language dailies. Public Understanding of Science, Vol. 9 (2000), pp. 123-140.

12. INS Press Handbook, 2005-2006, Allied Publishers Pvt Ltd, New Delhi

33

Energy Budget of Summer Crops in a Mountain Agro-Ecosystem in Kumaun Himalaya

KUNDAN SINGH, VIR SINGH, UMA MELKANIA AND J.P.N. RAI

Introduction

Energy, basically, is the capacity to do work. Every ecosystem starts functioning with an input of energy. The first form of energy is of an extra terrestrial source, and this is the form of energy which is the primary source of life on Earth. When solar radiation strikes the Earth it tends to be degraded into heat energy. On account of the mechanism of photosynthesis carried out by green plants, however, a fraction of the same energy is converted into potential or food energy. It is thanks to this wonderful mechanism operated by green plants that the entire energy received by the Earth is not wasted.

Every ecosystem has an energy flow as one of its essential features and as an inevitability for its functioning. The Lofty and fragile mountains of the Himalaya, as of Uttarakhand, attain some kind of ecological stability thanks to a variety of forests as the main natural resource of the region. An agro-ecosystem in the Himalayan mountains, as operating in the Kumaun region of Uttarakhand, is an interesting example of overlapping ecosystems. Forest/ grassland/ range-land, cropland, livestock and households' cluster and called are a village, make inseparable components of an agro-ecosystem. Thus, agro-ecosystem, as the one in the Himalayan mountains, is quite a heterogeneous and complex unit of nature managed by human beings. Here the major (rather exclusive) consumers are the human beings and, unlike in upland forest and open ocean ecosystems, it receives nature's subsidy either though a natural means or through human management. Thus an agro-ecosystem involves lot of energy for its functioning (Singh 2008).

Summer cropping (the *Kharif* season cropping) especially embraces a whole range of agro-biodiversity, including cropping patterns, large number of crops (cereals, millets, pseudo cereals, pulses, oil seeds, vegetables, etc.) and very large numbers of varieties of each crop. A strategy of agricultural development in mountain areas largely revolves round summer cropping. The summer cropping is interesting to understand in its totality including the energy flow between different components of the agro-ecosystem. Our current understanding of an energy flow is within a mountain agro-ecosystem and appears to be inadequate. This study is an attempt to fill this gap.

A Green Revolution type or modern agriculture as seen in favourable areas of the plains is cut off from its linkages with forests/ rangelands and livestock. The agro-ecosystem in which this type of production system is managed can be referred to as fossil fuel subsidized solar-powered ecosystem, unlike the one in the Himalayan mountains which can be referred to as a nature-subsidised solar-powered ecosystem. This classification emanates from the one presented by Mitchell (1979) based on Odum (Singh 2005).

The functioning of a mountain agro-ecosystem would require an energy input of various sorts, viz. i) direct solar energy which would be converted into potential energy by plants, ii) forest biomass in terms of manure through livestock, iii) seeds of a variety of crops, iv) animate energy of livestock and humans, v) crop residues as manure, and vi) livestock feeds, which are converted into draught power, milk and other products.

An agro-ecosystem would produce energy in various usable forms, viz., i) crop residues used in livestock feeding, ii) food grains/ fruits/ other plant parts, such as roots, shoots, leaves, etc. of food value, iii) green grass, weeds and tree/ shrub leaves used as green fodder for livestock, iv) dung/ manure, v) milk and other animal products, and vi) calf crop.

A quantitative estimate of the energy flow through a mountain agro-ecosystem would help to gauge the energetic efficiency of the food production system. The energy output-input ratio is one of the major indicators of the functioning and performance of an ecosystem. Determination of gross energetic efficiency, or the output-input ratio, is the logical and objective method of evaluating the productive performance of an agro-ecosystem as well as its linkages with various ecosystem components. Common denominator to which the input and output can be related with most accuracy is not economics, but energetics. The economic value of parameters involved tends to fluctuate with supply and demand. The calorific values remain relatively constant (Odend'hal, 1972).

Energetics being dependent on the inputs and outputs of an agro-ecosystem is also a useful indicator to suggest whether it is sustainable or not. This is, therefore, an important issue relating to the performance of an agro-ecosystem, which helps to understand the basic relationship of the production system with its environment (Singh and Sharma 1993; Singh 1998, Singh *et al.* 2003).

Materials and Methods

The Study Area

Uttarakhand, the 27th state of the Republic of India, lies in the Northern part of India amidst the magnificent Himalayan mountains. The state is bordering Himachal Pradesh in the North-West and Uttar Pradesh in the South and has international boundaries with Nepal and the Tibetan Autonomous Region.

The State has a total of 13 Districts which can be grouped into three distinct geographical regions, viz. the mountainous region, the mid-mountainous region and the Tarai region. The agro-ecological zones are upper Himalayas (snow-bound), lower Himalayas, and Shivaliks with great biodiversity.

The State is one of the most forested areas of India, which covers more than 50 percent forest area of its total land area and is known to be storehouse of biodiversity including a rich diversity of food crops. The agricultural system of the State is intimately associated with forest resource base and hence any increase in pressure on the agricultural system has a corresponding impact on the forest.

Agriculture, concentrated mainly between 1000-2000 m elevation belts, is the primary occupation of about 80 percent people of Uttarakhand. This belt is also called the populated or "problem zone" of the region as most of the human population is settled in this belt. Agriculture is practiced as a mixed farming system, where the crops, livestock and forests are the integral parts of an agro-ecosystem.

The study area is located in the Pithoragarh district in the Kumaun Division of Uttarakhand. This district extends over a geographical area of 468293 ha out of which 47.8% is under forests. Cultivated barren land occupies 10.8%, fallow land 2.6%, uncultivated land 4.6%, other than agriculture 1.8%, pastures 11.0% and orchard 8.8% area. The net sown area reported is 12.4%. Cropping intensity is 170% (Teli 2006).

Selection of Villages

Community land in mountain areas is shared by a cluster of villages called gramsabha, which is a local governing body. Selection of a Gramsabha thus is logical, for an agro-ecosystem receives energy and nutrients from a common uncultivated land which is looked after by the concerned Gramsabha. For this reason, Gramsabha Durlekh was purposely selected for the study. This Gramsabha has a cluster of five villages, viz., Bajani, Ajera, Hachila, Durlekh and Leemabhat. The selected agro-ecosystem is a typical case of mountain agriculture. The agro-ecosystem lies in the mid-altitude range (1400 to 2000). This is the altitude on which most of the mountain agriculture is concentrated.

Selection of Households

From each of the villages, ten farm families were selected randomly and required information was collected on the pre-structured format. In this way, a total of fifty households were selected for collecting detailed information.

Data Collection

The required information was collected from Government offices, Gramsabha office and from selected households. All agricultural outputs were recorded as per the estimates of the farmers, except for the amount of crop residues which was estimated based on the straw-grain ratio provided by Singh (1998).

Data Analysis

All the inputs and outputs used in the summer cropping in mountain agriculture were converted into their calorific values using standard table for calorific values and published research work.

Table 1: Energy values for different inputs and outputs

Item	kJ/kg.
Human foods	
Grains	16233.00
Pulses	17094.00
Leaf Vegetables (Fresh)	2839.20
Roots, tubers	3956.40
Vegetables (Fresh)	2410.80
Livestock Feed	
Green Fodder (Cultivated)	3956.40
Tree and Shrub Leaves	4204.20
Legume Hay	14985.60
Straw	13986.00
Hay	14557.20
Manure	7320.60
Fertilizer	30340.80
Pesticides (Insecticides)	148000.00

Source: Based on Mitchell (1979); value for pesticides (insecticides) based on Ralhan *et al.* (1991)

The energy values for different agricultural operations for cultivation of different crops have been taken from Singh (1998) and they have been converted into kilo joules (1 kWh = 860 kcal/h; 1 kcal = 4.186 kJ). The input and output values have been converted to energy by multiplying the quantities with standard values reported by Mitchell (1979) which have also been used by several workers, e.g. Pandey and Singh (1984), Singh *et al.* (1984), Negi *et al.* (1989), Ralhan *et al.* (1991), Singh (1991), Singh and Dankelman (1993) and Maikhuri (1996). These values, summarised in Table 1, are expressed on fresh weight basis.

Results and Discussion

Crop cultivation requires certain inputs in appropriate quantities. It also gives certain inputs of critical consumptive value to human society. Input estimation was based on farmers' own experiences, while outputs were mainly based on the productivity of individual crops, but the figures were agreeable by the farmers.

Input Estimation

Input energy used in crop production is: i) animate energy to be obtainable from human labour and bullock labour, and ii) energy to be provided by the other inputs, viz., seeds, manure, fertilizers, and pesticides. In animal production, the major input is feed and fodder. Outputs in crop production include food grains, straws, fruits and vegetables. In livestock production, outputs are milk, dung, and calf crop. The bullock and human work input depends upon the type of cropping system. Amongst the summer crops, the lowland rice demanded the highest number of bullock and human work hours to be invested in crop cultivation. On the other hand amaranth, a pseudo cereal, requires minimum number of bullock and human hours. The fruit crop, unlike food grain crops, did not need investment of bullock energy.

Table 2: Bullock and human hours invested in summer cropping in the mountain agro-ecosystem studied

Summer crops	Bullock hours	Human hours
Upland rice	144	639
Lowland rice	180	1158
Finger millet + pulses	158	502
Barnyard millet	131	467
Amaranth + kidney beans	79	343
Soybeans	167	798
Fruits	-	400
Vegetables	80	872

Source: Singh and Partap (2000)

Singh and Partap (2000) made a comprehensive study of bullock and human use in mountain agriculture. Their data for mid-altitude agriculture would also hold true in this case. We tried to testify their data while interviewing the farmers of the study area, which coincided with those already reported. So we opted to use the same data. The amount of different inputs during summer (kharif) cropping per ha cultivated land are presented in Table 2. An understanding of the input amount per unit area is essential to workout the energy value and finally energetics of the crop production.

Table 3: Amount of inputs (kg) per ha in summer cropping in study area

Crops	Seeds	Manure	Fertilizer	Pesticides
Upland rice	80	600	-	-
Lowland rice	60	300	120	5
Finger millet+pulses	30	350	-	-
Barnyard millet	25	350	-	-
Amaranth+kidney beans		12	100	-
Soybean	140	250	100	6
Fruits	-	500	-	10
Vegetables (fresh)	10	650	50	10
Roots and tubers	40	650	50	25

The rate of seed input for raising a particular crop is almost the same throughout the mountain region and in many cases it would be different from that in the plains. Farmers have their own perception of seed rate. Manure application, however, depends on many factors. Manure application rates are likely to vary from place to place depending upon livestock holding size, type of animals (i.e., kept on grazing or stall fed), family labour, etc. Fertilizer and pesticide use is seldom as per recommendations. Farmers do not apply these inputs under traditional systems. It is only the lowland rice, soybeans and vegetable cultivation where fertilizer and pesticides are used (Table 3).

Output Estimation

Average productivity of individual crops is presented in Table 4. Amount of straw was estimated according to the straw-grain ratio of mountain crops provided by Singh (1998). It was observed that amongst the food grain crops lowland rice produced maximum yields on per ha basis. Amongst the other food crops fruit production was highest on per ha basis.

Gross production of individual crops is presented in Table 5. The figures have been derived by multiplying productivity of individual crop by the total cultivated area devoted to that crop in individual villages.

Table 4: Average productivity (qha^{-1}) of summer crops at study site

Crops	Grain	Straw
Upland rice	24	31
Lowland rice	34	45
Finger millet+pulses	14+6	28+6
Barnyard millet	13	36
Amaranth+kidney beans	12+6	0+9
Soybean	18	0
Fruits	200	-
Vegetables (fresh)	50	-
Roots and tubers	120	-

Two outputs are of major human use, viz., main products that include food grains, vegetables, edible roots and tubers and fruits. The by-products serve as livestock feeds which are also of critical value. These feeds are eventually converted into milk and draught power by livestock. In addition, the by-products' energy is used in the maintenance and growth of livestock and a proportion of the feed voided as dung serves as a useful input for the soil.

Table 5: Total production (q) of different crops in the villages in study area

Crop	Bajani		Ajera		Hachila		Durlkeh		Leemabhat		Total		Average	
	MP	BP	MP	BP	MP	BP	MP	BP	MP	BP	MP	BP	MP	BP
Upland rice	384	496	1164	1503.5	600	775.5	432	558	744	961	4068	4294	664.8	858.8
Lowland rice	54.4	72	326.4	432	108.8	144	85	112.5	153	202.5	727.6	828	145.52	192.6
Finger	203	406	364	728	294	588	329	658	392	784	1582	3164	316.4	632.8
millet+pulses	87	87	156	156	126	126	141	141	168	168	678	678	135.6	135.6
Barnyard millet	45.5	126	52	144	143	396	175.5	486	188.5	522	6045	1674	120.9	334.8
Amaranth +	18	0	24	0	20.4	0	46.8	0	40.8	0	150	0	30	0
kidney beans	9	13.5	12	18	10.2	15.3	23.4	35.1	20.4	30.6	75	112.5	15	22.5
Soybean	81	0	99	0	72	0	117	0	140.4	0	509.4	0	101.88	0
Fruits	800	-	150	-	750	-	800	-	350	-	2850	-	570	-
Vegetables (fresh)	150	-	70	-	130	-	150	-	140	-	740	-	148	-
Roots and tubers	200	-	100	-	150	-	180	-	150	-	780	-	156	-

Energy Audit of Crop Production

Input Energy

Every kind of biomass has its specific energy value. These values have been determined by different workers. Values provided by Mitchell (1979) are frequently used wherever applicable. Values of bullock work for different kind of operations in mountain agriculture have been deduced by Singh (1998) and Singh and Partap (2000). Ralhan *et al.* (1991) has used energy values of pesticides. The animate energy invested in the cultivation of summer crops per ha of land in the mountain agro-ecosystem studied are shown in Table 6.

Table 6: Bullock and human energy invested in the cultivation of summer crops (x10^5 $kJha^{-1}$)

Summer crops	Bullock energy	Human energy
Upland rice	2.47	1.73
Lowland rice	3.09	3.13
Finger millet + pulses	2.12	1.36
Barnyard millet	1.79	1.26
Amaranth + kidney beans	1.24	0.93
Soybeans	2.59	2.15
Fruits	-	1.57
Vegetables	3.20	4.70
TOTAL	16.50	16.83

Source: Singh (1998)

The bullock energy and human energy input values are based on Singh (1998). These values have been found appropriate since they have been elicited by means of long-term experiments in mountain agro-ecosystems. Values of energy used through seeds, manure, fertilizer and pesticides have been presented in Table 7.

Table 7: Energy investment through various inputs in the cultivation of summer crops (x10^5 $kJha^{-1}$)

Crops	Seeds	Manure	Fertilizer	Pesticides	Total energy
Upland rice	12.99	43.92	-	-	56.91
Lowland rice	9.74	21.96	36.41	7.40	75.51
Finger millet+pulses	4.87	25.62	-	-	30.49
Barnyard millet	4.06	25.62	-	-	29.68
Amaranth+kidney beans		1.95	7.32	-	-
9.27					
Soybean	22.72	18.30	30.34	8.88	80.24
Fruits	-	36.60	-	14.80	51.40
Vegetables (fresh)	0.24	47.58	15.17	14.80	77.79
Roots and tubers	1.58	47.58	15.17	37.00	101.33
TOTAL ENERGY	58.15	274.50	97.09	82.88	512.62

Note: The values have been deduced by multiplying the input amount figures of Table 3 with figures in Table 1.

Maximum amount of energy (274.50x10^5 kJha^{-1}) was invested through manure (53.55 percent of the total energy), followed by chemical fertilizers (97.09x10^5 kJha^{-1}) and pesticides (82.88 x10^5 kJha^{-1}), and minimum through seeds (15.15x10^5 kJha^{-1}) during the production of summer crops in the mountain agro-ecosystem investigated during this study. The energy input in the form of chemical fertilizers and pesticides is an important one as these inputs are purchased from the market, unlike all other inputs which are produced within the system. Thus, the share of the imported energy input of the total energy input is considerable, 35.11 percent. Traditional agriculture tends to rely upon the inputs produced within the system. Many favourable areas of the mountains where agriculture has undergone considerable transformation, proportion of imported input energy is pretty high as has been revealed in a study of the Indian Central Himalaya (Singh 1998).

Output Energy

Various outputs of human use were converted into their respective energy values as indicated in Table 1. The energy values for the output of various summer crops raised on one ha land area in the mountain agro-ecosystem under study are shown in Table 8. In the form of useful outputs (grain and straw) gross energy value of 10260.95 x10^5kJha^{-1} was harvested during the summer cropping in the mountain agro-ecosystem under study. Out of this total energy, as much as 79 percent was harvested through grains only. Amongst the food grain crops, the largest share in total energy production was that of lowland rice. However, the largest proportion of energy was produced through fruit production (3246.60x10^5kJha^{-1}), which was about 32 percent of the gross energy produced in one ha cropland in summer cropping.

Table 8: Energy values (x10^5kJha^{-1}) of the useful outputs of summer crops in the mountains

Crops	Grain	Straw	Total energy
Upland rice	389.59	433.57	823.16
Lowland rice	551.92	629.37	1181.29
Finger millet+pulses	227.26+102.56	391.61+89.91	811.34
Barnyard millet	211.03	503.50	714.53
Amaranth+kidney beans	194.80+102.56	0+134.87	432.23
Soybean	292.19	0	292.19
Fruits	3246.60	-	3246.60
Vegetables (fresh)	811.65	-	811.65
Roots and tubers	1947.96	-	1947.96
TOTAL ENERGY	8078.12	2182.83	10260.95

Energetic Efficiency

The output-input ratio of a production system is an important indicator of its efficiency. The energy budget shown in Table 9 would be indicative of the efficiency of the individual crop cultivation as well as of the entire agro-ecosystem.

Ecosystems are the solar-powered machines in which the kinetic energy of sunlight is stored as organic molecules by green plants, which, in turn, can be used either for the vegetative growth of plant structures or for their maintenance (Mitchell 1979). In addition to the solar energy, the agro-ecosystems also use other forms of energy. The main forms of energy used in mountain agriculture can be broadly classified as direct energy (animate and biomass) and indirect energy (fertilizer and pesticide).

Table 9: Energetic efficiency of summer crops at the study site ($x10^5 kJha^{-1}$)

Crops	Output (grain)-Input Ratio	Output (straw)-Input Ratio	Output (biomass)-Input Ratio
Upland rice	6.375	7.094	13.470
Lowland rice	6.752	7.700	14.453
Finger millet+pulses	9.709	14.174	23.884
Barnyard millet	6.447	15.387	21.831
Amaranth+kidney beans	26.342	11.789	37.782
Soybean	3.438	-	3.438
Fruits	61.291	-	61.291
Vegetables (fresh)	9.471	-	9.471
Roots and tubers	17.833	-	17.833

Energy flow through different summer crops is depicted in Fig. 1. Gross energy output-input ratio of the tree-based fruit crops was the highest (61.291) of all the crops in the summer season. It is owing to the low input demand of tree crops. Deep-rooted trees fulfill virtually an entire input need the natural way. However, since the fruit crop is part of the socio-economic system, certain amounts of inputs, particularly in the form of human labour, are required. Manure on the floor of orchards is applied mainly for vegetable cultivation, but a considerable proportion is likely to be used by fruit crop. Root and tuber crops showed comparatively higher energetic efficiency (17.833) than other vegetables (9.471). Vegetables generally require chemical fertilizers and pesticides for their production, which narrow down their output-input ratios.

Amongst the food grain crops the maximum energetic efficiency (37.782) was recorded in case of amaranth-kidney bean mixed crop cultivation. The next crop combination in energetic efficiency was finger millet and pulses (23.884).

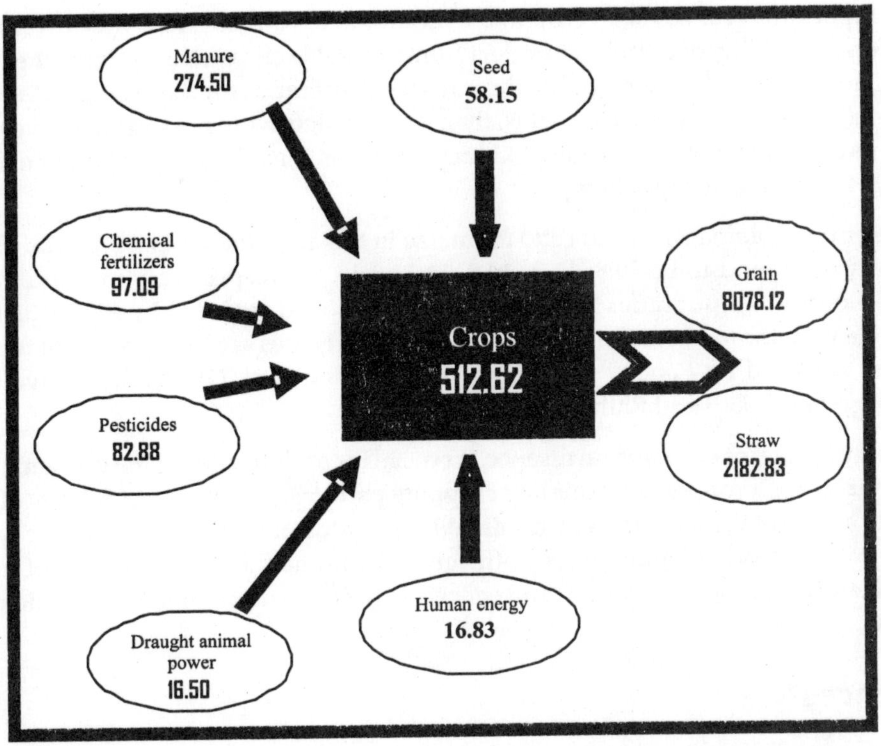

Fig. 1. Energy flow through summer crops in a mountain agriculture (all figures are x $10^5 kJha^{-1}$)

The poorest performance was recorded in case of soybean cultivation. Its narrow output-input ratio (3.438) was mainly because of very high amount of energy expended in soybean cultivation. This crop serves as a host of numerous insect pests which are controlled by the use of pesticides.

The overall energetic efficiency of summer cropping is comparable with the one reported by Singh (1998) from four types of agro-ecosystems in the Central Himalayas. According to his report, the output - input ratios in the hill, transformed, and high Himalayan agriculture were 2.98, 3.81 and 2.51 times lower, respectively, than in the traditional mountain agriculture. If only the main product (agronomic yields) is to be considered then these ratios, in the respective areas, were 3.53, 3.13 and 1.88 times lower than in the traditional agricultural areas. The high energy input compared with the energy output in the hills and transformed mountains and overall low energy output in the high Himalaya are attributable to the relatively lower energetic efficiency of the agriculture in these areas, according to Singh (1998).

The energetic efficiency of the crops in which a high amount of imported energy (chemical fertilizers and pesticides) is to be employed was lower than all other crops. For example, the output - input ratios for soybean crop in the mountain agro-ecosystem was lower than other crops. Non-availability of useful by-product from vegetable crops, as also from other crops, like soybean, further contributes to decline the output - input ratios for these crops. Amaranth too did not yield a useful by-product (that is, their by-products were not usable in livestock feeding), but its output together with that of kidney bean was considerably higher than these crops because of the saving on input energy.

The output (biomass) - input ratio for maize in Mexico, Guatemala, Nigeria, Philippines and India were found to be 80.8, 11.2, 29.8, 14.3 and 13.3, respectively in a study (Reijntjes 1992). The output - input ratios for selected crops in Nepal reported by Rijal *et al.* (1991) are comparable with those of ours in many cases. But our figures are higher (except for soybean) than those reported by Pandey and Singh (1984), Singh *et al.* (1984), Srivastava and Shah (1984), Negi *et al.* (1989) and Ralhan *et al.* (1991).

Energy budget of cropping, in essence, is crucial for understanding the energetic efficiency of individual crops, cropping systems and cropping patterns. The energy efficiency of cropping also serves as one of the indicators of sustainable agriculture. A crop or a cropping system or a cropping pattern having higher energy efficiency would help reduce amounts of inputs and hence is likely to be more economical and conceivably more sustainable than would otherwise be the case.

References

1. Maikhuri, R.K. 1996. Eco-energetic Analysis of Village Ecosystem of Different Traditional Societies of Northeast India. Energy , 21(12):1287-1297.
2. Mitchell, R. 1979. *The Analysis of Indian Agro-ecosystems.* Imprint, New Delhi.
3. Negi, G.C.S., Singh, V. and Singh, S.P. 1989. Energetics of Agricultural Systems: An Experimental Study from the Central Himalaya. In Pandey, D.C. and Tiwari P.C. (eds.) *Dimensions of Development Planning,* Vol 2. New Delhi, 459-476.
4. Odend'hal, S. 1972. Energetics of Indian Cattle in their Environment. *Human Ecology,* 1; 3-22
5. Odum, E.P. 1971. *Fundamentals of Ecology. 3rd edition.* Saunders, Philadelphia.
6. Pandey, U. and Singh, J.S. 1984. Energy-Flow Relationships between Agro- and Forest Ecosystems in Central Himalaya. *Environ. Cons.,* 11(1) : 45-53.
7. Ralhan, P.K.; Negi, G.C.S. and Singh, S.P. 1991. Structure and Function of the Agroforestry System in the Pithoragarh District of Central Himalaya: An Ecological View point. *Agriculture, Ecosystems and Environment,* 35, 283-296.
8. Reijntjes, C. 1992. "Fossil, Human or Bio-Energy". *ILEIA Newsletter,* 8(4) : 3-4.
9, Rijal, K.; Bansal, N.K. and Grover, P.D. 1991. Energy and Subsistence Nepalese Agriculture. *Bioresource Technology,* 36 (1991).

34

"Genetic Evaluation and Characterization of Bamboo Species"

M. S. BHANDARI, TRIPTI DAS AND S.K. TEWARI

Introduction

India possesses 25 per cent of the bamboo species found in the world and 43 per cent of the species found in Asia and have a rich species biodiversity. Bamboo occurs in all states of India except Jammu and Kashmir. It occurs naturally at heights ranging from sea level to over 3500 m in an extra- ordinary range of habitat in almost 10 million hectares of land and constitutes 12.8 per cent of the total forest area in India which yields about 4.5 million tonnes of bamboo per annum.

In India, bamboo provides almost the entire supply of long fibre pulping material for the pulp, paper, board and newsprint industry. About 70% of the pulp used in making paper in India come from bamboo, with an estimated annual production of 2,50,000 tonnes. Bamboo seeds are consumed as food in times of famine, with seeds of some species comparable to wheat in protein content and to rice in protein quality **(Sharma, 1982)**.

Bamboo has got a vast genetic diversity and most of it is not well documented so far. For a successful improvement programme in bamboo, the evaluation and characterization of these species is a prerequisite which will help in identifying and advocating a suitable species for a particular area. It will also help in formulating various propagation and conservation strategies.

Moreover, in distinguishing one species from the other, morphological and molecular markers play an important role. In cases where morphological markers are not effective in the characterization of different species, biochemical markers are found to be more accurate and reliable.

The development of bamboo as a multipurpose plant on a worldwide scale will critically depend on basic research. About the basic biology and genetics of bamboo very little is known and basic research is non-existent. Realizing the scope and importance of bamboo, the present study is proposed with the following objectives:

1. To evaluate the growth performance of a bamboo species in the field.
2. Characterization of a bamboo species on the basis of a biochemical basis.

Materials and Methods

The present work was carried out at the Agroforestry Research Centre, Govind Ballabh Pant University of Agriculture and Technology, Pantnagar, during *Kharif*-2005 (September). Experimental material of the present study comprised of eight species of bamboo planted in a Randomized Block Design with three replications and sixteen plants were maintained in each Block. Four plans from each block were randomly selected for recording observations. The details of the experimental material were presented in Table 1. The recommended packages of the cultural practices were followed during the experimentation for obtaining normal plantation growth. Phenotypic and genotypic coefficients of variation were obtained as the ratio of the respective standard deviation to the general mean of the characters and expressed in a percentage following **Burton** and **DeVane (1953)**, heritable observations in the broad sense following **Allard (1960)** and the expected genetic advance under a selection for different characters was estimated as suggested by **Allard (1960)**. Correlation coefficients were estimated following **Searle (1961)**.

Table 1. List of bamboo species used in the study

Treatment No.	Species	Distinct feature
T_1	*Dendrocalamus hamiltonii*	Clums hollow, leaves pubescent.
T_2	*Bambusa nutans*	Culm is dark green with whitish middle portion, white right below the node.
T_3	*Dendrocalamus asper*	Culm is green color, leaves covered with short shining hairs.
T_4	*Bambusa bambos*	Culm with sharp and stout spiny branches
T_5	*Bambusa balcooa*	Culms are hollow with tiny pseudo-spines.
T_6	*Dendrocalamus strictus*	Culms are semi-solid, leaves are glabrous.
T_7	*Bambusa tulda*	Culm is grayish green, basal portion of the culm has white strips.
T_8	*Bambusa vulgaris*	Culm is bright green or yellowish, green or yellow stripe, smooth and glossy.

Eight species were subjected to gel electrophoresis. New leaf samples were collected from species in the month of October-2006. Zonal electrophoresis in polyacrylamide gel electrophoresis (PAGE), simultaneously exploits differences in molecular size and charge for the purpose for the fraction. This technique, which is based on the mobility of the charged solutes in an applied electric field, is influenced not only by the charge, but also by voltage, distance between the electrodes, size and shape of the protein, temperature and time. Polypeptide chains studied by PAGE, in the presence of SDS migrate according to their molecular weights.

Table 2. Analysis of varince for different quantyitative character in bamboo during *Kharif*-2006

Source of variation	July			August			September		
	Number of shoot	Height	Diameter	Number of shoot	Height	Diameter	Number of shoot	Height	Diameter
Replication (2)	2.51	1773.26	0.15	8.16	531.23	0.19	14.11	11117.94	0.98
Genotype (7)	3.92*	28368.25**	2.31 **	12.09**	20465.17**	2.11 **	17.72**	38984.04**	4.18**
Error (14)	1.20	732.59	0.10	0.95	393.15	0.11	1.82	7148.61	0.44
CD at 5%	1.92	47.39	0.55	1.70	34.71	0.59	2.36	48.04	1.16

*, ** - Significance at 5% and 1 % probability lev,els, respectively.

Results and Discussion

The analysis of variance revealed a significant and positive treatment mean in squares for all the characters studied during July, August and September (Table 2). Mean performance of eight different species from the month of February 2006 to September 2006 at monthly intervals are shown by graphical representation (Graph 1, 2 and 3) for the three different characters studied during the experimental trial.

Analysis of mean for emergence of shoots showed that in the month of February, growth started in *Dendrocalamus hamiltonii, Bambusa nutans, Dendrocalamus asper, Bambusa bambos and Bambusa balcooa,* whereas no growth was observed in *Dendrocalamus strictus, Bambusa tulda* and *Bambusa vulgaris.* During April, May and June *Bambusa balcooa* and *Dendrocalamus asper* showed maximum growth whereas, *Bambusa tulda* showed highest shoot emergence in July, August and September followed by *Bambusa nutans* and *Bambusa tulda.*

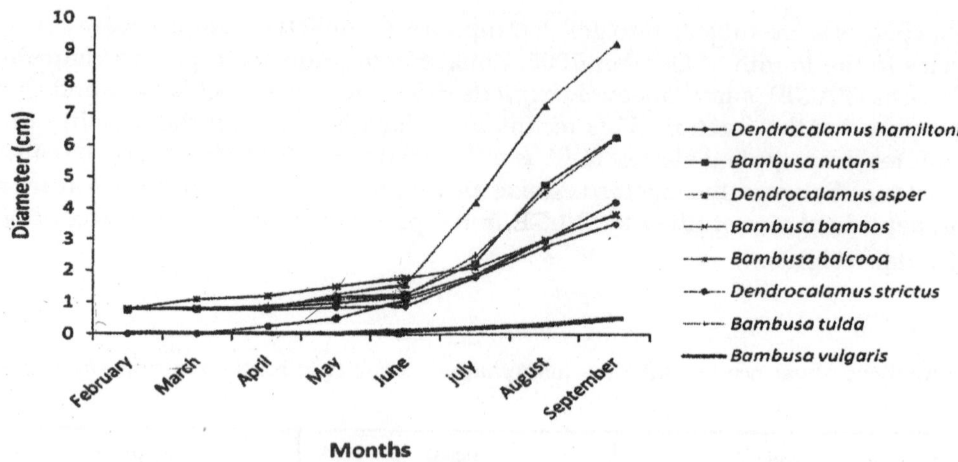

Graph 1. Emergence of new shoots in different months

The height of new shoots in different months showed variations among different species of bamboo. In February, March and April *Dendrocalamus hamiltonii, Bambusa nutans, Bambusa balcooa* and *Bambusa bambos* showed normal growth patterns for them plant height. During June, July and August all bamboo species except *Dendrocalamus asper, Bambusa tulda* and *Bambusa vulgaris* showed a radical increase in growth. Maximum plant height was observed in *Dendrocalamus hamiltonii* and *Bambusa balcooa.*

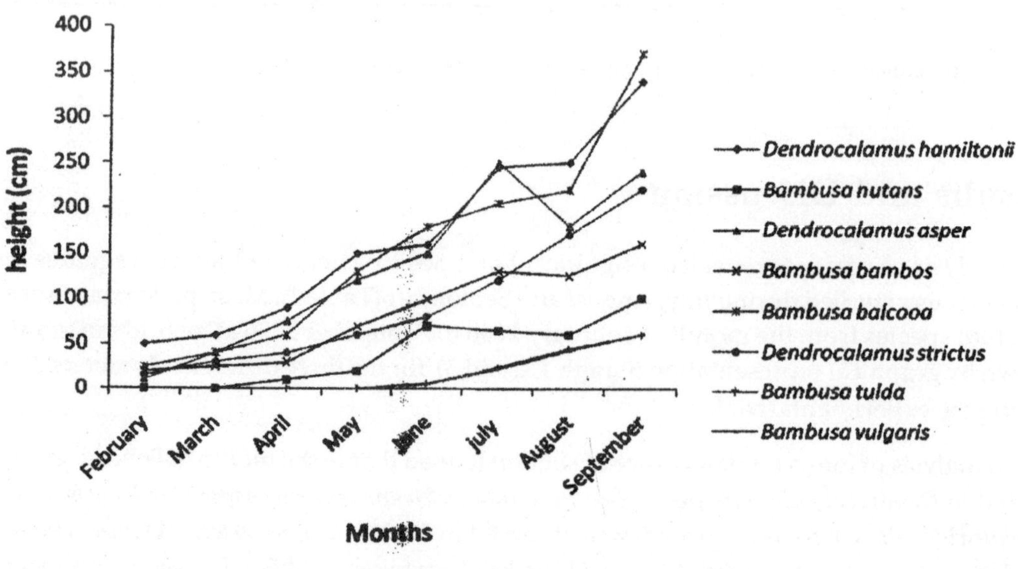

Graph 2. Height of new shoots in different months

Bambusa balcooa showed the highest growth in diameter of new shoots in February and depleted during March, April and May; but showed sufficient increment from June to September. *Dendrocalamus hamiltonii* showed a normal diameter growth from February to August and become maximum in the month of September. Other species followed the above growth pattern (in diameter) are *Bambusa tulda, Bambusa bambos* and *Dendrocalamus strictus.*

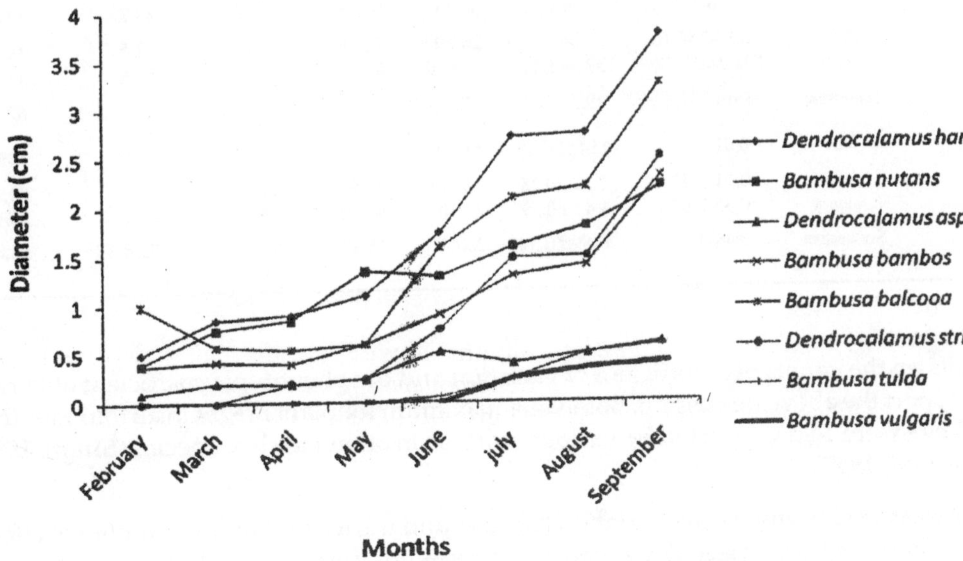

Graph 3. Diameter of new shoots in different months

The graphical representation clearly depicted that *Dendrocalamus hamiltonii, Bambusa tulda, Bambusa bambos* and *Bambusa balcooa* had positive effects in the rise in temperature, excess humidity, rainfall and other environmental conditions in pursuing a maximum growth rate in plant height, shoot emergence and diameter of new shoots. These growth patterns are in accordance with **Raina *et al.*, 1988** as observed in *Bambusa tulda.*

The heritage ability of any trait is an important parameter for any improvement programme. The estimate of this heritage ability and expected genetic advance provided the information on the total variation present in the population and simultaneously gave an indication of the influence of them environment on the character. These estimates help in choosing suitable selection time and method. The heritage ability estimates were analyzed for three characters in eight species of bamboo (Table 3). In June, July and August the plant height showed a maximum heritage ability whereas, in September the number of new shoot emergence showed the highest heritage ability. Similarly, genetic advance is maximum for plant height in June, July and August while in September it was maximum for the diameter of a new shoot. Therefore the selection is quite effective for characters such as plant height, number of new shoot emergences and the diameter of new shoots in the case of a bamboo species.

Table 3. Estimates of genetic parameters for agronomic traits in Bamboo during *Kharif*-2006

Number of shoot	June	0.08-2.00	1.21:!:O.25	36.46	56.70	67.42	70.7	98.34
	July	0.33-4.33	2.20::1:0.63	49.19	37.70	59.52	40.1	53.18
	August	3.00-7.23	3.70::1:0.56	25.73	50.94	57.10	79.5	95.67
	September	0.50-9.00	4.81::1:0.78	28.09	47.83	55.47	74.3	84.82
Plant height	June	3.16-179	93.68::1:1.23	31.24	69.75	76.43	83.2	131.15
	July	18.33-254.16	133.36:f:1.74	20.29	71.96	74.77	92.6	142.68
	August	50.00-25'0.06	139.59:f:2.01	14.20	58.59	60.28	94.4	117.30
	September	71.66-353.41	199.36:f:1.89	42.41	51.67	66.84	59.7	82.27
Diameter	June	0.02-1.73	0.94::1:0.23	43.00	66.82	79.39	70.8	119.14
	July	0.21-2.63	1.31:f:0.18	24.21'	65.59	69.77	33.0	89.31
	August	0.32-2.67	1.45:f:0.19	23.40	55.50	59.91	85.8	108.27
	September	0.48-3.66	1. 99:f:0.34	33.47	55.87	65.11	73.4	95.47

Results on the genotypic coefficient of variation and the phenotypic coefficient of variation also support the above discussion. Both were maximum for plant height in all four months. These findings were also supported by various workers in other bamboo species (**Singh, 1993** and **Indira *et al.*, 1997**).

Yield is a complex and highly variable character and it is a result of the cumulative effect of so many component characters along with the complex environmental role in bringing about the ultimate product i.e. yield. In formulating the suitable selection criteria, this complex genetic system of inter-character correlation is always given due consideration. In present investigation, inter-character correlation coefficients were estimated at phenotypic, genotypic and environmental levels. Plant height showed positive and significant correlation with diameters

Table 4. Correlation coefficients between agronomic traits in bamboo species

Characters		July			August			September		
		Number of shoot	Plant height	Diameter	Number of shoot	Plant height	Diameter	Number of shoot	Plant height	Diameter
Number of shoot	rg	1.00	0.01	-0.02	1.00	-0.25	-0.12	1.00	-0.07	-0.10
	rp		0.07	0.05		-0.21	-0.05		-0.05	-0.08
Diameter	rg			1.00			1.00			1.00
	rp									

rg = genotypic correlation; **rp** = phenotypic correlation;

*, ** - Significance at 5% and 1 % probability levels, respectively.

for all the three months (July, August and September) at both genotypic and phenotypic level respectively (Table 4). Similar findings were also reported by **Indira, 1998** and **Sagwal, 1987.** The environmental correlations include mainly the effect of soil heterogeneity, cultural irregularities and chance error in the experiment. Such factor causes a harmonic change in the plant behavior and may easily be interpreted in terms of physiological instruments. The environmental factors greatly influenced the expression of these attributes under field conditions.

Molecular markers can be useful tools at all taxonomic levels. PCR-based genetic markers are now well documented for cultivar identification. In order to find out the phylogenetic relationship among eight genotypes of bamboo, a protein markers was studied in the present study. Protein marker provided sufficient variation to allow a genetic analysis of bamboo genotypes. The use and advantages of electrophoresis as a tool in taxonomical and breeding work has been discussed by many investigators. (**Allard and Kahler, 1971**).

The methods that are usually used for identifying different clones/cultivar/species of plant are mainly based on morphological markers. Most of the morphological characters are quantitative in nature and governed by many genes. The contribution of these genes towards phenotypes is dependent upon the gene in their neighborhood and their interaction with them. Such morphological character is highly influenced by the environment. Under such a situation, in spite of inherent differences between or among species, the phenotypic expression of the characters could not be distinct. In these circumstances, the use of protein banding patterns for species identification is more reliable since the presence or absence of certain protein are expression of the genetic makeup of the plant and are little affected by the environment.

The comparison of electrophoreteic banding patterns of the species under study clearly indicated that the protein profile was specific to each species. By applying UPGMA (Unweighed Paired Group Mean Cluster Analysis) on 0-1 patterns, a dendrogram is formed by measuring the average distance between bamboo species. On analyzing the dendrogram, all the species were grouped into five clusters. Cluster I contained only one species i.e., *Bambusa tulda.* Cluster II contained *Dendrocalamus asper* and *Bambusa vulgaris. Bambusa nutans, Bambusa bambos* and

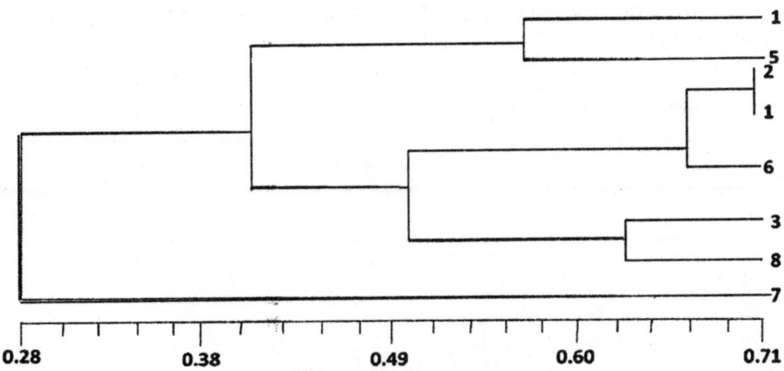

Figure 1. Dendrogram

Dendrocalamus strictus were placed in cluster III. Cluster IV contained *Bambusa balcooa* and Cluster V comprises of *Dendrocalamus* hamiltonii (Figure 1). **Biswas, 1997 and Boonsermuk *et al.*, 1992** reported similar results by conducting isoenzyme analysis on different populations of *Dendrocalamus strictus* and *Dendrocalamus asper* respectively.

Thus, on the basis of variation in number and intensity of bands, it was established that there were inherent differences among these species for leaf protein profile. The banding pattern revealed that there is a clear-cut diversity among species (Plate 1).

1. *Dendrocalamus hamiltonii*
2. *Bambusa nutans*
3. *Dendrocalamus asper*
4. *Bambusa bambos*
5. *Bambusa balcooa*
6. *Dendrocalamus strictus*
7. *Bambusa tulda*
8. *Bambusa vulgaris*

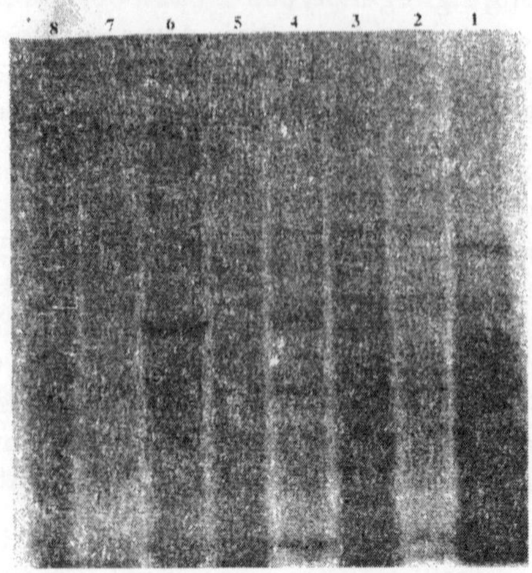

Plate 1. Gel banding pattern of leaf protein in Bamboo species

On the basis of a similarity index (Table 5), it was inferred that *Bambusa bambos* showed maximum affinity with *Bambusa nutans* (70%) followed by *Dendrocalamus strictus* (66%) and *Dendrocalamus asper* (57%). *Bambusa vulgaris* showed 61% and 60% similarity with *Dendrocalamus asper* and *Dendrocalamus strictus,* respectively. *Bambusa balcooa* showed 53% similarity with *Dendrocalamus hamiltonii.* The lowest similarity has been found between *Dendrocalamus asper* and *Bambusa vulgaris* (18%).

Table 5. Similarity Index between different bamboo species

	Similarity Index							
	T_1	T_2	T_3	T_4	T_5	T_6	T_7	T_8
T_1	1.00							
T_2	0.47	1.00						
T_3	0.52	0.35	1.00					
T_4	0.40	0.70	0.36	1.00				
T_5	0.53	0.31	0.57	0.26	1.00			
T_6	0.45	0.66	0.5	0.65	0.31	1.00		
T_7	0.21	0.38	0.18	0.31	0.30	0.29	1.00	
T_8	0.42	0.47	0.61	0.47	0.35	0.60	0.25	1.00

In the light of the results and discussion, it is to be concluded that the presence of variability gives selection of germplasm in a defined manner. Heritage as well as genetic advance clearly estimates the performance of progeny used for propagation. The distinction between species can be done with the help of a banding pattern obtained by the SDS-PAGE technique. Therefore genetic diversity conservation is helpful in (1) saving a plant from the infestation of pests and disease attack, (2) utilization for scientific studies, (3) the introduction of a cultivar to a new geographical area, (4) biotechnological programme as a source of genes and (5) used for aesthetic value e.g., ornamentals, lawn grasses and shrubs are for decorative purposes and are of great value in social life.

References

1. Allard, R.W. 1960. Principals of Plant Breeding. Jonn Wiley and Sons Inc. New York, 185 p.
2. Allard, R.W. and Kahler, A.L. 1971. Allozymes polymorphism in plant population. Stadler Symposium. 3: 9-24.
3. Biswas, S. 1997. Contribution to the isoenzyme studies on Indian bamboo, *Dendrocalamus strictus* (Roxb.). Nees with enphasis on diversity evaluation. *Annals of Forestry.* **5(2):** 168-172.
4. Boonsermuk, S.; Vongvjitra, R. and Suebka, A. 1992. Isozyme studies of *Dendrocalamus asper.* Bamboo abstracts. INBAR Bamboo Information Centre, China. Abstract 920006.
5. Burton, G.W. and E.M. Devane. 1953: Estimating heritability in tall *festuca* from replicated clonal material. *Agron. J.*, **45:** 478-481.
6. Indira, E.P. 1998. Correlation between growth characters and juvenile mature performance in *Bambusa bambos* (L.) Voss. *Journal of Tropical Science.* 14 (1): 1-4.
7. Indira, E.P. and Rugmini, P. 1997. Variability and heriatability in *Bambusa bambos* (L.) Voss. *Journal of Tree Science.* 16 (2): 120-122.
8. Raina, A.K.; Prasad, K.G.; Pharasi, S.C.; Kapoor, K.S. and Singh, S.B. 1988. Effects of nutrients on growth behavior of Bambusa Tulda in the nursery. *Indian Forester.* **114 (9):** 584-591.
9. Sagwal, S.S. 1987. Correlation studies in maggar bamboo (*Dendrocalamus hamiltonii* Nees & Arn.). *Journal of Tropical Forestry.* **3: 2**, 132-135; 7 ref.
10. Searle, S.R. 1961: Phenotypic, genotypic and environmental correlations. *Biometrics.* **17:** 474-480.
11. Sharma, Y.M.L. 1982. Some aspects of bamboos in Asia and the Pacific. Food and Agriculture Organization, Bangkok, Thailand.
12. Singh, N.B. 1993. Estimation of variance, heritability, genetic gain and correlations among some growth characters in *Bambusa pallida* Roxb. *Indian Journal of Forestry.* 1993, **16: 1**, 33-38; 10 ref.

35

Disaster Management through Remote Sensing and GIS

NEERAJ KUMAR SINGH

Introduction

It is well known fact that natural disasters strike countries, both developed and developing, causing enormous destruction and crating human suffering and producing negative impacts on national economics.

Due to diverse geo-climatic conditions prevalent in different part of the globe, different types of natural disaster like floods, drou8ghts, earthquakes, cyclone, landslide, volcanoes, etc., strikes according to the vulnerability of the area.

India is considered as the world's most disaster prone country. It has witnessed devastating natural disaster in recent past like drought, flood, cyclones, earthquakes, landslides, etc.

Natural Disaster in India

India is a large country and prone to a number of natural hazards. Among all the natural disaster that country faces, river floods are the most frequent and often devastating. The shortfall in the rainfall cause droughts or drought like situation in various part of the country. The country has faced some severe earth-quakes causing widespread damage to the life and property.

India has a coastline of about 8000 km which is prone to very severe cyclonic formation in the Arabian Sea and Bay of Bengal. Another major problem faced by the country is in the form of landslide and avalanches.

Remote Sensing

Remote sensing makes observation of any object from a distance without coming into actual contact. Remote sensing can gather data much faster than ground based observation, and can cover large area at one time to give a synoptic view. Remote sensing comprises Aerial Remote Sensing which is the process of recording information, such as photographs and images from sensor on aircraft and Satellite. Remote sensing which consists of several satellite remote sensing system which can be used to integrate natural hazard assessments into development planning studies. These are; Landsat, Sport Satellite Radar System, advanced very high resolution Radio. Satellite data is an indispensable tool for vulnerability assessment of human settlement and disaster management.

Geographic Information System, GIS

A GIS combines layers of information about a place to give you better understanding of place. What layers of information you combine depends on your purpose finding the best location for a new store, analyzing environmental damage, viewing similar crimes in city to detect a pattern, and so on. Unlike with a paper map where "What you see is what you get"; a GIS map can combine many layer of information.

GIS provides a tool for effective and efficient storage and manipulation of remotely sensed data and other spatial and non-spatial data types for both scientific management and policy oriented information. The can be used of facilitate measurement, mapping monitoring and modeling of variety of date types related to natural phenomenon.

The specific GID application in the field of Risk Assessment and – hazard mapping to show earthquake, landslides, flood or fire hazards. These map could be creator for cities, districts or even for the entire country and tropical cyclone. Threat maps are used by methodological departments to improve the quality of the tropical storm services and quickly communicate the risk to the people who are likely to get affected by the cyclone.

Manage disaster with GIS by using following steps –

- Assessing the location of risk and hazard in relation to population, property and natural resources.
- Integrating data and understanding the scope of an emergency of manage an incident.
- Recommending preventive and mitigating solutions.
- Determining how and where scarce resources should be assigned.
- Prioritizing for search and rescue tasks.
- Identifying staging area locations, operation of branches and divisions, and other important incident management needs.
- Assessing short or long-term recovery operation.

National Early Waning System for Tsunami and Storm Surges Background

Tsunami is a system of ocean gravity waves formed as a result of large-scale disturbance of the sea bed, mostly due to earthquakes (or volcanic eruptions or submarine landslides).

The Tsunami of December 26, 2004 one of the strongest in the world which left over 250,000 dead and resulted in an estimated 43 billions dollars of damages.

Tsunamigenic Zones in Indian Ocean

For a tsunami to hit Indian coast, it is necessary that a tsunamigenic earthquake occurs and its magnitude should be larger than M7, and the possible of such events are enclosed in blue circle and ellipse.

The Imperatives and System Design

Tsunami detection	-	End-to-end
The Imperative	-	System Design

Observation Network

- Network of 17 broadband seismic stations for real-time Earthquake detection.
- Network of 12 Deep Ocean Assessment and Reporting Systems (DOARS) for detection of Tsunami Waves.
- Network of 50 Automatic tide gauges of monitoring the progress of Tsunami Waves.
- Other complimentary observations include 5 Coastal Radars, 2 Current Meter moorings, 26 Surface Drifters, 2XBR Lines and other Surface, Met-Ocean observing platforms.

Modeling

Use of numerical models to simulate all possible scenarios with varying earthquake source parameters.

Results indicating travel time and run -up heights at coastal locations will be organized in a database from which the closest scenario will be extracted at the time to event.

High-resolution bathymetry as well as coastal topography being generated for use in modeling. Detailed inundation maps will be generated for delineation of evacuation routes and long-term planning in vulnerable coastal communities.

Early Warning Centre

State-of-art Information and Communication Technology infrastructure for data reception, storage, modeling, display and alert systems warming generation and dissemination.

Technical support facilities and Standard Operation Procedure for warming generation and dissemination to designated contacts.

Partners

India is the only country that is developing capability to detect tsunami generated in the two tsunamigenic zones that would affect Indian Ocean.

India has been elected a Chairman of International Coordination Group (ICG) set up by UNESCO/IOC for Indian Ocean Tsunami Warning and mitigation System, a network of 27 national systems.

India hosted the second session to ICG/IOTWS at Hyderabad during December 14-16, 2005

International Interface

Institutions from Ministry of Earth Sciences (MoES), Department of Space (DOS), Department of Science and Technology (DST), Center for Scientific and Industrial Research (CSIR), Ministry of Home Affairs (MHA), Ministry of External Affairs (MEA).